TRUST AND REPUTATION FOR SERVICE-ORIENTED ENVIRONMENTS

TRUST AND REPUTATION FOR SERVICE-ORIENTED ENVIRONMENTS

TECHNOLOGIES FOR BUILDING BUSINESS INTELLIGENCE AND CONSUMER CONFIDENCE

Elizabeth Chang

Curtin University of Technology, Perth, Australia

Tharam Dillon

University of Technology, Sydney, Australia

Farookh K. Hussain

Curtin University of Technology, Perth, Australia

John Wiley & Sons, Ltd

Email (for orders and customer service enquiries): cs-books@wiley.co.uk
Visit our Home Page on www.wiley.com

Other Wiley Editorial Offices

John Wiley & Sons Inc., 111 River Street, Hoboken, NJ 07030, USA

Jossey-Bass, 989 Market Street, San Francisco, CA 94103-1741, USA

Wiley-VCH Verlag GmbH, Boschstr. 12, D-69469 Weinheim, Germany

John Wiley & Sons Australia Ltd, 42 McDougall Street, Milton, Queensland 4064, Australia

John Wiley & Sons (Asia) Pte Ltd, 2 Clementi Loop #02-01, Jin Xing Distripark, Singapore 129809

John Wiley & Sons Canada Ltd, 22 Worcester Road, Etobicoke, Ontario, Canada M9W 1L1

Wiley also publishes its books in a variety of electronic formats. Some content that appears
in print may not be available in electronic books.

Library of Congress Cataloging-in-Publication Data:

Chang, Elizabeth.
 Trust and reputation for service-oriented environments : technologies for
building business intelligence and consumer confidence / Elizabeth Chang,
Tharam Dillon, Farookh K. Hussain.
  p. cm.
Includes bibliographical references and index.
ISBN 0-470-01547-0 (cloth : alk. paper)
1. Electronic commerce – Security measures. 2. Business
enterprises – Computer networks – Security measures. 3. Trust. 4. Consumer
confidence. 5. Consumer protection. I. Dillon, Tharam S., 1943- II.
Hussain, Farookh K. III. Title.
 HF5548.32.C47235 2006
 658.4'78 – dc22

            2005034992

British Library Cataloguing in Publication Data

A catalogue record for this book is available from the British Library

ISBN-13: 978-0-470-01547-6
ISBN-10: 0-470-01547-0

Typeset in 9/11pt Times by Laserwords Private Limited, Chennai, India

Contents

Preface

Trust has played a central role in human relationships, and hence has been the subject of study in many fields including business, law, social science, philosophy and psychology. It has played a pivotal role in forming contracts, carrying out businesses, and assisting people work together cooperatively and underpins many forms of collaboration. Closely related to this notion of trust is the concept of reputation within a community, with other peers or with society in general. This is frequently used as the basis of a judgement as to whether to trust an individual or organization, particularly in the absence of prior direct contact with them.

These two concepts of trust and reputation have assumed increasing importance in the information technology field.

We note that, increasingly in the networked economy, communication, collaboration, research, education, transactions, sales, marketing and industrial system controls are being conducted over the Internet. As a result, trust becomes crucial to the enterprise or corporation, business, service provider, government and consumer that is making a decision to carry out an interaction on the Internet. Lack of trust between communicating parties results in a situation that is popularly known as 'prisoner's dilemma', in which either of the two parties could resort to unfair practices. The two parties may be involved in transactions worth thousands of dollars, but it may not be possible to detect any unfair practices until the outcome of the transaction is known. Time is needed for the two communicating parties to develop trust in each other before the business transaction takes place. It has become a significant element of business intelligence.

This book aims to increase the level of understanding of the trust and reputation body of knowledge for service-oriented environments.

The scope of this book includes provision of the following:

(a) a detailed explanation of the concepts of trust, trust relationships, trustworthiness, trust value, trustworthiness value, trustworthiness measure, reputation, reputation value and reputation measure and trustworthiness prediction;
(b) a detailed analysis of the dynamic natures of trust and reputation, as well as the context-specific and time-dependent natures of trust and reputation;
(c) a clear understanding of trust and reputation ontologies;
(d) a detailed methodology for establishing trust and assigning trustworthiness through direct interaction or through third-party recommendation and reputation, and trustworthiness prediction through historical datasets and reputation values;
(e) a detailed trust modelling technique;
(f) advice on business intelligence through the use of trust and reputation technology and providing a clear distinction between existing business intelligence tools and this new class of technology in addition to identifying how they offer business intelligence and why;
(g) a clear distinction between web services and service-oriented environments, from both a technology perspective and from the business perspectives; and
(h) a clear distinction between trust and security, and how to build business intelligence and consumer confidence through trust.

The service-oriented network environment can be viewed in three types of networked environments. The first type is centralized networks such as client−server systems. The second type is decentralized networks such as peer-to-peer (P2P) or grid (a big networked virtual computer) or mobile networks. The third type is partially centralized and partially decentralized networks that are a combination of both client–server and server-to-server (peers) infrastructures such as an ad hoc network. The challenge, in the above three types of service-oriented environments, is due to the fact that the communicating parties or agents can carry out interactions in one of the three ways:

- Anonymous (identity is unknown during communication)
- Psuedo-anonymous (known by a masked identity during communication)
- Non-anonymous (identity is known during communication).

However, unlike the centralized client server, which enables non-anonymous communication (where trust is pre-established before communication), in distributed communication each server has its own trust management system.

It is difficult to establish trust between two communicating agents in a distributed network as most of the time these operate in an anonymous environment. Without trust relationships, an agent will be reluctant to carry out an interaction with another agent.

Service-oriented networked environments have assumed new significance with the advent of web services.

There is a common misconception that the studies of trust and security are one and the same.

Security in the computing field refers to the process of enabling secure communication between two communicating parties or machines in the service-oriented network environment, whereas, trust is the belief that the trusting agent has in another agent's willingness and capability to deliver the agreed quality of service, and it is context and time dependent. Trust is fundamental to business, commerce, education, politics and any form of collaboration. The more rationalized the trust we have, the lower the risk of interaction we achieve. Conversely, the lower the rationalized trust we have, the higher the risk we encounter. This book aims to integrate concepts from economics, computing, science, engineering, marketing, business, law, sociology and psychology to address the need for trust in service-oriented network environments.

The emerging trustworthiness technologies are reshaping the networked economy by providing business intelligence and consumer confidence through the creation of open, fair, convenient and transparent service-oriented environments for business transactions, quality of services, reputation of sellers and products, as well as loyalty of buyers and their accountability. The trustworthiness technology acts as follows:

- a quality assessor for businesses to fulfil their contractual obligations including quality of service and product delivery;
- a consumer watchdog for deterring unfair trading, dishonest dealing, defects in products and discriminatory processes;
- an interaction mediator that helps distributed network buyers, sellers and online users to communicate, build trust relationships and create an opportunity for business collaboration and competition;
- a strategic advisor for both online and off-line business providers to capture customer needs, competitors' operations, marketing trends, and so on;
- a brand advertiser for the trustworthy service providers to assist with maintaining their quality of service and retaining their customers; and
- a safeguard for all customers and all agents in the service-oriented network environment to remove untrusted agents, malicious service providers and to guard against their attacks.

This is a new-age, important, frontier technology for business intelligence. The study of this technology is applicable to computer science, computer engineering, software engineering and network engineering. It is also well suited for business, IT (Information Technology), IS (Information Systems) and e-Commerce (Electronic Commerce) courses. It is a booming and important area for research, teaching and commerce. Security is of paramount importance for most countries. However, trust plays a key role in security (in the broader sense) as it reduces the risk. We also see that the latest government funding has provided a huge amount of support for security and trusted platforms, programs and projects and many universities have newly created degrees in security. Despite this interest, there is not a lot of material available on trust and reputation, particularly in the newly emerging service-oriented environments. In the last couple of years, many worldwide prestigious conferences and workshops (IEEE and ACM) have incorporated 'trust' into their tracks or themes. Several groups run specific workshops or conferences specifically on trust. New subjects and fresh courses are appearing on topics such as 'Trust' and 'Reputation'. It is envisaged that trust management systems will become essential on any computer, server, network or business over the next 5 years.

This is the first book in the field that gives a detailed overview of the subject and provides methodologies to help solve problems in trust over networked communications, especially in peer-to-peer networks and service-oriented networks. The subject matter is cutting edge technology. The book has been reviewed by students to ensure that it is appropriate for its target audience.

The primary audience is third- and fourth-year undergraduate students and Masters students in degrees on information technology, information systems, e-commerce, security, marketing, commerce and management, computer science, computer engineering, network engineering, software engineering and computer engineering as well as MBA, business and law. This book can also be used by research students and computer professionals alike.

Secondary audiences include business practitioners, organizations with e-commerce applications, IT organizations, security firms, governments and consortiums, alliances and partnership SMEs (Small Medium Enterprises) for learning how trust may be built into their business operations, IT systems and network infrastructures.

Author Introduction

Prof Elizabeth Chang is a **Professor** in IT, Software Engineering and Project Management at Curtin University of Technology, Perth, Australia. She is also Director of the Research Centre for Extended Enterprise and Business Intelligence at Curtin. She has to her credit over 200 scientific conference and journal papers along with numerous invited Keynote papers at international conferences and has co-authored 40 publications on 'Trust'. All her research and development work is in the areas of software engineering, IT applications, logistics informatics and trust and information security.

Professor Tharam Dillon is a Professor in Computer Science. He is an expert in the fields of software engineering, data mining, XML-based systems, ontologies, trust, security and component-oriented system development along with trust and reputation. Professor Dillon has published five authored books and four co-edited books. He has also published over 500 scientific papers in refereed journals and conferences. He is Dean for the Faculty of IT at the University of Technology Sydney, Australia.

Mr. Farookh Hussain is an outstanding PhD student under the supervision of Prof Chang and Prof Dillon, who has become an expert in the field of trust. Over the last two years, he co-authored 40 papers on trust, together with Prof Chang and Prof Dillon. He obtained his MSc and BSc in Computer Science between 1998–2003. He is expecting to complete his PhD in 2006.

Acknowledgement

This book is an outcome of research carried out at the Centre for Extended Enterprise and Business Intelligence (CEEBI www.ceebi.curtin.edu.au) at Curtin Business School and the University of Technology Sydney's EXEL Lab (http://exel.it.uts.edu.au), funded by the Australian Research Council (ARC) and Curtin University of Technology's R&D Office.

The authors would like to thank our centre manager, Ms Sonya Rosbotham, who took over many duties including the organization of a major international conference to free us to concentrate on the book. She also helped proofread. Without her effort, we would not have been able to complete the book in such a timely manner.

The authors would like to thank Dr Patricia Thomson, who read the entire 14 chapters of the book and provided many valuable comments for content improvement. The authors would also like to give special thanks to our PhD and casual research assistants, Mr R. Rajugan and Mr William Gardner, for assistance with content development; Miss Pronpit Wongthongtham for validation of trust and reputation ontologies; Mr Gautam Mehra for evaluation of trust and reputation systems; Miss Jyoti Bhattacharjya, who carefully read through earlier drafts of CCCI methodology; and Mr Mark Rosbotham who proofread most chapters of the final draft. We would also like to thank our Masters and Undergraduate students, Sarah Pollard, Adnan Khan, Ketsiri Sekaravisuth, Wang Xin and Anshul Purohit, who helped review an initial draft of the book. Without the above staff and students' efforts, this book would not be of such a high quality.

We also thank Professor Dillon's secretary, Ms Samantha Carmen, and our research officer, Ms Lynne Harding, who helped share the Professors' load and coordinated the effort during the making of this book. Finally, we would like to thank all staff in the School of Information Systems and Management of Curtin R&D office and Curtin Business School for their support and encouragement, and the staff at the Faculty of IT at UTS for their encouragement of this endeavour.

1

Trust and Security in Service-oriented Environments

1.1 Introduction

The advent of the Internet and the Web provide connectivity and information richness over great distances at any time. This has created a dynamic, open and convenient environment for social and business development. It not only provides the opportunity for new entrepreneurial endeavours utilizing the Web, but also opens up new opportunities for old, static, closed, locally based businesses to adopt new business paradigms and new organizational forms. The Internet has also opened up modes of interaction and dynamic organizational configurations that were previously inconceivable within a wide array of human and business activities. However, these have also introduced challenges. One of the most pressing of these arises from the fact that in a business or social interaction on the Internet, we cannot rely on the usual physical, facial and verbal cues that we might have relied on to reach a judgement as to whether or not the other party will fulfil the service that they are promising. In addition, in the case of the purchase of physical goods over the Internet, we have no direct physical, sensory contact with the specific product and are reliant solely on the promise of the seller. We are being put in the position of 'buying a pig in a poke', rather than being able to 'squeeze the tomatoes' to determine their firmness. There could, in some cases, be difficulties ensuring that the purchaser pays for the goods. These factors and several others, when taken together, create the imperative for being able to make judgements within such an environment about the other parties' trustworthiness and capability to provide the service at a specific level of quality. Through adopting new *trust* technology in the service-oriented network environment, a platform for both consumers and businesses to learn from each other is created. Thus, real business value, increased consumer confidence, guaranteed quality of product and service could become a reality in the virtual word.

In this chapter, we study why trust is important and make clear distinctions between the concepts of trust and security. We also offer a detailed introduction to service-oriented environments, which are an integral part of a networked economy.

1.2 Why Trust?

In recent times, we have seen an increasing number of people carrying out a myriad of different activities on the Internet. These range from writing reports to looking at the news, from selling a car to joining a club, from the purchase of goods (e.g., [1]) to the purchase of services (e.g., Priceline.com for travel arrangements), from entertainment (music or games) to research and

Trust and Reputation for Service-Oriented Environments   Elizabeth Chang, Tharam Dillon and Farookh Hussain
© 2006 John Wiley & Sons, Ltd

development (information surfing), from private resource utilization (grid computing) to remote file sharing (peer-to-peer communication), from shopping at the mall (BizRate.com) to bargaining in virtual markets (e-Bay), from e-bill to e-pay, from the virtual community to virtual collaboration, from e-governance (e-administration) to mobile commerce (stock trading), from e-education (cyber-university) to e-learning (getting an MBA online), from e-manufacturers (remote control production) to e-factory (e-products), from offshore development (business expansion) to outsourcing (such as IT), from e-warehouse (warehouse space booking) to e-logistics (goods shipping orders), and limitless other possibilities.

Transactions have moved away from less face-to-face encounters to being more on the Internet. The infrastructure for the above business and information exchange activities could be client–server, peer-to-peer (P2P), or mobile networks. Most times, users on the network (the customer or business providers) carry out interactions in one of the following forms:

- Anonymous (no name is to be identified in the communication)
- Pseudo-anonymous (nicknames are used in the communication)
- Non-anonymous (real names are used in the communication).

In such distributed, open and often anonymous environments, *fraudulent* or *incomplete practice* could occur where the seller or business provider or buyer (the agents on the network) does not behave in a manner that is mutually agreed upon or understood, especially where terms and conditions exist. This could take several forms:

(a) The *seller* or *service provider* only delivers part of the service promised or is inconsistent in delivering the goods or services, for example, sometimes delivers and sometimes does not deliver or cannot deliver or never delivers what was promised or advertised.
(b) The *customer* or user may always be negative and disruptive of the business, or gives false or faulty credit details.
(c) The provider provides a *service*; however, it is not up to an acceptable standard.
(d) The seller's *product* is not of a good quality.

Trust and *trust technology* have come into the picture for the virtual environment recently to give an online user the sensation of being able to 'squeeze the tomatoes before you buy' or for providing opinions and assessments before a decision is made. It boosts consumer confidence and helps facilitate judgements about business reputations. In other words, you feel confident to pay for a service or product because you trust the seller's reputation or the quality of products (goods) or services. This helps mitigate the risk in the business transaction. On the other hand, *sellers* or *service providers* can learn about users and customers through trust technology so that they can improve on-demand service that meets customer needs better. Trust technology such as trustworthiness systems, or rating systems, or recommender systems already exist on the Web. For example e-Bay, Amazon, BizRate and CNet already have some rudimentary versions of trust technologies. Regardless of the fact that these examples of the use of the technology only provide some basic functions, trust technologies are becoming more and more popular and providing a convenient tool to simulate social trust and recommendation experience for online users.

Trust is a crucial ingredient in any mutual relationship and where transactions are carried out in an anonymous, pseudo-anonymous or non-anonymous distributed environment to provide the agreed to Quality of Service (QoS).

With trust technology–supported web-based e-business, one is able to respond to dynamic individual and business needs, thus achieving the targeted productivity improvement, lowered operational costs, and enhanced customer service. On the consumer side, such system support will allow greater confidence, leading to greater willingness to participate in transactions on the Web.

1.3 Trust and Security

Trust and security are not the same thing in the world of e-commerce. Unfortunately, a variety of uses, particularly of the term 'trust', could lead to some confusion. In this section, we clearly distinguish between trust and determining security and when they could be synonymous and when they are not. This will help develop a shared conceptual vocabulary, which will form the basis of the discussions in this book.

Connectivity and information richness through the Web have led to the possibility of convenient environments for social and business development. We move away from notions of static, closed, local and competitive environments to new interaction models, new business paradigms, new organizational forms and new economic environments that were previously inconceivable in many activities. Such an environment is able to respond dynamically to individual and business needs in a more timely fashion. Individuals and organizations have equal opportunity to utilize the Web to increase productivity, improve operational costs and enhance customer service. However, this new, open and flexible environment introduces two broad challenges, namely, *security* and *trust*.

Security focuses on protecting users and businesses from anonymous intrusions, attacks, vulnerabilities, and so on, while *trust* helps build consumer confidence and a stable environment for customers or businesses to carry out interactions and transactions with a reduction in the risk associated with doing these in a virtual world, thus allowing one to more fully reap the possible rewards of the increased connectivity, information richness and flexibility.

1.3.1 Security

The dynamic, open and convenient Web environment not only boosts business potential and the economy but also creates concerns of security, trust, privacy and risks. If these issues are not dealt with in a timely fashion, they could hamper business in utilizing the Web. Security issues can affect communication, infrastructure, servers, client browsers, e-products, e-services, software, hardware, electronic documents, business transactions and organizational back-end databases. We need to prevent hackers, attackers, unauthorized individuals and malicious users or servers from taking advantage of honest online users, from damaging private businesses and also from attacks on non-government and government organizations.

Security threats and attacks on the Internet include, but are not limited to, the following [2]:

- *Eavesdropping* – intercepting and reading messages intended for other users
- *Masquerading* – sending/receiving messages using another user's ID
- *Message tampering* – intercepting and altering messages intended for other users
- *Replaying* – using a previously sent message to gain another user's privileges
- *Infiltration* – abusing a user's authority in order to run hostile or malicious programs
- *Denial of service* – preventing authorized users from accessing various resources
- *Virus and worms* – micro virus or attachment virus, Morris worm, cert/cc, short for the Computer Emergency Response Team Coordination Centre, studied at Software Engineering Institute at Carnegie Mellon University examine Internet security vulnerabilities.

Security Technologies that are widely available to address these include the following:

- Encryption (RSA encryption, algorithms, keys, encryption standards, etc.)
- Cryptography (hiding messages in text)
- Steganography (hiding messages in pictures or media)
- Secret information sharing (algorithms, symmetric keys)
- Digital signatures and standards

- Authentication (digital certificates, verifying identities, public keys)
- Authorization (controlling access to particular information and resources)
- Data integrity (a receiver can detect if the content of a message has been altered or a receiver can detect it)
- Intrusion detection.

Currently, the above-mentioned security technologies are sufficiently mature for e-commerce, and most of the technologies are already standardized [2].

The field of security research is still very active in the following areas (though it is not limited to them):

- Electronic payment (electronic wallets, dual signatures, etc.)
- Digital money (blind signatures, coins, double spending, etc.)
- Web security (HTTP messages, header leaks, Secure Sockets Layer (SSL) tunnelling etc.)
- Server security (data and database security, copyright protection, etc.)
- Client security (privacy violation, anonymous communications)
- Mobile agent security (agent protection, malicious agents, attacking servers, sand box, cryptographic trace)
- Mobile commerce security (Global System for Mobile (GSM) security, subscriber ID authentication, etc.)
- Smart card security (SIM card, biometrics, etc.)
- Communication security (firewalls, security negotiations, virtual private networks, network layer securities)
- Data, database and information security (triple keys, Hippocratic databases)
- Security policies (international legislation and regulation, enforcement of security)
- Security management (infrastructure, network, application and database)
- Computer forensics (electronic evidence, expert witness, etc.)
- Risks and emergency responses
- Privacy (protecting the identity of individuals and their information and allowing them to control access to their information).

Security and trust are two distinct concepts. Security provides a safe environment and secure communication along with protecting end-user and business from intrusion. Trust is the belief or faith that a person or agent has in another person or agent with respect to certain activities at a given time. In order to acquire trust in another entity over the anonymous distributed network, security-establishing mechanisms may be necessary to provide sheltered communication or information protection.

1.3.2 Trust

In the networked virtual environment, we study trust along the following dimensions (though not necessarily limited to these):

- Trust, trust value, trust relationships, initiation, association
- Trustworthiness, trustworthiness value
- Reputation, reputation value
- Trust management
- Reputation management
- Dynamic nature of trust, implicitness, asymmetry, context specific, time dependency
- Trust ontology
- Trust and reputation modelling

- Trust relationship diagrams
- Establishing trust relationships, building and maintaining trust
- Trust management protocols
- Trustworthiness measurement
- Trustworthiness prediction
- Assigning trustworthiness
- Trustworthiness scales
- Trusted business, business value, consumer confidence
- Trustworthiness systems, rating systems, recommendation systems.

Most of these concepts will be discussed in this book.

Trust, trustworthiness and reputation are innovative technologies reshaping the world of e-commerce. Many of the largest commerce websites and organizations are already adopting these technologies albeit in an early and rudimentary form. They help business providers learn from their customers and help the customers find the best deals available and understand the risks associated with a transaction with a particular supplier.

1.3.3 Trust in Security Context

The concept of 'trusted computing' known as *Palladium technology* was initiated by Microsoft around 2001 as a combined software and hardware solution and a tamper-proof computer environment for secure communication (Figure 1.1). Microsoft's 'trusted computing' known as the *secure computation service* [4] claimed that 'this significant evolution of the personal computer platform will introduce a level of security that meets rising customer requirements for data protection, integrity and distributed collaboration' [4]. The significance of *trusted computing* (Palladium technology) is its potential to improve system integrity, personal privacy and data security. Reliability and security are achieved as the applications run in the protected communication environment provided by Palladium. This promising technology is available only in a beta version on the market, and its promises still need to be proved.

The concepts of Microsoft's 'trusted computing', 'trusted network', trusted communication', 'trusted agents', 'trusted. . . ', and so on, are here, related to security issues, security mechanisms, security technology and security services. All topics of security study and research are directed towards providing a secure and tamper-free environment, or network or communication. In this context, 'trust' is synonymous with 'secure', which is tied to 'security'. However, this is *not* the same as trust in the business paradigm, which is the subject of this book.

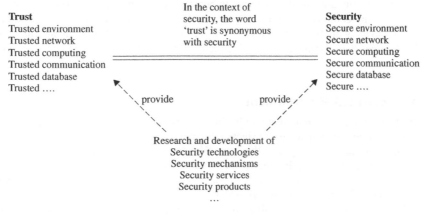

Figure 1.1   'Trust' in security context

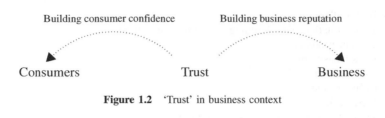

Figure 1.2   'Trust' in business context

Figure 1.3   'Trust' and 'security': complementary technologies

1.3.4 Trust in the Business Context

Trust is a belief of confidence or a feeling of certainty that one person has in another person or thing that he/she is interacting with. Everyone or every organization wants assurance, certainty and confidence about what they do and what they will receive. In the business world, trust is especially tailored for ensuring honest dealings and quality of products or services and that is usually related to mutual agreements and understandings (Figure 1.2).

When we discuss trust in a social or economic context, there is a limited relationship with security. The motivation behind trust technology is to help build business reputation, consumer confidence, fair trading and mutual relationships. This book is about trust in business and specifically focuses on trust technology, trust establishment, trust level measurement and prediction and trust relationship development.

Security can be used to support *trust*, through providing a secure trusted environment, secure network and secure communication so that trusted business transactions can take place.

However, building trust in social and economic environments also helps to reduce aspects of *security risk*.

Both trust and security are equally important in business, commerce and the world of technology. Trust and security are complementary to each other (Figure 1.3). In the field of security, the word 'trust' is synonymous with 'security'. However, in business and social contexts and their support through the Internet they mean different things and both require complex studies, research and development.

1.4 Service-oriented Environment

The advent of the Web and its intrusion into business, commerce, government and the health sector have led to Web-based e-commerce for business interactions and collaborations over great distances and at any time. In the last ten years, this new Web-based environment has enabled economic growth, industry development, technological innovation and resource sharing. This new business environment has led to the development of the *service-oriented environment* that *transcends* the previous *static, closed, competitive models* and has moved towards *flexible, open, collaborative, sharing* and *distributed environments* that are able to respond in a timely manner to consumer needs and business dynamics inherent in the networked economy.

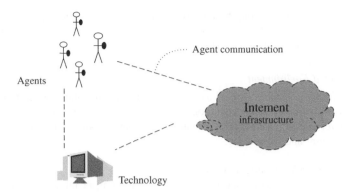

Figure 1.4   The definition of technology-based *environment*

1.4.1 Environment

Environment in social terms is defined as the external elements and conditions that influence the growth, development and survival of an individual or organization.

In technological terms, we define the *environment* as a networked virtual community in which there are agents, infrastructures and technologies which enable the interaction and operation of agents for the growth of the community, economic development and social evolution (Figure 1.4).

1.4.2 Essential Elements in the Technology-based Environment

1.4.2.1 Service

In technological terms, *services* often refer to *software applications, methods, operations* or *communications* between two computing objects, or the *interface* between two software components. These technological terms for *services* are influenced by the business community, as the aim of technology is not simply the development of new technologies but the development of real business solutions. It is important that we study technology not for its own sake but for its application to industry and business.

1.4.2.2 Agents

Agents in business terms are often referred to as intelligent representatives, acting on behalf of a company, business or individual and having the *power, authority* and *ability to make decisions*.

In technological terms, *agents* are referred to as *intelligent software agents*, or communication facilitators, or users of computer systems, and so on. The intelligence comes from a knowledge base, a rule base or ontology that we program or design or build into the software or hardware agents. Therefore, they can behave autonomously and perform business or services automatically without human intervention.

1.4.2.3 Agent Interaction

As in the physical world, where humans or enterprises need to interact to carry out business or services, in the technological environment *agents* need to communicate in order to carry out business activities. As a result of agent interaction, a business transaction or deal is accomplished.

1.4.2.4 Infrastructures

Infrastructure in a business sense is *a base* or *foundation* to provide basic facilities so that the primary business can be operated, and technology is a tool to support or facilitate the business.

The *infrastructure* in a technology-based environment includes basic support services such as networks, which incorporate servers and communication protocols, and so on. Technologies refer to the type of software or hardware tools used.

1.4.3 Service-oriented Environment

A Service-oriented Environment is defined as a collaborative, shared and open community in which agents utilize its infrastructure and technology to carry out business activities, such as product sales, service deliveries and information retrieval. It has at least four components: agents, business activities, infrastructure and technology.

- Agents (buyers, sellers/providers, users, websites or servers)
- Business activities (product sales, service deliveries, marketing or information sharing)
- Infrastructures (networks communications)
- Technologies (service publishing, discovery, binding and composition).

It is a *collaborative environment* because it is not a closed walled individual operation as carried out in a traditional business sense. Online users are often anonymous but help can be provided for each of them by posting and answering questions on the Web, and carrying out collaborative business, research or industrial processes.

It is a *shared environment* because agents share information on the Web about unknown agents, unknown products, unknown service providers, or merchants.

It is an *open community environment* because agents attach themselves to or leave the community as they need to rather than having a predefined set of agents listed within the community servers. Also, everything is on the Web, coupled with emerging trust technologies that ascertain the behaviour of sellers, producers, merchants, manufacturers including service providers using ratings by all kinds of online users, or other buyers, sellers, and so on.

Figure 1.5 depicts the service-oriented environment and its major entities.

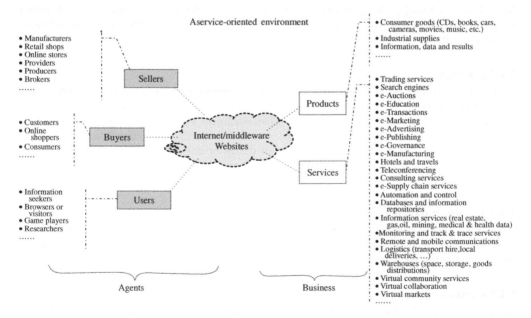

Figure 1.5   Major entities in service-oriented environments

The *Service-oriented Environment* has the following characteristics:

(a) multiple channels of sales, marketing, purchasing and information inquiries;
(b) a collaborative approach between sellers, buyers, users and service providers;
(c) high connectivity and electronic handling of information, data and documents;
(d) strong information infrastructure that extends beyond the *original physical* individuals and businesses;
(e) provision of products, services, end-user interaction and utilization of information services;
(f) self-organization and reconfiguration to meet dynamic business needs;
(g) capturing business intelligence through trust, reputation and smart information sharing; and
(h) value-added consumer relationships, customer service and strengthened small-medium businesses.

The *service-oriented environment* redefines the old industrial business models by means of new Internet-based technologies in order to maximize productivity, customer value and QoS.

1.4.4 Issues in Service-oriented Environments

Trust plays a crucial role in anonymous, remote, heterogeneous service-oriented environments and business communications, where the transactions and services are carried out. It involves several issues that need to be understood and addressed, such as the following:

- It is hard to assess the trustworthiness of remote unknown entities over the Internet.
- It is difficult to collect evidence regarding unknown transactions, unknown service providers, unknown consumers or unknown products.
- It is hard to distinguish between the high and the low quality of services on the Internet.
- It is possible to allow people to assess a much wider range of cues related to trustworthiness in the physical world than is currently possible through the Internet.
- Short-cut communication for which the information provided is not comprehensive and sometimes incomplete, for example, insufficient information about the service providers, goods or products as well as published services; and online users and consumers often have to take 'risks' that can leave them in a vulnerable position.
- An attractive website offers little evidence about the solidity of the organization behind it.
- The online user or consumer has no opportunity to see and try out the products, or 'squeeze the tomatoes before you buy'.
- The provider or seller, on the other hand, may not know whether he or she will get paid properly or on time by the customers or consumers.
- Trustworthiness systems have their own way of carrying out data entry or enquiry and can have different representations, interaction styles and trust rating scales. Some use five stars, some use four stars, some use a scale of $1-10$, some use percentages, and some use unbounded integers;
- The fewer the number of people participating in a trustworthiness rating system, the more inadequate the opinions or recommendations provided by the system.
- The effectiveness of any recommender system is dependent on many factors, not just the quality of the algorithm [5].
- The heterogeneity in the distributed network.
- Catering for anonymous, pseudo-anonymous and non-anonymous agents interaction in the public and open environment.

1.5 Agents in Service-oriented Environments

1.5.1 Agents in Service-oriented Environments

In *service-oriented environments*, we refer to the interacting parties as *agents*. An *agent* in a service-oriented environment is an *intelligent autonomous entity,* capable of making decisions. It

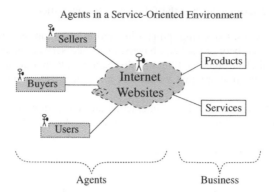

Figure 1.6 Agents carry out business (sell goods or provide services) in a service-oriented network environment

can be a buyer, a seller, a supplier, a merchant, a website, a service provider or an interacting peer, a software construct, or an intelligent server running behind the scenes on the network.

An agent on the network can be the trusting agent or trusted agent. For example, if a website is a virtual bookstore, when a customer (the trusting agent) ranks the trustworthiness of the bookstore, the bookstore is a *trusted agent*. However, the bookstore may have thousands of book suppliers, and the bookstore (trusting agent) can rank its supplier (trusted agent). The bookstore can also rank its customers; therefore, the customer becomes the *trusted recommending peer*. However, the context for which they measure or predict the trustworthiness is different in each of the above situations. Note that a particular agent can be both a trusted agent and a trusting agent in different contexts.

Figure 1.6 depicts the agents in a service-oriented environment.

1.5.2 Sellers

Sellers in the service-oriented environment are the agents making the sales. They could be *businessmen, brokers, dealers, merchants, retailers, auctioneers, online stores* or *shops, and salesmen or tradesmen*.

Note that *manufacturers* or *service providers*, and so on, can also be *sellers* if they want to sell their products or services and do not use intermediary agents. Many manufacturers do sell their products directly to the public. The most important thing here is that, over the Internet, we are primarily concerned with agents who interact with customers. From the customer or online user's point of view, the most important party is the one that is involved in the direct interaction with the customer. For example, if we want to buy a pair of running shoes, we are concerned more with the shoes and from whom we bought them, rather than where the factory that produced them might be. We seldom interact with the factory for most goods that we purchase.

1.5.3 Buyers

Buyers in the service-oriented environments are *purchasers, clients, consumers, customers* or *shoppers*. Buyers can also be business entities, such as in B2B e-commerce. The buyer is a very important entity in a service-oriented environment, because the buyer raises many of the issues that create business, infrastructure and technological needs that we are dealing with today such as security and trust and protecting consumer rights and privacy.

1.5.4 Users

Users in service-oriented environments are information *seekers, browsers, information requesters, game players, researchers, patrons* or *shoppers*. They could also be customers, buyers, sellers or providers. We separate users from buyers or sellers because the population of online users is much larger than buyers or sellers.

1.5.5 Websites

Finally, we define *websites* as agents. Websites are normally *service providers*, such as *business brokers, merchants, retailers, shopping centres* or *information databases* or *repositories*. They may provide a single service or multiple services. They, themselves, may have suppliers who provide goods for sale or they may be manufacturers who sell their own products, such as computers, or service providers who sell services, such as consulting services for e-learning.

This is because a *website* in a service-oriented environment is an intelligent entity on the Internet with a unique URL address. By incorporating and utilizing software technology and hardware infrastructure, it can carry out communication with other agents and achieve business objectives such as enabling sales and service deliveries.

Before concluding this section, we would like to summarize that an agent over the service-oriented network environment can be a trusting agent or a trusted agent, and an agent can measure or predict the trustworthiness of another agent or agents.

1.6 Business in a Service-oriented Environment

In a service-oriented environment, *business* refers to business activity and includes all the management processes and workflows that enable a company or resource provider to sell or deliver the products and services through the use of new technologies and infrastructure to maximize consumer confidence and business value.

From Figure 1.6, we note that there is a high-level decomposition of two types of activities in a service-oriented environment, namely, *product sales* and *service delivery*. These are discussed in the following text

1.6.1 Products

Products in a service-oriented environment are *goods* or *finished products* that are sold to or are for sale to consumers. A product could be any software or hardware, even an information kit, results, data sets, documents, experiment output and material products such as DVDs, cars or bags. A product could be purely information such as weather information or drug information.

We also define that some *information* retrieved from service-oriented networks are also products, regardless of whether they are free or not, because results, reports, documents or information are produced and may be the products of research or development, or an information provider's products. They may sell the information as a form of product. For example, IEEE has one of the world's largest scientific article collections, and they sell scientific papers as products.

The *quality of a product* can be evaluated or rated by consumers who use the product or customers who bought the product. They give the *quality of a product* rating based on their opinion(Figure 1.7).

1.6.2 Services

Services in a service-oriented environment are jobs, duties, tasks or activities that a business or a service provider offers to customers or consumers, such as logistics services (they have trucks or trains that can deliver goods or products for you; however, they do not own or produce the

Customer/User   Product

Figure 1.7   Customer or user gives an opinion on the quality of the product

goods), warehouses (such as space rental, refrigerated rooms or air-conditioned space), access to information databases, and so on.

From Figure 1.5, we see that services include (but are *not limited* to) the following: trading services, search engines, e-auctions, e-education, e-transactions, e-marketing, e-advertising, e-publishing, e-governance, e-manufacturing, hotels and travel, teleconferencing, consulting, e-supply chain services, automation and control, databases and information repositories, information services (real estate, gas, oil, mining, medical and health data), monitoring and track and trace services, remote and mobile communications, logistics (transport hire, local deliveries), warehouses (space, storage, goods distributions), virtual community services, virtual collaboration, and virtual markets.

From the above list, you may notice from the characteristics of services that they are not products; the service may or may not be free, and there may be a payment for using it; and service items, such as information services, may or may not be free.

1.6.3 Quality of Goods

In most countries, the consumer has to pay *Goods and Services* Tax on each transaction. This shows that sales of *goods and services* are key business activities. When you buy goods or services online, it is the same as if you buy them offline.

With the most popular products, their properties are easily ascertained even without extensive measurement. With some it may be necessary to use sophisticated measurements. However, with some others, such as software or buildings with hidden defects, it may be more ambiguous and require expert arbitration. In general, the room for the 'interpretation' of the quality of goods is much smaller.

Millions of product categories exist. Similarly, there are millions of types of services available. Nowadays, you can find or purchase most of these *goods and services* online.

1.6.4 Quality of Service

Quality of Service (QoS), in a service-oriented network environment, is defined as the *fulfilment* of the *service agreement* or *mutually agreed service.*

In a service contract or agreement, a service is defined by its context or functions, coupled with the terms and conditions, and is normally set by the agreement between the service requester and the service provider. In other words, a service agreement describes a mutually agreed service, and states that both the customer and the service provider have agreed upon all the terms and conditions. QoS can then be measured against the fulfilment of the mutually agreed service as specified in the service agreement. A service in the service agreement is clearly defined so as to have a clear context or functions and a set of terms and conditions that are tailored to the customer's requirements.

If there is no *service context* and there are no *terms and conditions* (or *requirements and constraints),* there would be no protection for either party. An *agreement* implies that the service context and *terms and conditions* are specified, understood and agreed to by both parties and especially that the customer has understood before signing the service agreement. It implies a mutually agreed service that contains obligations for both the parties, such as that the trusting agent has to pay the agreed amount within a given time and that the service provider has to deliver the service according to what has been agreed upon with the customer. Its purpose is to provide protection to both the parties from fraud or cheating in a service interaction, and in many situations, though the purpose of the service agreement appears as protection for the customer it fundamentally serves to protect the service provider. Consider the cases of, 'bank loan', 'house insurance', 'mobile phone service', 'car rental agreement', 'employee contract' or 'bugs in a SAP software module', and so on. Frequently, the customers may not get what they have understood or expected because of the hidden meaning or hidden costs contained in the terms and conditions, colloquially referred to as *fine print* in the agreement.

Customer   Service   Provider

Figure 1.8 *Quality* of service is measured by considering the customer's input as well as input from the service provider with respect to the level of service in the service agreement

1.6.5 Quality of Goods Measure

An *important* point to note, however, is that the quality measure of *goods and services* is quite different. Measurement of a goods' quality is generally much easier than measurement of the QoS. This is because when we give an incorrect measure or comment about the goods or a product, it is much easier to verify this against the specification. However, when we measure the quality of a service or give a rating to the quality of a service, because a *service* involves another intelligent entity, we cannot just accept one side of the judgement. Therefore, one side of the story may not count as much until both sides of the story are heard. This creates a *big challenge* in measuring or quantifying the *QoS* over the service-oriented environment.

1.6.6 Quality of Service Measure

A *service* always involves two parties: the service provider(s) and the service consumer(s). Therefore, it is always accompanied by a service agreement between the two parties. A service agreement may be very simple or very complex; this is dependent on the size of the job. *A service agreement* contains, among other things, *terms and conditions*. It contains both parties' responsibilities, for example, that the service customer should pay the right amount at the right time and that the service provider should deliver the QoS according to terms agreed.

QoS is determined in relation to the fulfilment of the service agreement. Normally, when a service contract is signed, it implies that both the parties understand the service agreement and mutually agree with all the terms and conditions in it. By default, the *service provider* should deliver the mutually agreed service to its *service customer*. The measurement of the QoS needs to take into consideration input from both the customer as well as the service provider, as the service agreement binds both the parties (Figure 1.8).

The judgement on delivery of QoSs and fulfilment of a service agreement is a much more complex issue. This is particularly true when the service agreement has complex terms and conditions that could sometimes even provide for a lower service level under circumstances beyond the service provider's control. An example of this could be a fixed delivery time of, say, two days for the delivery of the goods by a transporter. This could be softened a little in the contract in the event of inclement weather. There are also important issues related to the service receiver's expectations in relation to the level of service in the agreement and the service provider's interpretation of the service level.

1.7 Infrastructure in Service-oriented Environments

In service-oriented environments, the *infrastructure* mainly refers to the *Internet* and associated infrastructure. In the early days of e-commerce, business and transactions were mostly carried out over client–server environments. Since then, many other forms of network infrastructure have come into use. Even though not all the infrastructures are totally mature, they are regarded as the next generation of e-commerce or e-business platforms. The network infrastructures that can be used in service-oriented environments are as follows:

- Client–server network
- Peer-to-peer network

- Grid service network
- Ad hoc network
- Mobile network.

Any one of these network infrastructures, when provided within a service-oriented environment, permits e-Business (total e-solution), e-commerce (online transactions) and e-service to be carried out.

1.7.1 Client–Server Network

In the client–server network, one computer acts as the server and others act as clients. Thus, in the client–server network, there is only one server and the roles of communicating parties (either clients or server) are clearly defined from the beginning. For example, a client cannot be changed to a server for the same transactions once the infrastructure is set up.

1.7.2 Peer-to-Peer Network (P2P Network)

The Peer-to-Peer (P2P) network is totally different from the traditional client–server network. Each computer has the same roles and functions [6]. A P2P network distributes information among the nodes directly instead of interacting with a single server [7]. Moreover, a P2P network platform is independent. In other words, it supports heterogeneous systems [8]. Each node has its own repository for distribution to other nodes. There is no central repository in a P2P network as information is automatically spread in the network [6]. The number of nodes in a P2P network is dynamic. Nodes can enter and leave the network at all times. Napster, Gnutella, Kazaa and Freenet are among the most popular P2P applications [9]. To access the P2P network, a node can be anonymous (no user name or ID can be identified), non-anonymous (user name and ID can be identified), or pseudonymous (users use nickname(s) instead of real names). For an anonymous network, the identity of the node is unknown [10], while in a non-anonymous network, the identity of the node is known and the name of the node can be linked to a physical identity [10]. However, in most cases, nodes are pseudonymous. Nodes provide their identities at the beginning, and they use pseudonyms to connect with other nodes. As anonymity can lead to the breakdown of intellectual property [10], the anonymous network is not so common. Among the four most popular applications, as previously mentioned, *Freenet* provides anonymity in accessing the network [9].

P2P networks (such as [11]) are regarded as the next generation of the service-oriented networks. As a result, it changed the whole scenario from a centralized environment to a distributed environment. The main difference between P2P networks and the client–server environment is that P2P networks transfer the control from the servers back to the clients. In the P2P network, the users or business providers can carry out interaction in one of three forms:

- Anonymous (no name is to be identified in the communication)
- Pseudo-anonymous (nickname is used in the communication)
- Non-anonymous (real name is used in the communication).

1.7.3 Grid Network

The basic idea of a grid network is to assemble the existing components and information resources in order to be able to share them among the users [12]. The grid network provides the resource-sharing paradigm for clients. In particular, in the grid network, there is a collection of servers and clients working together [13]. Each node is autonomous. There is no central management. A grid network is similar in a few respects to P2P in that they both provide the sharing of resources and components among the nodes in the network. Unlike P2P, in the grid network, each node has a distinct role: either as a server or as a client. Both P2P and grid networks support heterogeneous systems. However, the heterogeneous nature of the resources is to some extent a distinct barrier for a grid network [14]. Even though, the grid network supports heterogeneous systems, to integrate enormous numbers of heterogeneous components and resources is expensive and with the current available technology this poses difficulties.

1.7.4 Ad hoc Network

An ad hoc network is a Local Area Network (LAN) or small network, where the connection is temporary. The communicating parties are in the network only for the period duration of a specific communication session [30]. An example of an ad hoc network is communication via Infrared transmission with mobile telephones.

1.7.5 Mobile Network

Nowadays, mobile devices play an important role in everyday life. In recent times, the number of mobile phones and wireless Internet users has increased very greatly. A mobile network is a kind of ad hoc network. Mobile networks provide users access to whatever they want without being tied to a fixed location PC, as they change their geographical location, using compact devices such as PDAs, smart phones and Internet appliances [15].

The *first generation mobile networks only provided* voice and data communication at low data rates while *second generation* (2G), digital multiple access technology provided enhanced features including paging and fax services [16]. For sending text messages over the mobile device, with Short Message Service (SMS) [17], 2G technology is used. The General Packet Radio Service (GPRS), a 2.5G technology, supports flexible data transmission rates, and is a radio technology for Global System for Mobile communications (GSM) [16]. *The third generation (3G)* technology seeks to connect users anytime, anywhere at high transmission rates and at low cost [18]. 3G technology is based on radio transmission. However, it is still developing towards the next generation in order to offer such superior services.

1.8 Technology in Service-oriented Environments

1.8.1 Service-oriented Architecture (SOA)

A typical Service-oriented Architecture (SOA) consists of the interactions between three roles, as shown in Figure 1.9, namely, service providers, service requesters and service brokers. It also involves three distinct activities, namely, (i) *description and publishing*, (ii) *finding and discovery* and (iii) *binding*. We explain these terms in more detail in the following text.

Service provider: The major task of the service provider is to have their services deliverable, *described* and *published*. Such publishing could be by registering with the service broker or making them available to service requesters directly [19].

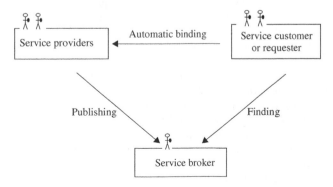

Figure 1.9 Service-oriented architecture

Service requester: Service customers or requesters may be either machines or humans. The service requesters *find* or *discover* the desired services from the service brokers or directly from service providers. Then they request those services, by sending messages, to either service brokers or the service providers directly. Once the service provider has been located, the service customers or requesters then *bind* to those services directly with the service provider.

Service broker: Service brokers provide central repositories or directories for *services* published by service providers. Basically, there are two types of service brokers: public brokers and private brokers. Public brokers are publicly available through the Internet or for open community, while private brokers are limited in access for only authorized groups or a closed community. A service broker has to be centralized in this regard, but service binding is distributed.

As mentioned above, *description and publishing, finding* and *discovery* and *binding* are the underlying principles of the SOA framework.

Description and publishing: Service providers have to provide a precise description of the services and the mechanisms by which they are accessed and publish their services with the broker or central registry.

Finding and discovery: A service requester or customer looking for a particular service does so by putting out a request for the service providing a precise description of the service. This could be responded to by the broker or central registry or alternatively by the service provider directly.

Binding: Calling or invoking a service such as *GetLastUKTradePrice*. This invocation will be transmitted over the network and bring the answer or delivery of the service to the customer or service requester.

1.8.2 Web Service

Traditionally, e-business is carried out through the use of a *Web application* to handle the communication between the provider and the consumer. Users need to provide some input to the Web application in order to get the result from the provider. There is no problem if one is working with only one provider. However, consumer requirements have become more and more complex and dynamic. Dynamic e-business has been emerging as a new phase of e-business development [20]. Essentially, dynamic e-business requires an infrastructure to integrate several providers in order to supply such a superior service [21]. In other words, we need a platform for automating both the provider and customer ends of the transaction [22].

In recent years, *Web services* have emerged with important underlying features, such as interoperability and loosely coupled models. Web services use a *service-oriented architecture*. Using a Web service, one can automatically invoke applications running in other businesses. Multiple applications communicate with each other regardless of the platforms on which they are. Therefore, by means of Web services, the company can effortlessly link its applications with those of its partners, customers and suppliers . However, Web services do not tend to replace the traditional infrastructures. Indeed they are complementary in order to offer the services. Thus, we clearly see that Web services are one of the key mechanisms that represent a significant advance in the e-business of the future [23].

The primary focus of this book is 'trust' and 'reputation', and not 'Web services'. However, there is a large amount of material available about Web services. Here, the authors would like to refer readers to a very easy to read and content-rich book on Web services [24], or to the W3 website at www.W3.org/TR/wsa-reqs, for more details on Web services.

Measuring or rating the quality of *goods and services* is important in modern service-oriented environments. It is reshaping the world of e-business by providing *trusted business* processes and the reputation of services and service providers. The next generation of the Internet must have some degree of control over business conduct in the Internet through measuring or rating goods and services and agents.

1.8.3 Web Service as Software Technology

Defining the term 'service' is the key to clarifying *Web services* in the technological sense.

Service in software or object-oriented systems is defined as a *method, process* or *communication* within an application or software. These 'services' or 'methods' are operations of the application that target some business need.

Web services, a specialized term, is defined as a middleware technology that offers standard communication interfaces that allow ease of communication between heterogeneous applications over the distributed network environment. Web services provide *inter-application* operability, inter-organization collaboration and business integration to achieve wide commercial objectives.

There are many technical definitions of Web services in the literature, not all of which are consistent with each other. A good working definition is provided by Web Service Activity Group of the W3C, which defined a Web service as '*a software application identified by a URI, whose interfaces and binding are capable of being defined, described and discovered as XML artefacts. A Web service supports direct interactions with other software agents using XML-based messages exchanged via Internet-based protocols*' [25]. There are several key features here, namely, that the application is open and capable of being defined, described, discovered and interacted with by external software agents; and that it uses standards-based interfaces that rely on XML messaging. Currently, these include XML, SOAP, WSDL and UDDI [29] standards. These are briefly described in the following text:

XML provides declarative semantics to the data through the use of tags for the data.

SOAP specifies the communication message format that defines a uniform way of passing XML-encoded data [26]. The SOAP messages support '*publish, bind* and *find*' operations for a services-oriented architecture. SOAP lets an agent or application invoke another agent or application by using an XML message over the Internet [27].

WSDL is known as *Web Services Description Language*. It is used for describing the services available in the broker or in the service registry (for service publishing and finding). It gives the location of a *Web service*, the functions and operations of the service and their data type information. From the *service provider's* perspective, they use WSDL to describe what services they can provide in order for other people to find their services and use them. From the *service requester's* point of view, they use WSDL to describe what service the requesting agent is looking for.

UDDI is known as *Universal Description, Discovery and Integration* [28] (for service publishing and finding). It provides three main and crucial functions:

1. It collects all the services available on the network to the *service broker* or *central registry*.
2. It then acts like *Yellow Pages* in that it clusters similar services under similar headings for the clusters and lists them individually.
3. Then, it provides publish/find functions for service providers and service requesters/customers.

Therefore, UDDI provides universal common description of services and allows search, query, discovery or the ability to locate the services and enables run-time or automated dynamic binding or integration between the service requester and service provider.

Services are carried out by sending messages and communication between the agents. A tailored Web service protocol is known as *Web Service Protocol Stack*, and this has four functional layers that offer the following (Figure 1.10):

1. Sending messages between agents or applications via HTTP. This layer is called the *Transport Layer*.
2. Encoding messages into XML format so that the messages can be understood by all agents in the *Web Services* environment. The commonly used protocol is Simple Object Access Protocol (SOAP) . The tailored name for this layer is called *XML messaging*.

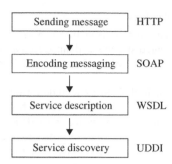

Figure 1.10   Web service protocol

3. Describing the services available in the service registry with WSDL in XML vocabulary. This layer is called the *Description Layer*.
4. Querying the broker or registry and obtaining service information, support interaction and negotiation between the agents and application and the facilitation of dynamic binding, through UDDI. This layer is called the *Discovery Layer*.

The above use of the term 'service' appears to be borrowed from the business term. Fundamentally, the purpose of IT is for business, not for IT itself. In companies, we often have business units telling us 'We do not want technologies, we want business solutions'. This viewpoint should impact on how we perceive Web services. Should we stop at the technology level, or lift its potential to incorporate the business sense? If so, Web services and the SOA will have more influence in the business world.

Despite the popularity of the *SOA, Web services* have not been widely adopted in inter-organizational situations. We believe one factor is the limitation of its use in technology environments rather than in business environments.

It is important to note that *not* all the *Web services* need to use the Web. For example, in the intra-organization communication situation, most companies have heterogeneous software components within the organization that *do not* talk to each other, or *do not easily* talk to each other. By adopting Web service technology, old legacy systems are wrapped with a Web service interface (standard interfaces). These *heterogeneous* software components now become *homogeneous* software components. For example, all of them use XML as the communication medium. Intra-organization applications can now be seamlessly integrated. In this situation, the organization uses Web-service technology without the Web. It is also conceivable that not *all* services (business) on the Web are based on Web services (technology). Because of the specialized definition of Web services, we hope the technology will evolve and the two separate terms designating services (business and technology terms) will progressively coalesce because they both ultimately should serve business and commercial needs.

1.8.4 Web Service as a Business Solution

Since the introduction of the term *Web services*, we must begin to distinguish between two kinds of services that exist within the service-oriented environment. One is 'service' in a real business sense, and the other one is 'service' from a technology point of view. However, if we think about what technology is used for, it is not hard to integrate the two kinds of service perspectives into one, because the purpose of the technology is not for technology itself, but for business. Therefore, it is wise to understand the linkage between the technology term of 'Web services' and real-world business service.

Web services provide a state-of-the-art middleware service that enables different business applications from different business organizations across the Internet to work together to achieve seamless information exchange and business transactions over the distributed environment. It provides dynamic and automated application integration service via the following:

- Standard homogeneous interfaces
- Standard computing languages
- Standard communication protocols.

These allow an organizational application to invoke or to be invoked or to be discoverable and accessible across the Web to carry out business services.

Even though, the service broker is centralized (Figure 1.11(a and b)), the inter-company communication is distributed and works in a peer-to-peer communication fashion (Figure 1.11c). Figure 1.11 (a–c) show how Web services achieve a service-oriented architectural framework for business.

We can view Web services as business solutions because they utilize an SOA. Web services are loosely coupled, cost beneficial, provide ease of use, privacy and security protection, help in the creation of a trusted community, automation and intelligence as well as reshaping of IT application development methodologies.

Loosely coupled: Traditional middleware is centralized. Only one of the business organizations involved needs to install it. For Web services, only the broker is centralized, while the actual binding of services is distributed and there is no need for a fixed connection (Figures 1.11(a–c))

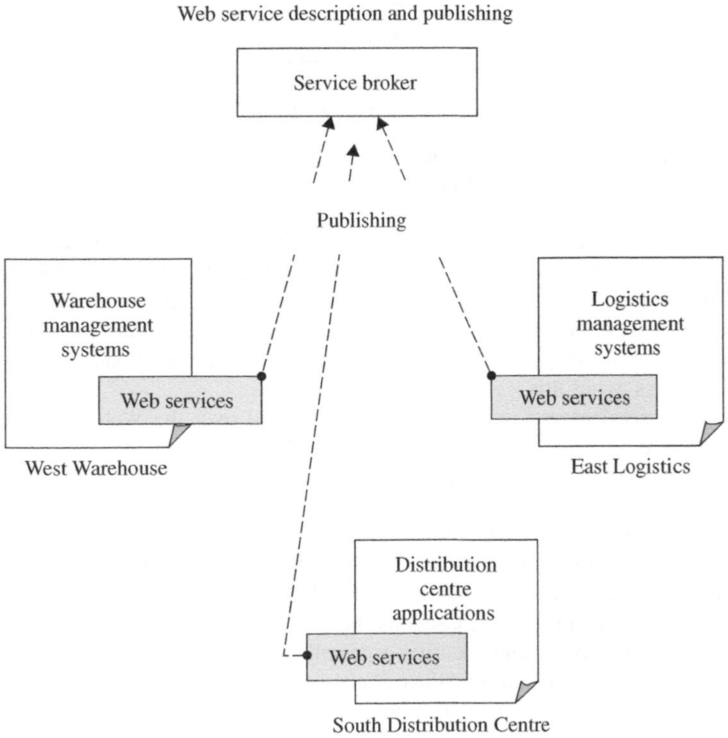

Figure 1.11(a)   Service providers can publish their *business services* through *Web service*s (a standard interface) to the service broker

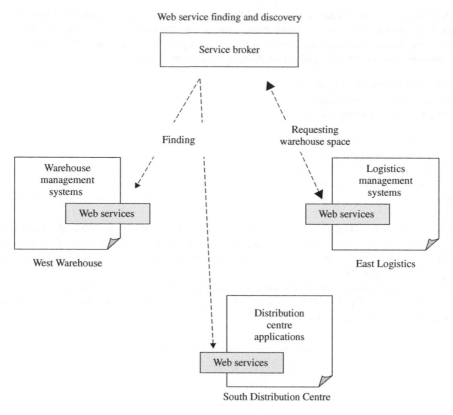

Figure 1.11(b)   Service requester (East Logistics) sends a *request* for special services to the broker and the broker helps *find* the suitable service providers

Cost beneficial: Historically, the distributed application integration is achieved by using different middleware or different application brokers. This is acceptable for a closed community, where few organizations are partners, and business is carried out only between them. However, since the Internet provides much wider business opportunities, one organization may have to use a number of different middleware to integrate with another organization's applications and the development is costly. Web services avoid this.

Ease of use: Web services are standardized and published over the Web and all organizations have an equal chance to learn and use them. This is different from the use of other kinds of middleware, as here you only need to learn one.

Privacy and security protection: Web services provide multiple standardized middleware interfaces, and every organization has one. When communication takes place there is no centralized control. Unlike traditional middleware (e.g., a B2B broker that is centralized and several organizations use one middleware), whoever has it will be able to tap into the middleware and monitor the system-wide communication and transactions (Figure 1.12(a) and b)). This creates a concern for organizations about their privacy and security, especially data security and their protection.

The creation of a trusted community: Since Web services are loosely coupled and there is no central control, it permits the use of the P2P communication model. There is no concern about where a middleware has to be located or who should have the middleware with the potential advantage that this might bring. It creates trust between companies because Web services are not centralized middleware.

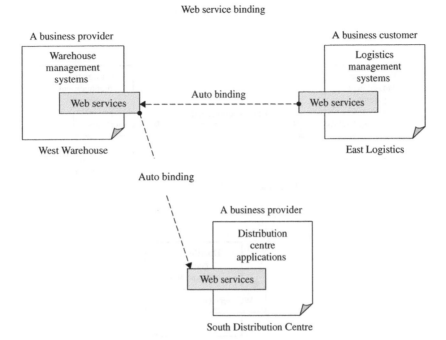

Figure 1.11(c)   Binding *business services* via *Web services*

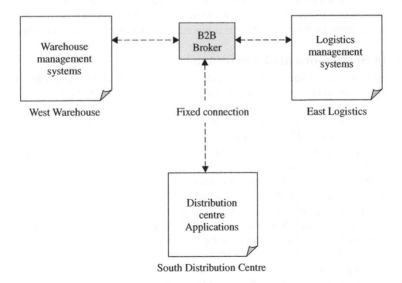

Figure 1.12(a)   Traditional middleware technology only one middleware, and location of it has to be decided

Web services middleware technology

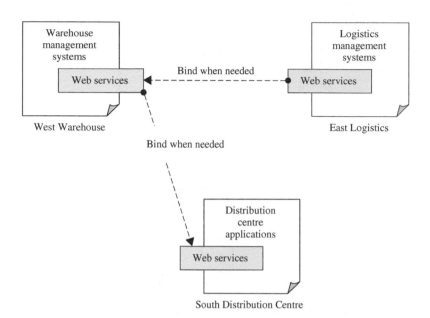

Figure 1.12(b)   Web service middleware technology with *multiple* homogeneous interfaces

Automation and intelligence: Each service is an autonomous agent and is able to respond to enquiries and accomplish the task automatically and independently.

Reshaping the IT application development methodologies: The advances in SOA are reshaping IT application development and design by considering everything as services, rather than as objects or entities or procedures or functions. This is a big step towards business and IT integration.

1.9 Trust in Service-oriented Environments

A Web services broker has a strong client–server flavour. The Web service requester and provider, on the other hand, can communicate in a P2P communication fashion. Alonso *et al.* [24] state that *'Peer-to-peer interactions are likely to be the accepted interoperability paradigm for business in the future. Web services are indeed a natural solution to the standardization problem. Web services enable a company to open its IT infrastructure to external partners. Web service technology is still evolving and the ultimate goal is to provide the infrastructure for business transactions, sales and services to take place over the Web'.*

The goal of Web services is to allow automated clients to browse the service registries, find adequate services and service producers, and automatically discover how to interact with the service, and finally invoke the service. All of this is done automatically without human intervention. Simple Object Access Protocol (SOAP), UDDI and Web Services Description Language (WSDL) form the basic infrastructure. Through continuing development and enhancements of Web service technology, it should become a reliable platform offering all the necessary functionalities for electronic commerce and e-business. Through a rich and structured descriptive language, the use of a business ontology (ontology is agreed and shared knowledge) and yet additional layers of software, it should be possible to entirely automate the discovery and interaction among Web services and business operations [24].

The service-oriented environment is an unstructured open community environment. The trend has resulted in a move from a small closed community involving direct interaction and static binding towards open indirect and dynamic interaction. In such an environment, trust and QoS are major concerns. Introduction of trust and trustworthiness technology including recommendation and rating systems provides some degree of control and QoS. These rating systems address the QoS and give buyers aggregated information about the provider, and provide some degree of reliability and assurance. Measuring trust, prediction of trust and rating and certifying services as well as sellers or providers facilitates the dynamic service selection process. The actual service is directly invoked by calling the selected service provider.

1.10 Chapter Summary

In this chapter, we discussed the following:

- Why trust is important to e-business
- The difference between trust and security
- What is a service-oriented network environment?
- Issues in service-oriented environments
- The components in service-oriented environments
- The technology in service-oriented environments
- The infrastructure in service-oriented environments
- What are agents, buyers, sellers, customers, users, website, products and services?
- What is Quality of Service (QoS) and quality of products?
- The service-oriented architecture
- Web service as a technology.

Unlike the traditional e-commerce platform, the new set of *trustworthiness technologies* and *systems* for service-oriented environments is reshaping business intelligence through building trust relationships with end-users and consumers, learning from customers and competitors and improving customer service and business performance through creating trust, reputation and consumer confidence. The new set of technologies provides a platform for sellers, service providers, websites, manufacturers, business partners, customers, buyers and end-users to learn from each other about their trustworthiness and reputation.

From a service provider's point of view, such an automated system would assist business in finding out what customers/consumers really want, finding both their likes and dislikes, and taking customer's/consumer's input as an opportunity to improve the relationships with them.

From a consumer's point of view, such an automated system provides for online sensations such as being able to 'squeeze the tomatoes before you buy' or enables them to seek other opinions before they undertake a deal or make a transaction.

The adoption of such a technology and systems becomes more and more important in today's e-business, as many large sites have already started utilizing a portion of the technology. It is such an important technology that every online business and service provider will have to take advantage of it to maintain his or her competitiveness. This is because it may provide the necessary technology to improve customer service, to boost consumer confidence and to help with a business's reputation. It may also enhance consumer learning and facilitate them to seek the best value online. In this chapter, we have provided an introduction to the service-oriented network environment and some of the issues it brings up in terms of trust, trustworthiness and reputation of agents, products and services. We also distinguish these ideas from the more familiar one of security.

In the next chapter, we introduce the basic concepts and model for trust.

References

[1] amazon.com. (2004) *Web Service*, Available: [http://www.amazon.com/gp/browse.html/103-6530699-8373443?node=3435361] (Aug 24, 2004).

[2] Hassler V. (2001) *Security Fundamentals for E-commerce*, Artech House.

[3] Algesheimer J., Cachin C., Camenisch J. & Karjoth G. (2000) *Cryptographic Security for Mobile Code*, IBM Research, Zurich, Switzerland.

[4] Carrol A., Juarez M., Polk J. & Leinger T. (2002) *'Microsoft Palladium': A Business Overview*, Microsoft whitepaper, June 2002. http://www.microsoft.com/PressPass/features/2002/jul02/0724palladiumwp.asp.

[5] Swearingen K. & Sinha R. (2002) 'Interaction design for recommender systems'. In *DIS 2002*, ACM Press.

[6] Tomoya K. & Shigeki Y. (2003) 'Application of P2P (Peer-to-Peer) Technology to Marketing', in *Proceeding of International Conference on Cyberworlds (CW 2003)*, Singapore pp. 1–9.

[7] Parameswaran M., Susarla A. & Whinston A.B. (2001) *P2P Networking: An Information-Sharing Alternative*, *http://crec.mccombs.utexas.edu/works/articles*.

[8] Schneider J. (2001) *Convergence of Peer and Web Services*, Available: [http://www.openp2p.com/lpt/a/1047] (Oct 6, 2004).

[9] Tsaparas P. (2004) *P2P Search*, Available: [www.cs.unibo.it/biss2004/slides/tsaparas-myP2P.pdf] (3/10/,2004).

[10] Foster I., Iamnitchi A., (2003) On death, taxes, and the convergence of Peer-to-Peer and grid computing. *Proceedings of 2nd International Workshop on Peer-to-Peer Systems (IPTPS'03)* Springer-Verlag LNCS 2735, pp. 118–128.

[11] Napster. (2004) http://www.napster.com/ntg.html.

[12] Roure D., (2003) Semantic Grid and Pervasive Computing *GGF9 Semantic. Grid Workshop*, Chicago, pp. 70–76.

[13] Berman F., Fox G. & Hey T. (2003) *Grid Computing – making the global infrastructure a reality, (eds book)*, Wiley, pp. 1000

[14] Gannon D., Ananthakrishnan R., Krishnan S., Govindaraju M., Ramakrishnan L. & Slominski A. (2003) Grid web services and application factories in Berman F., Hey A. & Fox G. (eds.) *Grid Computing – Making the Global Infrastructure a Reality*, John Wiley & Sons, pp. 251–264.

[15] Weisman C. (2002) *The Essential Guide to RF and Wireless*, 2nd ed, Prentice Hall PTR.

[16] Toh C.K. (2001) *Ad Hoc Mobile Wireless Networks: Protocols and Systems*, Prentice Hall PTR.

[17] ITU. (2003), http://www.itu.int/home/.

[18] Alcatel, (2004), *Mobile Network Evolution: From 3G Onwards'*, http://www.bitpipe.com/detail/RES/1074104543_898.html.

[19] Roy J. & Ramanujan A. (2001) Understanding web services, *IT Professional* vol. 3, no. 6, pp. 69–73.

[20] Xiao Feng J., Junhua X., Hua Z. & Zuzhao L. (2003) A realizable intelligent agent model applied in dynamic e-business. *Proceedings of the 2003 IEEE International Conference on Information Reuse and Integration, IRI 2003*, Las Vegas, NV, USA http://ieeexplore.ieee.org/xpl.

[21] Huy H.P., Kawamura T. & Hasegawa T. (2004) Web services gateway – a step forward to e-business, in *Proceedings of the IEEE International Conference on Web Services (ICWS'04)* San Diego, California, USA.

[22] Aissi S., Malu P., Srinivasan K (2002) E-Business process modeling: the next big step, *IEEE Comput* **35**(5), pp. 55–62

[23] Cruz S.M.S., Campos M.L.M., Pires P.F. & Campos L.M. (2004) Monitoring e-business web services usage through a log based architecture, in *Proceedings of the 2004 IEEE International Conference on Web Services (ICWS'04)*. San Diego, July 6–9, pp. 61–69.

[24] Alonso G., Casati F., Kuno H. & Machiraju V. (2004) *Web Services: Concepts, Architectures and Applications*, Springer, Berlin.

[25] W3C. Web Services Architecture Requirements Oct 2002, http://www.w3.org/TR/wsa-reqs.

[26] Cerami E. (2002) *Web Services Essentials*, O'Reilly & Associates, Inc, Sebastopol.

[27] Hess D. (2002) *Simple Object Access Protocol (SOAP) and Web Services: An Introduction*, *Gartner*.

[28] Universal Description, Discovery and Integration. *(UDDI)* 2000 (2004) Available: [http://publib.boulder.ibm.com/infocenter/wsphelp/index.jsp?topic=/com.ibm.etools.webservice.consumption.doc/concepts/cuddi.htm] (Sep 10, 2004).

[29] Alston J., Hess D. & Ruggieo R., (2002) *Universal Description, Discovery, and Integration (UDDI) 2000*, Gartner.

[30] *What is ad-hoc network*. (2003) Available: [http://whatis.techtarget.com/definition/0,289893,sid9_gci213462,00.html] (7/10/2004).

2

Trust Concepts and Trust Model

2.1 Introduction

The concept of trust can be found in a number of different fields including sociology, business and computing. This chapter introduces the primary concepts associated with trust in the networked environment. In the networked environment, the communicating parties are called *communicating agents*. The agent can be a user, a computer or a software agent. Trust in the agent environment is belief; it involves a trusting agent and a trusted agent, and it is dependent on the time and context. The association formed between a trusting agent and a trusted agent is called a *trust relationship*. Each relationship has a trust value associated with it. Initiation of the relationship refers to the type of initial introduction that results in the trust relationship. In this chapter, we introduce the basic set of trust concepts, and these concepts are essential to help understand the trust model to be introduced at the end of this chapter, the trustworthiness to be introduced in Chapter 3 and the trust ontology for service-oriented environments to be introduced in Chapter 4.

2.2 Trust Environments

One can distinguish between two environments in which trust can exist:

- The physical trust environment
- The virtual trust environment.

2.2.1 The Physical Trust Environment

The *physical trust environment* is one in which trust is often established between two parties who know each other through personal interaction. In such environments, we often see that the parties get involved in a trust relationship through face-to-face interactions or by introductions or recommendations.

2.2.2 The Virtual Trust Environment

In a *virtual trust environment*, a trust relationship is established between two parties who normally have never met or may never meet and where communication takes place through a virtual interaction medium. They normally do not know each other on a personal level or by introduction or recommendation but there is a trust relationship established between them. An example of this is when you buy a book from Amazon using your credit card. This environment is termed a *virtual trust environment*. In a virtual environment, the communicating parties are referred to as *agents*.

Trust and Reputation for Service-Oriented Environments   Elizabeth Chang, Tharam Dillon and Farookh Hussain
© 2006 John Wiley & Sons, Ltd

Client–server or peer-to-peer (P2P) networks allow individual users, hosts (agents) or clients to carry out transactions, share files or distribute information over the Internet. These networks are free and open. This is both a strength and a weakness. The strength lies in the wide range of business transactions that these networks enable. The weakness lies in the fact that there are opportunities for malicious agents to cheat, disrupt and even attack the harmony of the networked world. For example, let us consider a situation where a quotation for the delivery of goods over the Internet was $2000. However, when the goods arrived, the price went up to $6000. In order to prevent such a situation, each user or agent establishes a trust protocol and validates each transaction or business behaviour, identifying the context of the transaction and automatically keeping a record of it. This record may be used by the trusting agent for his future interactions with the trusted agent or for making a recommendation within the organization or to others. However, as the network is so open and *ad hoc*, establishing trust is not an easy task.

2.3 Trust Definitions in Literature

Many of us use the word *TRUST* in our daily lives in different contexts and at different times. It is generally understood to mean an assured reliance on the character, ability or truth of someone or something. Additionally, sources from sociology, psychology, business and law and computing have their own definitions of trust.

2.3.1 Trust Definition in Sociology, Psychology, Business and Law

From a psychological point of view, *trust* could be understood as when a person has faith in the trustworthiness of another person. From a social perspective, *trust* is defined as a firm reliance on the integrity, ability or character of a person or thing. In the legal system, an example of *trust* would be a legal title to property (real estate, money etc.) held by one party for the benefit of another (the trustee, the beneficiary of the property). Other examples of trust include the willingness to make an investment or participate in a pension scheme. In a commercial situation, *trust* between business partners could result in less competition and lower prices.

Trust has also been categorized into different types by various people. Hartman [1] has classified trust into three types.

- *Blue trust* (or competence trust) is placed in individuals or entities that the trustee feels are competent enough for a given job. It answers the question: 'Can you do this job?'
- *Yellow trust* (or Integrity trust) is placed in individuals or entities that the trustee feels will constantly look after his/her interests. Yellow trust answers the question: 'Will you constantly look after my interests?'
- *Red trust* (or Intuitive trust) answers the question: 'Does this feel right?'

Ratnasingham [2] classifies trust into four types, namely:

- *Deterrence-based trust*: relates to a threat of punishment and is seen as a negative factor
- *Calculus-based trust*: a positive trust from a relationship because of fulfiling the required actions
- *Knowledge-based trust*: linked to knowledge of the other trading partner (the trustee)
- *Identification-based trust*: linked via empathy and common values with the other trading partner's desires and intentions to the point where one trading partner is able to act on, or as an agent for, the other with the evolution of time.

The above classifications do not quantify trust into degrees or provide a method for the validation of the degree of trust assigned in a trust relationship.

2.3.2 Trust Definition in Computing

In computing literature, Marsh [3] was the first person to introduce the concept of *trust* in distributed artificial intelligence. Marsh [3], Rahman [4, 5] and several other researchers in the area of computing use the definition given by Gambetta [6], who defines trust as follows:

> '... trust (or, symmetrically, distrust) is a particular level of the subjective probability with which an Agent will perform a particular action, both before [we] can monitor such action (or independently of his capacity of ever to be able to monitor it) and in a context in which it affects [our] own actions'.

The above definition classifies trust as a probability. However, one may note that trust is also a belief or confidence and sometimes we do explicitly know what we trust, in a particular context and at a particular time. For example, 'I trust my mother to manage my bank account', 'I trust her' or 'I trust that we will have a nice holiday' could be explicit, context dependent and time dependent.

Wang [7] defines *trust* as '... *an Agent's belief in another Agent's capabilities, honesty and reliability based on its own direct experiences'*.

The above definition ties trust to direct interaction only. However, it must be noted that we can have the concept of trust even without direct experiences. For example, we may trust a bank and invest our money with them because we have been told either by our parents or by friends that this bank can be trusted with our money. In fact, we often invest with or borrow money from banks, credit unions or building societies with which we have no direct experience.

Note that most of the above definitions concentrate on the action or behavioural aspects of trust while some cover the context-dependent nature of trust [8, 9]. However, there are many other aspects related to the concept of trust. These include the dynamic nature of trust (the value of trust changes as time passes [10]) and the anticipated behaviour of the trusted party that would influence the trust. There are also psychological factors for the trusting party and the trusted party as well as the agent's calibre (knowledge, capability and professional qualities) that need to be taken into consideration. Another important factor that affects trust is the *association type* (how the relationship is formed) of the trust relationship that needs to be determined.

2.4 Advanced Trust Concepts

In this section, we introduce an improved definition of *trust* that can be used in the agent communication.

2.4.1 Trust

Definition: Trust is defined as *the belief* the *trusting agent* has in the *trusted agent's willingness* and *capability to deliver a mutually agreed service* in a given *context* and in a given *time slot*.

The terms 'belief', 'trusting agent' and 'trusted agent', 'willingness', 'capability', 'delivery', 'mutually agreed service', 'context', and 'time slot' are essential when defining trust. These terms are the building blocks of the trust concept. These terms are explained in the following text.

2.4.2 Trusting Agent and Trusted Agent

Trusting agent

We define a *Trusting agent* as an *entity* who has faith or belief in another entity in a given context and at a given time slot.

Figure 2.1 Trusting agent

We define a *Trusted agent* as an *entity* in whom faith or belief has been placed by another entity in a given context and at a given time slot.

Trusted agent

Figure 2.2   Trusted agent

In the following sub-sections, we explain the other seven terms that form the building blocks of the concept of trust.

2.4.3 Belief

Trust corresponds to the *belief/faith* the trusting agent has in the trusted agent.

The *belief/faith* the trusting agent has in the trusted agent communicates the trust the trusting agent has in the trusted agent. From Figure 2.3, we observe that Alice is of the opinion or belief that if she lends her car to Bob, Bob will not damage the car.

This belief that Alice has in Bob communicates the trust that Alice has in Bob. In other words, Alice *trusts* that Bob will not abuse her trust and will take care of her car. This situation could be true especially in relation to family or friends.

In the above example, Alice places her faith or trust in another entity, namely, Bob. Alice takes the role of the trusting agent. Since trust is placed in Bob, Bob plays the role of the trusted agent.

2.4.4 Context

The term *context* defines the nature of the service or service functions, and each *Context* has a name, a type and a functional specification, such as 'rent a car' or 'buy a book' or 'repair a bathroom'. Context can also be defined as *an object* or *an entity* or a *situation* or a *scenario*. This can be explained by following examples.

Example 1: **Context = 'rent a car'**

Context name	'Rent a car' or 'borrow a car'
Context type, also known as *service type*	'Car rental'
Functional specification	'Bob wants to rent a small, automatic car'

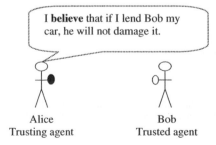

Figure 2.3   Alice *trusts* Bob with her car

Example 2: **Context = 'buy a book'**

Context name	'Buy a book' or 'order a book'
Context type, also known as *service type*	'Bookshop'
Functional specification	'Alice wants to buy a book on Trust'

Example 3: **Context = 'repair a bathroom'**

Context name	'Bathroom repairs'
Context type, also known as *service type*	'Handyman service'
Functional specification	'Alice looks for a tradesman to fix bath wall'

Example 4: **Context = 'digital camera'**

Context name	'Digital camera'
Context type, also known as *service type*	'Household product'
Functional specification	'Alice wants a camera with a long battery life'

We state that trust is context dependent because the belief the trusting agent has in the trusted agent, in a given context, will not necessarily be the same in another context.

Consider the two contexts in Figure 2.4. The trust that Alice has in Bob in the context of his borrowing her credit card may or may not be the same as the trust that Alice has in him in the context of his borrowing her car.

2.4.5 Willingness

The term *willingness* captures and symbolises the trusted agent's will to act or be in readiness to act honestly, truthfully, reliably and sincerely in delivering on the mutually agreed service (Figure 2.5).

If the trusting agent does not have belief in the trusted agent, then it signifies that the trusting agent believes that the trusted agent is not willing to deliver on the mutually agreed behaviour.

In contrast, if the trusting agent has belief in the trusted agent, then it signifies that the trusting agent believes that the trusted agent is willing to deliver on the mutually agreed service.

Figure 2.4   Alice's *trust* in Bob is *context* dependent

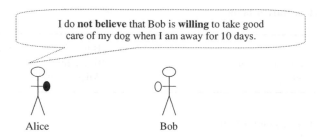

Figure 2.5   Alice does *not believe* that Bob is *willing* to do her the favour

The willingness of a trusted agent to deliver on the mutually agreed service is one of the two characteristics that the trusting agent can make a qualitative inference about from the actual behaviour of the trusted agent in its interaction. The other characteristic that the trusting agent can make a similar inference about is the *capability* of the trusted agent.

2.4.6 Capability

The term *capability* captures the skills, talent, competence, aptitude, and ability of the trusted agent in delivering on the mutually agreed behaviour (Figure 2.6). If the trusting agent has low trust in the trusted agent, it may signify that the trusting agent believes that the trusted agent does not have the capability to deliver on the mutually agreed service.

In contrast, if the trusting agent has a high level of trust, then it signifies that the trusting agent believes that the trusted agent has the capability to deliver on the mutually agreed behaviour.

2.4.7 Time Space, Time Slot and Time Spot

A *time spot* is defined as a particular time at which an entity interacted with another entity and subsequently assigned a trustworthiness value to it. A *time spot* refers to a specific time, such as 15/1/2005.

A *Time slot* is a period of time between two time spots, and is defined as the breadth or duration of time over which the trust value is collected.

The *time space* is defined as the total duration of time over which the behaviour of the trusted entity will be analysed and the trustworthiness measure and prediction will be carried out. Within a time space such as between 2000 and 2005, we can define many time spots, and between two time spots, there is a time slot. In each time slot, there is a unique trust value. Within a *time space*, there could be a trust value; however, this trust value is an average or aggregated trust value from all the time slots (Figure 2.7).

Figure 2.6   Alice *believes* that Bob has the *capability* to deliver the mutually agreed objectives

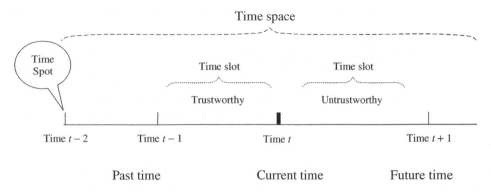

Figure 2.7   Time space, time slot and time spot for trust modelling

In order to analyse the dynamic behaviour of trust values [10], a time space consists of a number of non-overlapping time slots. An entity will have a trust value for each time slot. The trust value or the repute values (from recommendation) are aggregated and used for predicting future trust values, also known as a *trustworthiness value*.

Trust is dynamic and as such the amount of trust changes as time passes. This dynamic nature is due to the following three reasons:

1. The trusting agent can get a better idea of the capability and willingness of the trusted agent to deliver on the mutually agreed service in a given context by engaging in *further dealings* with the trusted agent.

   The trusting agent, upon *further interactions* and transactions with the trusted agent, might gain a better understanding of the capability and willingness of the trusted agent to deliver the expected service. As a result of getting a better idea of the trusted agent's willingness and capability, in a given context, than he had before, the trust or belief that the trusting agent has in the trusted agent in that context is altered (Figure 2.8).
2. The capability or the willingness of the trusted agent to deliver on the mutually agreed service or obligations in a given context *may vary over time*.

   It is possible that in any type of trust relationship, with the *passage of time*, the capability or the willingness of the trusted agent to deliver on the mutually agreed service in a given context may *increase, decrease* or *remain constant*. The reasons behind the change in the level of trust depend on both the parties and their relationship.

   In order to reflect this change in the willingness or capability of the trusted agent, the trust that the trusting agent has in the trusted agent will change correspondingly in the same direction.

Figure 2.8   Alice changes her *trust* in Bob in a given context upon *further interactions* with him

For example, if the willingness of the trusted agent increases, the level of trust that the trusting agent has in the trusted agent will increase.

3. Getting recommendations from other agents about the trusted agent in a given context may have an impact on the trust that the trusting agent has in that context. Upon querying other agents about the trusted agent, the trusting agent can get a better idea of the willingness and capability of the trusted agent to deliver on the mutually agreed behaviour in a given context. This may result in a change in the trust that the trusting agent has in the trusted agent.

The term *a given time slot* in the definition of trust captures the dynamic nature of the trust.

2.4.8 Delivery

The term *delivery* captures the actual service by the trusted agent or the actual behaviour of the trusted agent in the interaction. The trusted agent in the interaction may or may not deliver the mutually agreed goals or services. Actual service delivered by a trusted agent is also known as the *conduct* of a trusted agent in an interaction.

The trusting agent believes that the actual service delivered by the trusted agent in the interaction will match the mutually agreed service. The amount of trust that the trusting agent has in the trusted agent denotes the extent to which the trusting agent believes that the trusted agent will deliver on the *mutually agreed service.*

Let us consider the interaction between two supply chain companies [11] East Logistics (in China) and West Warehouse (in France). East Logistics will store some goods in West Warehouse only if the amount of trust placed by East Logistics (the trusting agent) in West Warehouse (the trusted agent) is high. This trust signifies that East Logistics believes that West Warehouse has the willingness and capability to deliver on the mutually agreed services (such as basic warehouse operations, goods handling, goods check in, goods check out and goods transfer operations), at the time under consideration.

2.4.9 Mutually Agreed Service

The term *mutually agreed service* implies the *commitment* of the trusted agent and states the expectations for the trusting agent. It is understood by the trusting agent that the trusted agent is committed to providing the agreed service and meeting all the *terms and conditions* set out in the agreement. It can also be interpreted that the trusted agent will deliver the *defined service* stated in the contract or service agreement.

> *Deliver mutually agreed service* implies the *fulfilment of mutually agreed obligations, or deliver the objectives or goals that are mutually agreed and understood by both parties.*

For example, in the case of a loan from a bank or mortgage broker, the service agreement is frequently drawn up by the service provider. In this agreement, the customer, the borrower (the trusting agent), trusts the bank and signs. The document could be so lengthy and full of legal jargon, that the customer often decides to sign it without reading because he or she trusts the bank. However, later on when he/she compares the terms of the loan with friends or colleagues who have the same-sized loan and he/she finds their monthly repayments are lower because his/her bank charges a monthly account keeping fee and repayment insurance fees, and so on. This might be written obscurely and appear in an insignificant place in the document. He/she is unhappy about this and decides to terminate the loaded agreement; however, the cost of doing this is much greater and the customer cannot just withdraw from the deal even if the agreement permits withdrawal at anytime.

Let us consider the example of the two supply chain companies again. East Logistics wants to store some of its consignment of goods in West Warehouse. It sends a request to West Warehouse asking for warehouse space of 6000 sq feet for a period of 2 months. West Warehouse quickly agrees to allow East Logistics to store the goods in its warehouse at a low cost and without many conditions. It is the *commitment* of the trusted agent, West Warehouse, which motivates the trusting agent, East Logistics, to place its trust in West Warehouse now and in the future.

The mutually agreed service is based on the following *conditions*:

— Warehouse space is available
— 6000 sq feet to rent out
— A discount rate with minimum conditions
— Allocate the space for 2 months from a specified date
— Space will be taken up within a week of agreement.

If the trusted agent fails to deliver on their commitment in a given context during its service delivery with the trusting agent in a given Time slot, then the trusting agent will not trust West Warehouse in a subsequent situation.

2.5 Trust Relationships

2.5.1 Trust Relationships and Trust Values

Trust is realized by the concept of a *trust relationship*. Without a relationship, trust has no meaning. However, a relationship is conditioned by the parties. Without the involvement of parties, there can be no trust relationship.

For the purpose of this discussion, we will define a trust relationship as a *bond* or *association* between a trusting agent and a trusted agent. Each relationship that the trusting agent has with the trusted agent is coupled with a numeric value that denotes the *strength* of the trust relationship in a particular context (Figure 2.9).

The above figure shows the realization of trust through a *trust relationship*. In each relationship, there must be a certain *degree of trust* that can be represented by a *trust value*.

2.5.2 Unidirection in Trust Relationship

The trust relationship between two agents is always *unidirectional*. If we assume the trusting agent is A and the trusted agent is B, the trust value assigned to the relationship is from Agent A to Agent B. The trust value is assigned in *one direction* from the trusting agent to the trusted agent on a scale of 0–5. This is due to the fact that, in a given context, for example, borrowing a credit card, the level of trust that Agent A has in Agent B may be different from the level of trust that Agent B has in Agent A. Let us suppose that Agent A (the trusting agent) assigns a *trustworthiness value* of 5 to Agent B, since he considers Agent B to be very trustworthy. From Agent B's point of view, however, he is the trusting agent and Agent A is the trusted agent. Agent B might assign a trustworthiness value of 3 to Agent A if he considers Agent A to be partially trustworthy. In the service-oriented business world, it is very important to recognize the fact that the trust measure is unidirectional from the trusting agent to the trusted agent.

Figure 2.9   Each trust relationship has a trust value

2.5.3 Multi-Context, Multi-Trust, Multi-Relationships

A trusting agent may have several trusted agents; this will result in multiple trust relationships. They may be in the same context or in different contexts. However, each individual relationship will result in an individual trust value.

At any given time slot, multiple trust relationships can exist between multiple agents (Figure 2.10) or between the same agents where an additional association exists between them (Figure 2.11). Also for the same *context*, multiple relationships may be formed between multiple agents (Figure 2.12).

In Figure 2.10, Alice is involved in three trust relationships. Alice (the trusting agent) trusts Bob (the trusted agent) to write a book. She trusts Budi (the trusted agent) to borrow her car and she trusts Sarah (the trusted agent) to build her house. Budi is involved in another trust relationship. Budi (the trusting agent) trusts Liz (the trusted agent) to borrow his credit card.

In Figure 2.11, Alice regards Bob as very honest and reliable and assigns a trustworthiness value of 5 to Bob in the context of 'lending her house'. Alice assigns a trustworthiness value of 2 to Bob in the context of 'lending her car'. This signifies that Alice does not regard Bob to be completely reliable with her car. Therefore, we can clearly infer the intensity of the trust relationship is context dependent.

Alice is involved in two relationships in the same context, such as a local delivery job, for example, delivering pizzas to her customers.

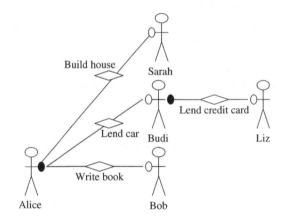

Figure 2.10   Multiple contexts, multiple trust relationships and multiple agents

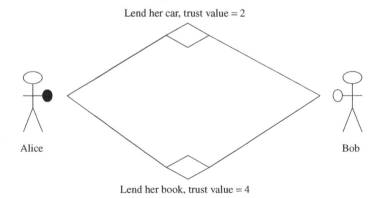

Figure 2.11   Multiple contexts and multiple trust relationships between the same agent

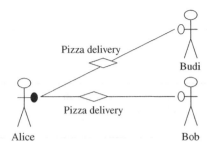

Figure 2.12   Multiple relationships with multiple agents in the same context

When we speak about trust in a network communication, it always implies *a trust relationship* and a specific context and a specific time slot, even though we understand that multiple trust relationships may exist between the same agents in *multiple* contexts and *multiple* relationships may exist between *multiple* agents in the same context. Therefore, when we talk about trust in a business situation, we have to qualify it with *a single* trust relationship between a trusting agent and a trusted agent.

2.6 Trust Relationship Diagram

A trust relationship diagram is a pictorial representation of the entities and attributes in a trust relationship. We will present a detailed discussion of this diagram in the following sub-sections.

2.6.1 Many-Many-to-One Trust Relationship

Figure 2.13 (a) shows that in a relationship between a trusting agent and a trusted agent there is always a *trust value* that expresses the strength of the relationship (or the degree of trust) from the trusting agent to the trusted agent. Figure 2.13 (b) illustrates the representation of trust concepts with cardinality.

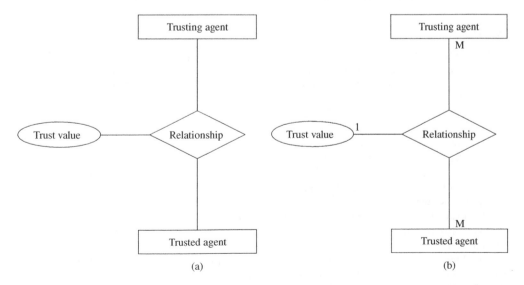

(a)   (b)

Figure 2.13   (a) Representation of the trust concepts, (b) Representation of the trust concepts with *cardinality*

Table 2.1 Trusting agent, trusted agent, trust value and M:M:1 relationship

#	Trusting agent	Trusted agent	Trust value on a user-defined scale of 0–5
Relationship 1	Alice	Sarah	2
Relationship 2	Alice	Bob	4
Relationship 3	Tom	Bob	3

> You cannot assign more than 1 trust value in 1 relationship. For example, it is not possible for Alice to give Bob two trust values 2 and 5 in a given context and in a particular timeslot. It is not possible for one to distrust and trust another at the same time, in the same context.

We know that each trusting agent can have many trusted agents. For instance, Alice may trust Sarah as well as Bob. Each trusted agent can have many trusting agents. Bob, for example, may be trusted by Alice as well as by Tom.

There is also a particular trust value associated with each relationship. This is illustrated in Table 2.1.

From the Table 2.1, we can observe the following:

(i) There is a M:M (Many-to-Many) relationship between the trusting agent and the trusted agent.
(ii) The trust value is unique in each of the trusted relationships.
(iii) For a given trusting agent and a given trusted agent engaged in a given trust relationship there can be *only one trust value*. Therefore, there is a M:M:1 (Many-to-Many-to-One) relationship between the trusting agent, the trusted agent and the trust value (Figure 2.13 (b)).

2.6.2 Contexts and Time slots

We know trust is both context and time dependent. For example, Alice trusts Bob in different contexts, such as 'writing an assignment', 'building a house', or 'managing an IT project'. Alice has a different level of trust (trust value) in Bob associated with each of these contexts. On the other hand, for each of the 'contexts', as time goes on, the level of trust may change. For example, Alice may trust Bob to manage an IT project now. After three months, Alice might not trust Bob as much, and after another three months, Alice might completely distrust Bob.

Therefore, we extend the diagram of the trust concept in Figure 2.14 (a) to show the *context* and *time* dependencies of a trust relationship.

A trust relationship is determined by a particular *time* and in a given *context*. Each of the relationships has to be associated with a trust value to reflect the level of the trust.

You may ask why *context* and *time slot* in the above figure are represented as circles and linked to a diamond (the diamond represents the relationship). This is because they are attributes of the *trust relationship*. These are important attributes that determine the meaning of the relationship.

Context and *time* are multi-valued attributes. This is illustrated with the help of the Table 2.2.

From Table 2.2, we see that Alice (the trusting agent) assigned Bob (the trusted agent) a trust value of 4 (on a user-defined scale of 0–5) in the context of lending her car to him in the Time slot 2002–2003. However, in a different context and in the same Time slot, Alice may assign Bob a

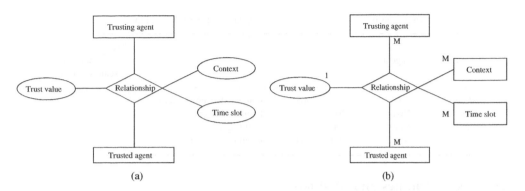

Figure 2.14 (a) Context and time dependence in a trust relationship, (b) Context and time dependence in a Trust relationship with cardinality

Table 2.2 Context and time slot as multi-valued attributes

#	Trusting agent	Trusted agent	Time slot	Context	Trust value on a user-defined scale of 0–5
R1	Alice	Bob	2002–2003	Lending her car	4
R2	Alice	Bob	2002–2003	Lending her credit card	1
R3	Alice	Bob	2003–2004	Lending her car	2

completely different trust value in the context of lending her credit card in the Time slot 2002–2003. Alice assigned Bob a trust value of 1. In the same context, 'lending her car', but in a different Time slot, Alice assigned Bob a trust value of 2.

This example shows that there are three relationships, R1, R2 and R3, and we see that the *Time slot* and *Context* are multi-valued attributes with regard to Alice and Bob's relationship, and the combined attributes of *trusting peer, trusted peer, time slot* and *context* are unique and the trust value is unique in each of the relationships R1, R2 and R3.

2.6.2.1 Entity Representation of Multi-valued Attributes

In section 2.6.1 and in the preceding text, we saw that Time slot and Context are multi-valued attributes. Instead of using circles to represent multi-valued attributes we can use rectangles. This is in line with 'Entity modelling' theory, where Ullman [12] states that '*a multi-valued attribute forms an Entity*' and an entity can be represented by a rectangle. In Figure 2.13 (b), Multi-valued Attributes 'Context' and 'Time slot' are represented in Entity notation or in other words, they are represented as boxes rather than as ovals.

Given a particular trusting agent and a particular trusted agent engaged in a trust relationship in a given context and in a given time slot, there can be *only one trust value* assigned in that relationship. This is illustrated in Table 2.3.

From Table 2.3, we see that given a particular trust relationship between a given trusting agent (e.g., Alice) and a given trusted agent (e.g., Tom), in a given Time slot (2001–2002) and in a specific context (lending her credit card), there can be *only one trust value* assigned by the trusting agent (Alice) to the trusted agent (Tom).

Table 2.3   There can be *only one trust value* assigned to a trust relationship

Trusting agent	Trusted agent	Time slot	Context	Trust value on a user-defined scale of 0–5
Alice	Tom	2001–2002	Lending her credit card	1
Tom	Budi	2001–2002	Lending his car	3
Liz	Bob	2002–2003	Building her house	5
Diana	Alice	2003–2004	Building her house	2

2.7 Trust Attributes and Methods

We can also treat the concepts of trust as objects in a trust relationship diagram. There are both *attributes* and *methods* associated with these concepts.

2.7.1 Trust Attributes

Each trust concept has a unique definition, attribute and value. The attributes and values of the concepts are illustrated in Figure 2.15(a).

2.7.2 Values for Trust Attributes

Example values for the attribute Context could be lending a car, lending a book, or any business process such as booking a warehouse space, or the local delivery of goods.

Example values for the attribute Time slot could be 2001, 2002–2003, or Aug 04 or Jan 2005 onwards.

A trust value could be any numerical value in a user-defined range, for example, a numerical value in the range 0–5, based on the context and time.

Example values for trusting or trusted agents are the name of an agent, name of an organization, an address, the nature of business, and so on.

2.7.3 Trust Methods

Each one of the trust concepts has associated methods. Methods carry out operations that determine the values of attributes for each concept. These methods are illustrated in Figure 2.15 (b).

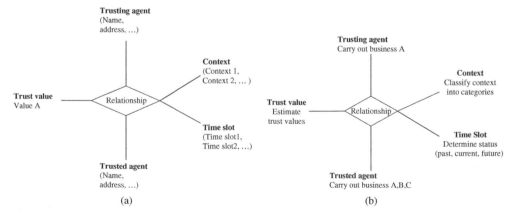

Figure 2.15   (a) Each trust concept has *attributes* and *values*, (b)Each trust concept has *methods*

An example of a method for context would be one that classifies the context into categories or knowledge areas so that the trust can be validated. An example of a method for Time slot would be one that determines the status of each Time slot within the whole time space, so that the trust value (whether it be a past, current or future trust value) can be determined. An example of a method for trust value would be one that calculates current Trust values or estimates future Trust values. An example of a method for trusting agent or trusted agent is one that carries out business or a friendship or a partnership. We have introduced methods and trust relationship diagrams since we need to utilize them in the rest of the book.

2.8 Initiation of the Relationship

In this section, we introduce the *initiation* concept in the context of a trust relationship. The initiation of the trust relationship results in the actual association or involvement of the trusting agent in an interaction with the trusted agent.

2.8.1 Initiation

We define the *initiation of the trust relationship* as a type of initial introduction that results in an association or a relationship and *provides methodologies* for *calculating or deriving* the trust value. There are *mainly three* different types of initiation as mentioned above and depicted in Figure 2.16:

1. Direct interaction
2. Reputation
3. History.

2.8.2 Direct Interaction

Initiation of the relationship by direct interaction: This is started by direct contact between the agents without any mediator or without the parties knowing each other upfront or from any

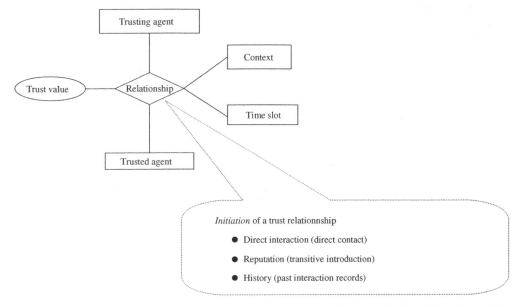

Figure 2.16 The trust relationship is complicated as you need to consider the *initiation* of the relationship

recommendation. The relationship generally begins from a mutual sense of requirement, for example, 'she became his new boss', or a random meeting, 'they met in the train', or direct agent to agent interaction, 'she was buying a footy ticket from him'. The *direct interaction* can result in trust and assigning of a trust value from the trusting agent to the trusted agent.

2.8.3 Recommendation

Initiation of the relationship by recommendation (also known as *introduction or obtaining reputation*): This relationship is begun by a third-party mediator who provides an introduction or recommendation. For example, Alice knows Liz through Budi. However, this relationship may not start with mutual trust. It depends on what Budi tells Alice about Liz and how much Alice believes Budi. This initiation method is known as *recommendation. Recommendation* is a method that helps to form a trust relationship by deriving the initial trust value, also known as *reputation* based on references or recommendations collected from other parties. The *aggregated* recommendation value is called the *reputation of the agent.*

2.8.4 History Review

Historical (or past knowledge) review or looking at past records may result in a new or renewed trust relationship. Historical data could be obtained from the trusting agent's own history repository (past personal interaction data). Therefore, historical review can be part of 'personal interaction'. Historical review only gives you trust values of the past. This relationship may or may not start from mutual respect if one party knows the history of the other party upfront. Over the e-service network, sometimes we interact with a service provider only a few times, and the more we interact with this service provider, the more precise our valuation of the trustworthiness of the service provider.

2.8.5 Initiation of the Relationship and Notations

Once an initiation of a relationship is complete, an association of the relationship is started and it lasts as long as the agents have to associate with each other. In fact, the type of association is not really important, but only the *initiation of the relationship* is very important, because it starts a relationship and it determines how much trust a trusting agent should have in the trusted agent at the beginning of the relationship or for a specific purpose and this is especially important in the business world.

The notation in Figure 2.17 will be used for the representation of initiation of trust relationship.

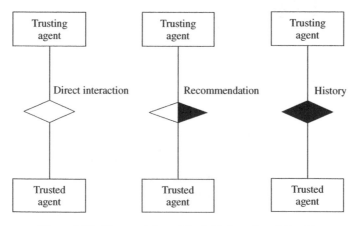

Figure 2.17   Representation of the Initiation of a relationship

In the next section, we shall put all these important concepts together in order to describe a trust model.

2.9 The Trust Model

The purpose of introduction of the trust concepts in the previous sections is to create the groundwork for a discussion of the trust model. The trust model can be used to help determine trust relationships and trust values, help build the trust relationship, gain trust and maintain trust values.

2.9.1 Existing Trust Models

Some work related to 'recommendation' or 'transitive introduction' can be found in existing literature on trust models by Wang *et al.* [13], Dragovic *et al.* [14], Rahman *et al.* [4, 5, 15], Aberer *et al.* [16], Xiong *et al.* [17], Kamvar *et al.* [18], Burton *et al.* [19], Singh *et al.* [20], Ooi [21], Cornelli [22] and Chen [23]. In their work, sometimes the concepts of *trust, trust values* and *trustworthiness* are used interchangeably and there is a lack of explicit definition of the *model* in an e-service environment such as client–server or *ad hoc* networks and in P2P communication networks.

2.9.2 Trust Model and Trust-based Decision

The trust model is used to help in the collection of trust values from the interactions of agents or from the recommendations or history of agents. The trust model will enable the trusting agent to determine the trustworthiness of the trusted agent. This is known as 'making a trust-based decision', that is, a decision about whether the agent should go ahead and do business with the trusted agent. This trust model will also be used for modelling, designing and implementing the trust data repository (or trust database), the trust network and virtual collaborative systems.

2.9.3 New Trust Models

We use a discrete rating to represent the level of trust between the agents (see Chapter 3). We define the trust value of an agent as *continuous numeric values that represent the trustworthiness (Chapter 6)*. Each trusted agent in a trust relationship is associated with a single trust value that captures the extent to which it can be regarded as trustworthy in that context and in that time slot.

Figures 2.18 (a–c) show representations of the trust model for different types of initiations of trust relationships.

The trust model can be programmed. This helps in the automatic assigning of a trust value to a trusted agent after an interaction. For example, the agents will be able to automatically assign *an initial trustworthiness value* to the trusted agent if they have direct interaction. The more interactions they have with the trusted agent, the more accurate the determination of the trust value. The trusting agent after assigning the trustworthiness value to the trusted agent can communicate this trustworthiness value to other agents in order to help them to determine the reputation of the trusted agent. This can help make a trust-based decision.

2.10 Chapter Summary

The following important concepts have been defined and explained in this chapter:

- Trust can exist in two environments: virtual and physical.
- A virtual trust environment is one in which a trust relationship is established between two parties who may or may not have met and interaction takes place via communication over a virtual interaction medium. A physical trust environment is one in which trust is established

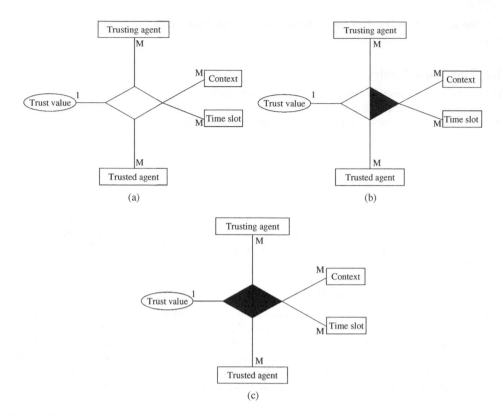

Figure 2.18   Trust model in which the initiation of the relationship is (a) *Direct interaction*, (b) *Recommendation*, (c) *history review*

between two parties who have the opportunity to know each other through personal interactions or introduction.

- A trusting agent is an entity that has faith or belief in another entity in a given context and in a particular time. The entity in which trust has been placed in a given context and in a particular time is known as the *trusted agent.*
- Trust is defined as the belief the trusting agent has in the trusted agent's willingness and capability to deliver a mutually agreed service in a given context and in a given Time slot, as expected by the trusting agent. The key concepts in this definition are as follows:
  - The *belief* that the trusting agent has in the trusted agent captures the trust that the trusting agent has in the trusted agent.
  - The term *willingness* symbolises the will exhibited by the trusted agent to act, or be in readiness to act gladly, honestly, truthfully, reliably and sincerely.
  - The term *capability* captures the skills, talent, competence, aptitude and ability of the trusted agent.
  - The *context* is the scenario or the situation in which trust is placed in the trusted agent by the trusting agent.
  - The term *Time slot* captures the dynamic nature of trust.
  - The term *delivery* represents the actual service delivered by the trusted agent in its interaction with the trusted agent.
  - The term *mutually agreed service* captures the quality of delivery by the trusted agent.

- A trust relationship is the bond or association between a trusting agent and the trusted agent.
- The trust relationship is unidirectional since the trust value is assigned in one direction, from the trusting agent to the trusted agent.
- Each trust concept has a unique definition and has attributes and methods.
- The initiation of a trust relationship is defined as a type of initial introduction that results in a trust relationship. Initiation can take place in three ways: by direct interaction, by transitive introduction, or by history.
- A trust model represents the trust concepts in pictorial form and helps in trust modelling.

This chapter introduced several primary concepts that are useful for Chapter 3 (Trustworthiness) and 4 (Trust ontology for service-oriented environment). The intention of studying these concepts is to help develop trust management systems for building trust, assigning trust values, maintaining the trust and for trust system design and implementation.

In the next chapter, we will discuss the concept of trustworthiness and the difference between trust, trustworthiness and trust values.

References

[1] Hartman F. (2003) 'The role of trust in successful system development and deployment', *Proceedings of IEEE Conference on Industrial Informatics 2003*, Banff, Canada.

[2] Ratnasingham P. (1998) 'The importance of trust in the digit network economy', *Electronic Networking Applications and Policy,* vol. 8, pp. 313–321

[3] Marsh S. (1994) *Formalizing Trust as a Computional Concept*, Ph.D., University of Sterling.

[4] Rahman A.A & Hailes S. (2003) *Relying On Trust To Find Reliable Information*, Available: http://www.cs. ucl.ac.uk/staff/F.AbdulRahman/docs/dwacos99.pdf.

[5] Rahman A.A. & Hailes S. (2003) *Supporting Trust in Virtual Communities*, Available: http://citeseer.nj.nec. com/cache/papers/cs/10496/http:zSzzSzwww-dept.cs.ucl.ac.ukzSzcgi-binzSzstaffzSzF.AbdulRahmanzS zpapers.plzQzhicss33.pdf/abdul-rahman00supporting.pdf.

[6] Gambetta D. (1990) *Can we Trust?* Available: http://www.sociology.ox.ac.uk/papers/gambetta213-237.pdf.

[7] Wang Y. & Vassileva J. (2003) *Trust and Reputation Model in Agent-to-Agent Networks*, Available: www.cs.usask.ca/grads/yaw181/publications/120_wang_y.pdf.

[8] Hussain F., Chang E. & Dillon T.S. (2004) 'Taxonomy of Trust Relationships in Peer-to-Peer (P2P) Communication', *Proceedings of the Second International Workshop on Security in Information Systems*. Porto, Portugal, pp. 99–103.

[9] Hussain F., Chang E. & Dillon T.S. (2004) 'Factors of Trust that influence Trustworthiness in Peer-to-Peer (P2P) based e-commerce', *Proceedings of the International Workshop of Business and Information*, Taipei, Taiwan.

[10] Dillon T.S., Chang E. & Hussain F.K. (2004) 'Managing the dynamic nature of trust', *IEEE Transaction of Intelligent Systems*, vol. 19, no. 5, pp. 77–88, Sept/Oct 2004.

[11] Chang E., Gardner W., Talevski A. & Kapnoullas T. (2003) 'A virtual logistics network and an e-hub as a competitive approach for small to medium size companies', *Proceedings of the International Conference on Web and Communication Technologies and Human.Society@Internet*, Seoul, Korea, pp. 265–271.

[12] Ullman JD. (1988) *'Principles of database and knowledge-based systems'*, vol. 1, ComSci Press.

[13] Wang Y. & Vassileva J. (2003) *Bayesian Network Trust Model in Agent-to-Agent Networks*, Available: http://bistrica.usask.ca/madmuc/Pubs/yao880.pdf.

[14] Dragovic B., Kotsovinos E., Hand S. & Pietzuch P. (2003) 'Xeno trust: Event based distributed trust management', *Proceedings of DEXA'03*, 1st ed, IEEE, Los Alamitos, California, Prague, Czech Republic, pp. 410–414.

[15] Rahman A.A. & Hailes S. (2003) *A Distributed Trust Model*, Available: http://citeseer.nj.nec.com/cache/ papers/cs/882/http:zSzzSzwww-dept.cs.ucl.ac.ukzSzcgi-binzSzstaffzSzF.AbdulRahmanzSzpapers. plzQznspw97.pdf/abdul-rahman97distributed.pdf (5/09/2003).

[16] Aberer K. & Despotovic Z. (2003) *Managing Trust in an Agent-2-Agent Information System*, Available: http://citeseer.nj.nec.com/aberer01managing.html.

[17] Xiong L. & Liu L. (2003) *A Reputation-Based Trust Model for Agent-to-Agent eCommerce Communities*, Available: http://citeseer.nj.nec.com/xiong03reputationbased.html.

[18] Kamvar S.D., Schlosser M.T. & Garcia-Molina H. (2003) *The EigenRep Algorithm for Reputation Management in P2P Networks*, Available: http://citeseer.nj.nec.com/kamvar03eigentrust.html.

[19] Burton K.A. (2002) *Design of the OpenPrivacy Distributed Reputation System*, Available: http://www.Agentfear.org/papers/openprivacy-reputation.pdf.

[20] Singh A. & Liu L. (2003) *TrustMe: Anonymous Management of Trust Relationships in Decentralized P2P systems*, Available: http://www.cc.gatech.edu/~aameek/publications/trustme-p2p03.pdf (11/10/2003).

[21] Ooi B.C., Liau C.Y. & Tan K.L. (2003) *Managing Trust in Agent-to-Agent Systems Using Reputation-Based Techniques*, Available: http://citeseer.nj.nec.com/cache/papers/cs/30109/http:zSzzSzwww.comp.nus.edu.sgzSz~ooibczSzwaim03.pdf/managing-trust-in-Agent.pdf.

[22] Cornelli F., Damiani E., Vimercati S., De Capitani di V., Paraboschi S. & Samarati P. (2003) *Choosing Reputable Servents in a P2P Network*, Available: http://citeseer.nj.nec.com/cache/papers/cs/26951/http:zSzzSzseclab.crema.unimi.itzSzPaperszSzwww02.pdf/choosing-reputable-servents-in.pdf (20/9/2003).

[23] Chen R. & Poblano Y.W. (2003) *A distributed Trust Model for Agent-to-Agent Networks*, Available: http://www.jxta.org/docs/trust.pdf (20/9/2003).

3

Trustworthiness

3.1 Introduction

In the previous chapter, we introduced the concepts of trust, trust values and trust relationships. However, we did not discuss how trust values may be quantified and their related semantics. This issue is addressed by the *trustworthiness measuring system*. *Trustworthiness* determines the trust level by quantifying the trust values utilizing a well-defined *trustworthiness scale*.

We define seven discrete levels of trustworthiness and discuss the *semantics* and *postulates* associated with each level. We also distinguish between trust, trust values, trustworthiness and trustworthiness values and discuss both positive trust and negative trust in relation to the trustworthiness measure.

3.2 Trustworthiness in Literature

3.2.1 Existing Definitions of Trustworthiness

In the existing literature, there has been no clear distinction made between the concepts of trust, trust values and trustworthiness. There is also no clear semantics or definitions that have been provided for these concepts. There has been a lack of systematic methodical description of how to determine the trustworthiness level of an agent. Moreover, there is a lack of a standardized trustworthiness scale that can be unified to represent the trust levels. For example, a system with three levels such as good, average and poor is better than a system with two levels: good and poor. This is especially important in open and often anonymous, virtual collaborative environments such as service-oriented networks.

3.2.2 Existing Trustworthiness Scales

There are different proposals for trustworthiness scale systems. Wang [1, 2] has proposed a Bayesian network–based model for determining trust, where each root node has two values: 'satisfying', denoted by 0, and 'unsatisfying', denoted by 1. Aberer *et al.* [3] have proposed the use of a decision function for determining trustworthiness, where 'trust' is represented by 1 and 'mistrust' is represented by −1. Xiong *et al.* [4] have developed a trust metric based on a number of parameters that include the amount of agent satisfaction and the credibility of agent feedback. Kamvar *et al.* [5] define normalized, local trust values, that is, trust values between 0 and 1. Yu and Singh [6] proposed a system in which the trustworthiness of an agent is the expectation of the cooperativeness, modelled as a probability of the agent.

Chen *et al.* [7] proposed a scale of −1 to 4. Rahman [8] proposes four different levels of trustworthiness ratings: 'very trustworthy', 'trustworthy', 'untrustworthy' and 'very untrustworthy'.

Trust and Reputation for Service-Oriented Environments   Elizabeth Chang, Tharam Dillon and Farookh Hussain
© 2006 John Wiley & Sons, Ltd

Cornelli [9] proposed a non-numeric rating expressed using stars with each additional star denoting a higher rating. The highest possible rating in their proposed method is five stars and the lowest possible rating is 1 star. However, there is a lack of semantic explanation of the different trustworthiness levels and a lack of coverage of all possible trustworthiness levels. For example, if 1−5 stars denote positive trust, how is negative trust to be denoted? There is also a lack of precise meaning and clear definition for a given trustworthiness level.

3.3 Advanced Trustworthiness Definition

3.3.1 Trustworthiness

Definition: Trustworthiness is defined as a measure of the level of trust that the trusting agent has in the trusted agent. The trustworthiness is measured against the trustworthiness scale.

A scale system provides the reference standard for trustworthiness measurement and trustworthiness prediction. It quantifies the trust values and scales the trust into different degrees or levels.

The terms 'a measure', 'the level of trust', 'a scale system', 'trustworthiness measure' and 'quantification of trust values' are essential when defining trustworthiness. These terms are important concepts in the definition of trustworthiness. These terms are explained below in the context of the definition of trustworthiness.

3.3.2 A Measure

The term '*a measure*' refers to trustworthiness, which gives *an estimate* of the level or the degree of trust. *An estimate* is the result of *a tentative measure*. A tentative measure could be in the form of an expert opinion or appraisal or it is a scientific judgement or prediction. *A measure* gives an approximate estimate against some scale or standard, and often the result is an amount or a value.

A trustworthiness value represents a measure, or a value, that depicts the level of trust that the trusting agent has in the trusted agent, in a particular context and in a particular time slot, through the use of a predefined trustworthiness scale system.

3.3.3 The Level of Trust

The term '*the level* of trust' determines the *amount* of trust that the trusting agent has in the trusted agent. It can be represented numerically or non-numerically.

If the trusting agent has a high degree of trust in the trusted agent, this implies that the trusted agent's trustworthiness level is high, that is, the *amount* of trust that the trusting agent has assigned to the trusted agent is high on the trustworthiness scale. Conversely, if the assigned trustworthiness level is low, it means that the trusting agent has very little trust in the trusted agent.

The level of trust represented by the trustworthiness is *unidirectional* from the trusting agent to the trusted agent, and it depends on the context and time, as was the case with trust as explained in the last chapter.

For example, let us assume that Alice trusts Bob to borrow her house for a month. However, this does not imply that Alice trusts Bob to drive her car. It is possible that Bob does not know how to drive or that he is a bad driver and, as a result, Alice does not trust him with her car. Therefore, the trust that Alice has in Bob in the context of lending her house will not be the same as that in the context of lending her car. Hence, the trustworthiness level or value assigned by Alice to Bob is high in the context of 'borrow her house', but it is low in the context of 'borrow her car'. The trustworthiness rating reflects a single scenario or context. When we speak about the trustworthiness of the trusted agent, we qualify it with the context in which it is applicable.

The *level of trust* changes from time to time. If the strength of the trust relationship changes, then the level of trust will change and the trustworthiness representation will also change.

As explained earlier, *the level of trust* changes from one *time slot* to another within a time space. Continuing with the previous example of Alice and Bob, let us assume that Alice extends her contract with Bob and lends her house to Bob for another 2 months. Upon further interaction with Bob, Alice finds out that Bob has made a number of holes in the walls of her house. Alice's trust in Bob in the context of 'lending her house' reduces. The trust that Alice has in Bob has decreased and thus the trustworthiness value will decrease in this time slot.

Further, let us extend the example to a decentralized peer-to-peer (P2P) service-oriented network, where multiple identities exist. Alice has interactions with both James and Jon, and she has assigned very high trustworthiness values to both of them. Later, she comes to realize that James was impersonating Jon. Alice then assigns the lowest possible trustworthiness values to both James and Jon. The trustworthiness values corresponding to the trust relationships that Alice shares with James and Jon decrease in this time slot. Hence, when we speak about the trustworthiness of the trusted agent, we qualify it with the time slot in which it is applicable.

3.3.4 Quantifies the Trust Values

In Chapter 2, we have stated that a *trust value* is a numeric value that depicts the amount of trust in the trust relationship that the trusting agent has in the trusted agent in a given context and in a given time slot. However, we did not discuss the following:

- How do we represent trust numerically? Do we use fractions, integers, percentages or some other representation?
- What is the range of possible values for a trust value?
- What meaning can we assign to a particular value?

These questions are answered by using a trustworthiness scale.

3.3.5 Trustworthiness Scale

A trustworthiness scale provides a simple metric that helps to *determine* the *amount of trust* that the trusting agent has in the trusted agent. Here, we provide a seven-level trustworthiness scale system, which helps to quantify the trust values. *Quantify*, here, means to calculate the trust value in order to determine the corresponding trustworthiness levels.

'*A scale system*' is defined as a measurement system that can be used to determine the level of trust. The scale system can have either numeric measures or non-numeric measures. We define the *numeric measure* of a trust level as an assessment of a trust relationship expressed in terms of an integer or a real number. We define the non-numeric measure of a trust level as a valuation of a trust level expressed neither in terms of an integer nor in terms of real numbers, but as lexicons such as 'very trustworthy' or 'untrustworthy'.

Normally, in the service-oriented network environment, a trustworthiness scale system is also known as a ranking system or rating system for the trust, such as that shown in Figure 3.1.

The *trustworthiness scale* in a service-oriented network environment includes seven levels and associated numerical and non-numerical measures.

A scale is normally defined as an instrument with ordered markings at fixed intervals and is used as a standard of reference for measurement. A ruler with markings in inches or centimetres is a type of scale. The trustworthiness scale provides a standard measuring system that allows us to measure the amount of trust that the trusting agent has in the trusted agent or helps the trusting agent to assign trust levels to the trusted agent.

The *numeric scale* of trustworthiness can represent *a measure* of trust by ascertaining *a value* and expressing it in terms of *an integer* or *a real number* (e.g., 5, 8.9, 100 %).

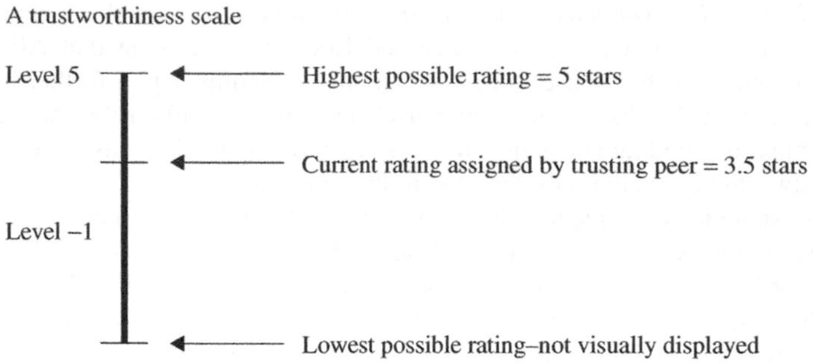

A trustworthiness scale

Level 5 ─┬─ ◄─────── Highest possible rating = 5 stars

         ─┼─ ◄─────── Current rating assigned by trusting peer = 3.5 stars

Level −1

         ─┴─ ◄─────── Lowest possible rating–not visually displayed

Figure 3.1 *A trustworthiness scale* that depicts a trustworthiness level with respect to the highest and the lowest possible ratings

The *non-numeric scale* of trustworthiness can represent *a rating* of trust by ascertaining levels, grades or rankings and expressing these not in terms of integers nor in terms of real numbers, but in *categorical terms* such as *very trustworthy* or *five stars*.

Trustworthiness is determined by numerically *quantifying* the trust values and *qualifying* the trust levels non-numerically. The term *qualify* means to give a specific meaning to the level that is derived.

In other words, the trustworthiness measurement system provides a *trust rating*, which is a non-numerical description of trust levels and if the *amount of trust* is high (numerical rating) then the *trust rating* is also high (non-numerical rating).

Scale the trust or rating the trust is the process of using the trustworthiness scale (numerical and non-numerical ratings) to qualify the trust level to the trusted agents in the network.

Scale the trust (or rating the trust) for a trusted agent is an important practice that a trusting agent in a service-oriented network needs to undertake before any business can take place. This is also true in the physical world because one does not want to conduct business with untrustworthy people. However, in the physical world, face-to-face contact makes it much easier for one to determine the *trust level* that one can assign to an individual. If anything goes wrong, one has the feeling that one can locate and confront the wrongdoer in person. However, if something goes wrong in the virtual world, where one may carry out a transaction with an agent one has met or may never meet, one may either lose money or spend a lot of time and effort recovering losses. Therefore, we see that many well-established sites use trust rating systems that automatically assign trustworthiness to all agents they interact with (both providers and customers). Examples of such sites include Amazon, e-Bay and Yahoo. More recently, third-party trust brokers that can help online customers to recognize good or bad service providers are starting to emerge on the Internet. Examples of trust broker sites include sites that carry information for travellers regarding accommodation, hotels, logistics services. Such sites provide a lot of valuable information.

3.4 Seven Levels of the Trustworthiness

In this section, we define a seven-level trustworthiness with both numeric and non-numeric measures for the evaluation of the trustworthiness. We define each level and discuss the semantics (linguistic definitions), postulates and visual representations. The *linguistic definition* of each level provides the meaning of the confidence or the trust that the trusting agent has in the trusted agent.

We represent the levels by using a system of *stars* (★) and half stars. The highest trustworthiness level is represented by five stars.

The higher the trustworthiness value, the higher the corresponding trustworthiness level and the higher the number of stars representing the level. For example, if Alice (the trusting agent) assigns

Bob (the trusted agent) the highest trustworthiness level, the visual representation of Alice's trust in Bob is *five stars*. According to the semantic representation for this level, Bob can be described as a *very trustworthy agent*.

3.4.1 Seven Trustworthiness Levels

The trustworthiness scale is an ordinal scale with seven discrete levels and corresponding semantics (linguistic definitions). In order to help explain the significance of each of the trustworthiness levels, we map the approximate ranges on a user-defined interval scale to the levels on the ordinal scale. Table 3.1 illustrates the seven levels of trustworthiness and semantics with corresponding user-defined interval ranges.

You may ask why we have proposed exactly seven levels. Could we perhaps have used fewer or more than seven levels? From the practical point of view, five levels are not enough to quantify the trust levels, and ten levels are possible but not necessary. We note that Level 2 spans 50 % on the interval scale, that is, 'partially trustworthy' has elements of being 'trustworthy' and 'not trustworthy'. Thus, it both falls below 50 % as a level and above 50 % as a level. Noting this, we see that below 50 % on the interval scale we only see three levels, and above 50 %, there are four levels. We could add another level between 0 and 50 %, and define it as 'very untrustworthy' or 'minimally trustworthy'. This may be necessary in the social environment, but in a business network environment, very little worthwhile distinction can be made between 'very untrustworthy', 'untrustworthy', 'barely trustworthy' and 'minimally trustworthy'. No one would want to conduct business with such service providers or business partners. However, it is good for a consumer or business partner to learn that this is the case.

3.4.2 Semantics or Linguistic Representations

In Table 3.1, we can see the semantics or linguistic definitions of the seven levels of trustworthiness. A fuzzy linguistic definition gives an approximate meaning expressed in natural language to each level of trustworthiness. It is fuzzy because we cannot give an *explicit definition* of each of the trustworthiness levels.

It is quite common to use linguistic terms to explain a concept, a measure or a value. For example, if we say Faro is tall, 'tall' itself is a fuzzy linguistic term, unless we use a metric or a scale to say a person has to be at least 6 ft to be described as tall. However, suppose that Mark has a height of 5.9999 ft. Should we call him short? In order to solve this problem, we create more categories, such as, 'average', 'short', 'very short', and 'very tall'. An alternate approach would be to use fuzzy sets to represent the notion of 'tall' [10].

Table 3.1 *Seven levels* of trustworthiness and corresponding semantics, linguistic definitions and approximate user-defined interval ranges

Trustworthiness scale (ordinal scale)	Semantics (linguistic definitions)	Percentage intervals (user defined) (%)	Trustworthiness value (user defined)
Level −1	Unknown agent	n/a	$x = -1$
Level 0	Very untrustworthy	0–19	$x = 0$
Level 1	Untrustworthy	20–39	$0 < x \le 1$
Level 2	Partially trustworthy	40–59	$1 < x \le 2$
Level 3	Largely trustworthy	60–79	$2 < x \le 3$
Level 4	Trustworthy	80–90	$3 < x \le 4$
Level 5	Very trustworthy	90–100	$4 < x \le 5$

Let us consider another example. When we say Liz is *friendly* and Joe is *not*, what measure or scale can we use to quantify the fact that Liz is *friendlier than Joe*? In order to describe who is *friendly* and who is *not*, we need to develop a scale for friendliness (composed of levels or markings like a ruler). In order to express *the levels of friendliness*, we may translate *friendly* and *not friendly* into a set of positive and negative actions, and map the actions to a set of numbers or percentages for easy calculation. Then we examine Liz and Joe's actions and count the right actions and the wrong ones. We end up with some values or estimates. This is followed by checking these values against a predefined friendliness scale to justify the fact that Liz is friendlier than Joe.

The above discussion applies to trust and trustworthiness as well, because terms such as 'trust', 'trustworthy', or '*un*trustworthy' are all fuzzy linguistic terms. It is very important to develop a *scale system* so that linguistic concepts or ideas can be quantified and qualified to a certain degree. This approach is a frequently accepted scientific approach [10].

The actual representation can be calibrated and predicted using fuzzy inference systems and fuzzy neural networks (Roger Jang [11] and Takagi–Sugeuo [12]) using training data. A detailed fuzzy calibration system and discussion can be found in [11].

3.4.3 Ordinal Scale

The trustworthiness scale is an ordinal scale. The term 'ordinal scale' means that the set of numbers are sequentially ordered from high to low or vice versa, but the difference between the numbers may not be equal. For example, on our trustworthiness scale, the difference between Level 2 and Level 3 may not be the same as the difference between Level 3 and Level 4.

There are many examples in our daily lives where the *ordinal scale* is used. For example, in a teacher's class survey we often see 'strongly agree', 'agree', 'not sure' and 'disagree'. We know that *strongly agree* is higher in value than *agree*, and *agree* is higher in value than *not sure*, and it is much higher in value than *disagree*; however, the intervals in between the categories may or may not be the same. Another example is the scale of 'excellent', 'very good', 'good', 'average' and 'poor', which is widely used in surveys and references.

3.4.4 Percentage Interval Scale

In order to explain the trustworthiness ordinal scale, in Table 3.1, we have mapped the ordinal scale to a user-defined percentage interval scale of [0–100 %]. Using this representation, if we were to describe our confidence in the three agents A, B and C as 40, 60 and 85 %, then these values would map to Levels 2, 3 and 4 respectively. Note that the difference between the value at Level 2 and the value at Level 3 is not the same as the difference between the value at Level 3 and Level 4. The actual percentage represents an interval scale. On the other hand, the levels constitute an ordinal scale. This is known as a user-defined scale derived from the user's interpretation of the domain.

3.4.5 Trustworthiness Value

Trustworthiness is an estimate of the level of trust, and this *estimate* can be represented numerically or non-numerically. Table 3.1 illustrates *seven levels* of trustworthiness, and a corresponding non-numerical rating (semantic level) and a numeric rating (trustworthiness values, there, two possible values are represented).

For *non-numerical representation*s of trustworthiness, we use Level 0 through Level 5 together with their predefined semantics.

For numerical representations of trustworthiness, we use defined real numbers and ranges that correspond to the Levels as shown in Table 3.2. As described in the previous two sections, the exact value (e.g., 5 or 90 %) is not as important as the meaning and the Level that it represents. This is similar to the grading system used in schools in many western countries. The grades A, B, C, D, F are Levels in the grading system with the corresponding semantics 'high distinction',

Table 3.2 *Seven levels* of trustworthiness and a corresponding visual representation as well as corresponding QoS (Quality of Service) ratings

Trustworthiness level	Trustworthiness value (user defined)	Visual representation (star rating system)	QoS rating (linguistic definitions)
Level −1	$x = -1$	Not displayed	New agent
Level 0	$x = 0$	Not displayed	Terrible
Level 1	$0 < x \leq 1$	From ★ to ⯪	Bad
Level 2	$1 < x \leq 2$	From ★⯪ to ★★	Average
Level 3	$2 < x \leq 3$	From ★★⯪ to ★★★	Good
Level 4	$3 < x \leq 4$	From ★★★⯪ to ★★★★	Very good
Level 5	$4 < x \leq 5$	From ★★★★⯪ to ★★★★★	Super or excellent

'distinction', 'pass', 'deferred' and 'fail'. The marks (e.g., '79' or '80') are normally not presented in the academic record but are used internally within the academic institutions for ease of evaluation. These marks can be mapped to grades for ease of communication to external users.

We note that the trustworthiness value can be expressed as any numerical value in a range [0, 5] or [0, 100 %]. Different value ranges (intervals) correspond to different trustworthiness levels. The trustworthiness levels are *discrete* and constitute an *ordinal scale*. We can associate clear semantics with each level. However, the trustworthiness values must be represented by a continuous variable that can take any value within the defined intervals. For example, any values between 80–90 %, such as 82 % or 89.9 % are valid trustworthiness values that can be mapped to Level 5 (*very trustworthy*).

3.4.6 Postulates

We define each trustworthiness level and provide the *postulates* associated with each trustworthiness level in Section 3.5. *Postulates* represent additional information, informal rules and conditions that further extend or clarify the definition of the trustworthiness concepts. A postulate as a concept is defined as something assumed without proof as being self-evident or generally accepted, especially when used as a basis for an argument.

3.4.7 Star Rating System

We can represent the *seven discrete trustworthiness levels* and their semantics visually using a system of stars and half stars. Table 3.2 illustrates the visual scale.

From Table 3.2, we note that Level 0 and Level −1 are labelled 'not displayed' or 'normally not displayed'. It is recommended that these levels should not be displayed via service-oriented networks. This is because customers or consumers would only be interested in doing business transactions with a business provider that can be trusted to some degree. On the other hand, from the business provider's point of view, it is a waste of web space or time for them to advertise businesses or business services that are unknown or untrustworthy.

For example, if we were searching for a hotel in Hawaii, we would normally want to find one that is reasonably safe and trustworthy. We would not look for an untrustworthy hotel to live in. Therefore, online service systems in Hawaii may not list a lot of hotels that are either not trustworthy or are ranked badly.

Let us consider another example. When we buy books from dotcom sites, the sites usually suggest reliable, low-cost and fast logistics services for deliveries, so that we do not end up paying too much for postage or shipping or do not experience excessive delays in the receipt of goods or loss of goods. There is no reason for a dotcom site to list a lot of bad logistics providers because the customers will not use them anyway. Additionally, by making bad referrals, one risks one's own reputation. Customers would not want to waste their time browsing through the names of bad logistics providers.

3.4.8 QoS Rating

The purpose of the *trustworthiness measurement system* is to present the Quality of Service (QoS) or the quality of business providers to the open networked environment and to help end-users choose the service or business providers. It also helps businesses build their reputations, business values and consumer confidence. In Table 3.2, we present the mapping of trustworthiness levels to QoS.

Trustworthiness concepts are frequently used to describe a person, a business or a company. These concepts are useful as there are millions of for-profit and not-for-profit organizations, such as dotcoms, e-organizations and websites on the Internet that provide various types of services.

When trustworthiness concepts are used in the context of service providers, they often refer to the quality of service that a dotcom or a business organization provides. For example, we could say that '*the service* from West Warehouse is *very good*', instead of saying that 'the service from West Warehouse is trustworthy'. We have often seen that such ratings for dotcoms or websites are very popular with end users.

3.4.9 Trust Rating, Positive and Negative Trust

From Table 3.2, we see that the seven levels of trustworthiness can be divided into two groups, positive trust and negative trust. This is an additional description to further clarify the trustworthiness levels.

A negative trust rating is given to a trusted agent that delivers bad service or behaves in a contradictory manner to the mutual agreement that the trusting agent has with trusted agent. Here, *negative trust* is only represented with a rating that is less than 1. We could represent negative trust with a finer granularity of levels, but these representations would not be significant in the virtual world. However, in some specific cases in the physical world they may be useful. For example, in a school grading system, the negative grade 'F' (fail) represents marks below 50 %. However, if we were required to make a decision regarding whether to ask a student to leave the school because he or she has not received a passing grade in all the subjects, a grading system with more than one negative level might help in the analysis of the student's marks. For example, there could be separate grades corresponding to 'did not submit one assignment' or 'never came to school'.

A positive trust rating is assigned to a trusted agent that is willing to or has the capability required to deliver some, if not all, of the services that are expected by the trusting agent. Here, we present 4 Levels. Level 2 (*partially trustworthy*) is around half way on the trustworthiness scale, and it could be positive or negative; the customer has to decide whether to go ahead with the service provider. They need to know that there is an approximate 50 % possibility that the service provider will not deliver the quality of service and that the customer knows the rating and wants to take the risk of using such a service. It also applies, for example, to the situation in a hotel booking, where a one-star hotel might be cheaper but the hotel might be located in a remote area or in a bad neighbourhood.

In the next few sections of this chapter, we will provide detailed descriptions of the seven Levels of trustworthiness.

3.5 Semantics Representation and Postulates for Trustworthiness Levels

3.5.1 Level-1: Unknown Trustworthiness

(a) Definition

Level-1 (Unknown trustworthiness) is defined as the trustworthiness level assigned to a trusted agent by the trusting agent when the trusting agent is unable to ascertain an estimate or carry out a measurement of trust in the trusted agent in a given context and time slot and for a given initiation of the association.

Let us consider the following example. Alice needs to employ Sonya as a financial controller (context) for a period of three months (time slot), and Alice has received Sonya's curriculum vitae (initiation of the association). However, Alice is unable to make a decision about Sonya's ability to manage funds, as there is no mention in Sonya's curriculum vitae of any prior experience related to financial management. In this case, she cannot ascertain whether Sonya is trustworthy or untrustworthy. If the initiation was by 'direct contact', if Alice knew Sonya because they play in a band together, Alice might still be unable to assign a trustworthiness level to Sonya in the *context* of financial management for a period of three months.

(b) Semantics

In a service-oriented network environment, if a trusting agent cannot determine the trustworthiness of the trusted agent, it might be due to the fact that the trusting agent does not know the trusted agent and did not have any direct contact with the trusted agent, in that particular context and time slot. The trusting agent, in this situation, cannot assume that the trusted agent is trustworthy nor can he assume that the trusted agent is untrustworthy. Therefore, the trusting agent cannot make an informed decision regarding the trustworthiness of the trusted agent. In this situation, Level −1 (*unknown trustworthiness*) should be assigned to the trusted agent.

(c) Trustworthiness value

We propose that a trustworthiness value of −1 be assigned to the trusted agent corresponding to Level −1 of trustworthiness. A trustworthiness value of −1 implies that the trusting agent has no idea about the trustworthiness of the trusted agent.

It should be noted that this does not imply distrust. All new agents in the network begin with this value of trustworthiness. The trustworthiness value of −1 is assigned to an agent when there is no precedent or no previous interaction that can aid the trusting agent in determining the trustworthiness of the trusted agent.

It should also be noted that if we do not know a person or have never heard of a person, we generally do not trust them at all. Therefore, this level would also have within it the notion of negative trust.

This level of trustworthiness is normally not displayed visually because it is useless for Internet users and service providers. In other words, we are normally only interested in those service providers that are good enough to provide the required services. This does not mean that the reputation of such an agent will be ignored. If, for example, we were searching for a hotel in Hawaii, we would normally want to know the trustworthiness level of the hotel we are interested in. Online service systems in Hawaii, however, might not list hotels that have been ranked badly or not ranked at all.

(d) Postulates

A trusted agent can be assigned a trustworthiness value of −1 under the following conditions:

(i) The trusted agent is new or has recently joined the network.
(ii) The trusting agent has not had any previous interaction with the trusted agent and is not in a position to determine the trustworthiness of the trusted agent.
(iii) All the other agents that the trusting agent inquires of for obtaining recommendation regarding the trusted agent, have not carried out any interaction with the trusted agent previously and have no knowledge of the trustworthiness of the trusted agent. Therefore, the trusting agent cannot associate a trustworthiness value with the trusted agent by using reputation values.

3.5.2 Level 0: Very Untrustworthy

(a) Definition

Level 0 (Very untrustworthy) is defined as the trustworthiness level that the trusting agent assigns to the trusted agent when the trusting agent cannot trust the trusted agent at all in the specific context and in that particular time slot.

(b) Semantics

If the trusting agent assigns Level 0 of trustworthiness to a trusted agent, it means that in that specific context and specific time slot, the trusted agent is exceptionally unreliable in performing a given action.

We term this kind of trust of the trusting agent in the trusted agent 'the highest possible level of negative trust'. We refer to such a trusted agent as 'untrustworthy'.

(c) Trustworthiness value

We propose that a trustworthiness value 0 be assigned to the trusted agent corresponding to Level 0 of trustworthiness. It can be anticipated that an agent that is assigned a trustworthiness value 0 will, in future interactions, behave in contradiction to what the trusting agent expects.

This level of trustworthiness is normally not displayed visually because it is useless for Internet users and service providers. Consider, for example, a dotcom site that sells books. The site may also provide suggestions for some fast and low-cost logistics services for delivery of the purchased books to customers. There is no reason for the site to list a lot of bad logistics providers because the customers will not use them. However, it may be useful for the dotcom site to have this information in case this bad logistic provider asks to be listed on their site or a user suggests that they be included in the list.

(d) Postulates

The following are the conditions under which a trusted agent can be assigned a trustworthiness value 0:

 (i) Either the majority of, or all, the other agents that the trusting agent inquires of to provide recommendations have assigned a trustworthiness value 0 to the trusted agent.
 (ii) The trusted agent was involved in a fraudulent deal with the trusting agent in a situation where the trusting agent had previously communicated the entire basis on which the trustworthiness of the trusted agent would be evaluated to the trusted agent in clear and explicit terms.

3.5.3 Level 1: Untrustworthy

(a) Definition

Level 1 (Untrustworthy) is defined as the trustworthiness level that the trusting agent assigns to the trusted agent when the trusting agent has very little trust in the trusted agent in a given context and in a given time slot and for a given initiation of the association.

(b) Semantics

Trustworthiness level 1 marks the beginning of the negative trust rating. Agents that are assigned trustworthiness level 1 are referred to as 'untrustworthy'.

If the trusting agent assigns trustworthiness level 1 to a trusted agent, it means that in a specific context and in a specific interval of time, the trusted agent is not reliable in performing a specific action.

(c) Trustworthiness value

We propose a trustworthiness value $0 < x \leq 1$ corresponding to Level 1. It can be anticipated that an agent that is assigned a trustworthiness value $0 < x \leq 1$ could, in future interactions, behave in nearly, but not exactly, the opposite way to what the trusting agent expects.

This trustworthiness level is visually represented by a half (✩) to one (★) star. Note that many sites do not display this level of service.

(d) Postulates

The following are conditions under which a trusted agent can be assigned a trustworthiness value $0 < x \leq 1$:

(i) A majority of the other agents that the trusting agent enquires of to obtain recommendations has assigned a trustworthiness level of '1' to the trusted agent.
(ii) The trusted agent was unable to deliver the committed service or fulfil the terms of the service agreement with the trusting agent, or the trusting agent has been misled by the trusted agent in the service agreement.

3.5.4 Level 2: Partially Trustworthy

(a) Definition

Level 2 (Partially trustworthy) is defined as the trustworthiness level that the trusting agent assigns to the trusted agent when the trusting agent has around 50 % *confidence* Table 3.1) in the trusted agent in a given context, time slot and initiation of the association.

(b) Semantics

If the trusting agent assigns Level 2 (partially trustworthy) to a trusted agent, it means that in a given context and in a given interval of time, the reliability of a trusted agent in performing a given action is uncertain. Broadly speaking, we term this type of trust in another agent as *halfway* trust. Assignment of trustworthiness Level 2 to a trusted agent in an interaction would communicate that the behaviour of the trusted agent was such that it cannot be given a negative trust rating. However, the trusted agent does not act in a consistent manner with respect to the trusting agent's expectations. This could be because it does not fully provide the required quality of service or it only intermittently provides the quality of service.

Trustworthiness Level 2 marks the beginning of the positive trust rating. We term such agents as 'partially trustworthy'.

(c) Trustworthiness value

We propose a trustworthiness value $1 < x \leq 2$ for trustworthiness Level 2. It can be anticipated that an agent that is assigned a trustworthiness value $1 < x \leq 2$ will behave in a less than consistent manner with regard to the trusting agent's expectations in future interactions.

This trustworthiness level is visually represented by one and a half (★ ✩) to two (★ ★) stars. The criteria for the rating could include price, commitment to service and fulfilment of service or consistency in fulfilment of service. The method for assigning the trustworthiness values and stars will be described in Chapter 6.

(d) Postulates

The following are the conditions under which a trusted agent can be assigned a trustworthiness value $1 < x \leq 2$:

(i) A majority of the other agents that the trusting agent enquires of to obtain recommendations have assigned a trustworthiness value $1 < x \leq 2$ to the trusted agent.

(ii) The trusted agent acted neither in a fraudulent manner (as represented by trustworthiness levels 0 and 1) nor in a good or very good manner (as represented by trustworthiness Levels 3, 4 and 5). For a given interaction, the behaviour of the trusted agent can neither be regarded as very bad nor as very good, and the trusting agent communicated ALL the bases on which the trustworthiness of the trusted agent was to be evaluated to the trusted agent in clear terms prior to the interaction. It could also be that the trusted agent performs in a good manner only about half the time.

3.5.5 Level 3: Largely Trustworthy

(a) Definition

Level 3 (Largely trustworthy) is defined as the trustworthiness level that the trusting agent assigns to the trusted agent when the trusting agent has basically positive trust or a confidence of around 70 % (Table 3.1) in the trusted agent in a given context and time slot and for the given initiation of the association.

(b) Semantics

It can be anticipated that an agent that is assigned trustworthiness Level 3 will behave as reasonably well and in a largely positive manner in a future interaction also with the trusting agent. In other words, it will generally behave as the trusting agent expects. However, there could be a few occasions when it did not fully meet expectations. We refer to such agents that are almost trustworthy as 'largely trustworthy'.

(c) Trustworthiness value

We propose that a trustworthiness value for a *largely trustworthy* agent be $2 < x \leq 3$. The trustworthiness level is visually represented by two and a half (★★⯪) to three (★★★) stars.

(d) Postulates

The following are conditions under which a trusted agent can be assigned a trustworthiness value $2 < x \leq 3$:

 (i) The majority of agents that the trusting agent enquires of to offer recommendations have assigned a trustworthiness value $2 < x \leq 3$ to the trusted agent.
(ii) A trustworthiness value $2 < x \leq 3$ (Level 3) indicates that the trusted agent acted in a good or positive manner most of the time in a given context and in a specific time slot for a given initiation of a trust relationship.

3.5.6 Level 4: Trustworthy

(a) Definition

Level 4 (trustworthy) is defined as the trustworthiness level that the trusting agent assigns to the trusted agent when the trusting agent has *a confidence of over 80 %* (Table 3.1) in the trusted agent in a given context and time slot and for a given initiation of the association. The trusted agent satisfying this criterion is also referred to as trustworthy and will, in an interaction, behave almost exactly as the trusting agent expected.

(b) Semantics

If the trusting agent assigns trustworthiness Level 4 to a trusted agent, it means that in a given context and in a given interval of time, the trusted agent can be relied upon to perform a given action.

We term this kind of trust that the trusting agent has in the trusted agent as 'positive trust'. A trusted agent that has been assigned this trustworthiness level can be relied upon in a given context to perform an action. It is the second highest possible Level of trustworthiness that a trusting agent can have in a trusted agent. We refer to such a trusted agent as 'trustworthy'.

(c) Trustworthiness value

We propose that the trustworthiness value for this Level be $3 < x \leq 4$. If the trusting agent assigns a trustworthiness value $3 < x \leq 4$ to a trusted agent, it means that in a given context and in a given interval of time the trusted agent can be relied upon to perform a given action. It can be anticipated that an agent that is assigned this trustworthiness level will, in future interactions, be highly regarded by the trusting agent.

This trustworthiness level is visually represented by three and a half (★ ★ ★ ⯪) to four (★ ★ ★ ★) stars.

(d) Postulates

The following are conditions under which a trusted agent can be assigned a trustworthiness value $3 < x \leq 4$:

(i) A majority of agents that the trusting agent inquires of to obtain recommendations have assigned a trustworthiness value $3 < x \leq 4$ to the trusted agent.
(ii) The trusted agent fulfils most of what was expected from it by the trusting agent in their interaction.

3.5.7 Level 5: Very Trustworthy

(a) Definition

Level 5 (Very trustworthy) is defined as the trustworthiness level that a trusting agent assigns to a trusted agent that is completely trustworthy, and the assignment of Level 5 indicates a very *positive trust or a confidence of over 90%* (Table 3.1) of the trusting agent in the trusted agent in a given context, time slot and for a given initiation of the trust relationship. A very trustworthy agent is also referred to as completely trustworthy, and in any given situation, the trusted agent is certain to behave exactly as the trusting agent expects.

(b) Semantics

If the trusting agent assigns trustworthiness Level 5 to a trusted agent, it means that in a given context and in a given interval of time the trusted agent can be fully depended upon to perform a given action. This is the highest possible level of positive trust that a trusting agent can have in the trusted agent.

It can be anticipated that a trusted agent that is assigned trustworthiness Level 5 will, in future interactions, behave exactly as expected by the trusting agent. We term an agent that is assigned this level of trustworthiness as 'very trustworthy'.

(c) Trustworthiness value

We propose a trustworthiness value $4 < x \leq 5$ for this Level of trustworthiness. If the trusting agent assigns a trustworthiness value $4 < x \leq 5$ to a trusted agent, it means that in a given context and in a given interval of time the trusted agent can be fully depended upon to perform a given action on every occasion.

This trustworthiness level is visually represented by four and a half (★ ★ ★ ★ ⯪) to five stars (★ ★ ★ ★ ★).

(d) Postulates

The following are the conditions under which a trusted agent can be assigned a trustworthiness value $4 < x \leq 5$:

(i) The trusted agent fulfils all that is expected of it by the trusting agent in an interaction. We explain mathematically what we mean by the term 'all' in chapter 6.

(ii) Most, if not all, of the other agents that the trusting agent enquires of to obtain recommendations have assigned a trustworthiness value $4 < x \leq 5$ to the trusting agent.

3.6 Trustworthiness Measure and Prediction

Trustworthiness measures result in trust values based on direct interactions. Given a series of interactions from the past, one gets a series of trust values. Only when we have had a set of previous trust values are we able to carry out trustworthiness prediction.

3.6.1 Trustworthiness Measure

Trustworthiness measurement is carried out by a trusting agent after a direct interaction or direct experience with the trusted agent by using a method to derive a trust value. This trust value could be for a trusted agent, or a service or a product. The result of the trustworthiness measure gives us a better idea whether we should interact again with the trusted agent. For example, if Charlie said 'Tobi gave me two bad haircuts', should I have a haircut from Tobi again?

Trustworthiness measures provide information for the following:

- Influencing the decision of whether to trust the trusted agent in the near future;
- Determining the attributes of trust for the trusted agent;
- Determining the attributes of the trust relationship;
- Determining trustworthiness of the trusted agent;
- Determining the trustworthiness of the trust relationship.

We can measure trust for the trusted agents, services or products. A more difficult measure in a service-oriented environment is the measurement of QoS.

There are three types of measures: from direct interaction, from reputation or from history. However, only direct interaction results in a trustworthiness value. Reputation gives third-party opinions that are not from trusting agents themselves and history reviews given past trust values.

In Chapter 5, we shall introduce the challenges in trust measures, which give us an appreciation of the methods provided in Chapters 6 through to 10.

'Trustworthiness measure' is defined as an estimate of the level of trust or the trustworthiness value assigned to the trusted agent *AFTER* a business service interaction over the distributed service-oriented environment.

In order to understand this definition, let us consider the following real-world scenario where Liz has hired Sonya as a financial controller on the basis of the initial trust value that Liz has assigned to Sonya. This initial trust value is based on Sonya's reputation, and it is determined by Liz after inquiring from other agents that Liz trusts, and that have previously interacted with Sonya. Let us suppose this initial trust value is 4. On the basis of Liz's interaction with Sonya, Liz may assign a trustworthiness value to Sonya that may be higher or lower than the initial trust value.

Therefore, the trustworthiness measure or trustworthiness value refers to the trust value assigned by the trusting agent to the trusted agent *after* the interaction.

3.6.2 Trustworthiness Prediction

'*Trustworthiness prediction*' is defined as the trust value assigned to the trusted agent for time slots $N + 1$ and $N + M$ into the future where N is the current time slot and M is M time slots into the future for which prediction is carried out. These predictions are an initial assignment for these time slots. In some cases, this may be BEFORE any interactions with the trusted agent, in which case, the trusting agent would have to rely on a reputation measure.

It is important to note that a *trustworthiness value* may be reassigned to a trusted agent by a trusting agent after an interaction. However, before the interaction takes place, the trusting agent can only use a predicted trust value. In the previous example, since Sonya has never worked as a financial controller for Liz previously, Liz can only assign Sonya an initial trust value based on Sonya's reputation. Note that, even if Liz knew Sonya in a different context (e.g., business marketing), she would be able to assign Sonya only an initial trust value or a predicted value in the context of her abilities as a financial controller before she actually interacted with her in this context.

Before we can carry out any prediction, we must have some historical data or reputation data (data that is recommended by third-party agents) that we can use to predict future trust. In the real world, this method has not often been used in forms of trust measurement or prediction.

However, in the electronic context of business in the computer and networked world, agents are relatively autonomous. In many of these exchanges, the threat of malicious agents exists in the network. If there is historical data or reputation and recommendation data related to these threats, there should be the ability to carry out trustworthiness predictions.

Historical data can be collected from experiences or from other sources such as third party's opinions (reputation recommendation data).

Note that trustworthiness measure and prediction would take into account dynamic trends rather than purely snapshots at a particular instant of time.

The trust value or trustworthiness value is only used by the trusting agent, and may not necessarily be shared with any other agents on the network, and if the measure or prediction of trust is wrong, trusting agents have to take sole responsibility for that information. However, if they decided to pass the wrong trust information to the public or to other agents on the network, these trusting agents have nothing to lose; however, it puts the trusted Agent in a vulnerable situation. Therefore, it is important to note that when a trusting agent submits its trustworthiness opinion about the trusted agent, it should be honest, and from the point of view of the trustworthiness system (or customer feedback system), it should make sure that the opinion is fair, truthful and honest. In later chapters, we shall learn how the trustworthiness system ensures the fair evaluation of trust and reputation.

3.7 Challenges in Trustworthiness Measure and Prediction

One of the significant challenges in trustworthiness measures and predictions is being able to examine different concepts of the trust definition and the relationship between the concepts. In Tables 3.3 and 3.4, important aspects of the nexus between trust concepts and relationship are considered.

If we tie each aspect of the trust definition to agents and relationships, we find the connections are as illustrated in Table 3.3:

Table 3.4 extends this consideration of the relationship from trust concepts to both internal and external factors of agents in the trust relationship.

By the term 'internal factors' of an agent we mean factors that no other agent can see, feel or hear directly. Moreover, they are implicit. Therefore, in order to more fully understand these

Table 3.3 Trust definition and their connections to concepts in the trust model

Concepts of the trust definition	Concept relationship
Willingness, capability	Peers
Context, time, trust	Relationship

Table 3.4 Further analysis of the trust concept

Concepts of the trust definition	Relates to
Willingness, capability	*Internal* factors of agents
Quality of service	*External* factors of agents

factors, some measurement needs to be applied. These factors can only be determined by some correlation analysis.

'External factors' of an agent are factors that other agents can see, feel, hear or sense directly. The context and time aspects of the trust definition relate to the trust relationship only, that is, the trust relationship only exists in a specific time slot and in a specific context.

The measurement of the Quality of Service can only be done by correlating the behaviour or fulfilment of the service, in a given context, in a particular time slot. The measurement is carried out by evaluating the actual delivered service against the committed service.

Trust values change over time. This is known as the dynamic nature of trust. It is most accurately defined as the change in the trust value assigned to an entity by a given trusting entity with the passage of time in different time slots within the time space.

We note that in Table 3.5

 (i) since capability and willingness are by and large *not* directly observable,
 (ii) we must arrive at an estimation of these by utilizing the external factors, and
(iii) undertake this within the relationship characteristic.

By using common sense, we understand that *dynamism* means changing with time. Some changes can be predicted and some cannot. If the changes can be predicted, then the changes can be managed by humans or machines. If the changes cannot be predicted, or there is no good method to predict them, then this dynamism, or change, is hard to manage.

We note that change can be caused by external factors as well as internal factors. In real life, both factors could cause the change. However, internal factors are hard to predict, even with great scientific studies. This is unlike external factors, where one can measure them, predict them and consequently

Table 3.5 Some trust concepts are unable to be measured directly. These create the challenge in trustworthiness measures

#	Concepts	Association	Specifiable, observable, measurable
(I)	Willingness, capability	Internal factors of peers	No
(II)	Actual behaviour, expected behaviour	External factor of peers	Yes
(III)	Context, time, initiation relationship	trust relationship	Yes

Table 3.6 Aspects of the trust model and their impact on trust measurement and prediction

Concepts in the trust model	Factors	Dynamism
Willingness, capability	Internal factors of agents	These factors are the internal factors of peers. Internal factors are very hard to capture. They could be changed by both internal influences and external influences. For example, more education or an circumstances could change the capability of a person, or psychological advice could affect the level of willingness. However, no one can predict how much they will change with time. Therefore, they are classified as 'dynamic' in the trust model.
Behaviour (actual behaviour), expectations (expected behaviour)	External factors of agents	External factors can be determined by other agents.
Context, time slot, initiation method	They are part of the relationship, and not peers themselves.	These factors are not dynamic.

try to manage them. Therefore, the internal factors that cause the dynamism or changes are the factors that humans or machines cannot manage. Humans or machines can manage the external factors that cause the dynamism or changes because these factors can be observed and thus measured.

In Table 3.6, we note that the proposed trustworthiness measure cannot be used to quantify the willingness and capability aspects of trust, which are shown in Tables 3.4 and 3.5. This is because these aspects are internal and not predictable. We also note that no one has yet developed a methodology for measuring willingness and capability in humans. This has to be done for humans before this can be done for machines.

The purpose of this analysis is to help the reader appreciate trust, trustworthiness technologies and methodologies that are to be introduced in the rest of the book.

In the following chapter, we devote the entire discussion to developing a methodology for the measurement of trust that can overcome the difficulties associated with an agent's willingness and capability in different contexts.

3.8 Chapter Summary

Trust, trust value and trustworthiness are three distinct concepts. *Trust* is the belief the trusting agent has in the trusted agent. A *trust value* represents the level of trust, or belief, that the trusting agent has in the trusted agent. *Trustworthiness* provides a measure for the level of trust that the trusting agent has in the trusted agent in *a trust relationship* in a given *context* and in a given *time slot*. Trustworthiness quantifies the *trust values* and rates the level of *trust* against a *trustworthiness scale*. In this chapter, we have given an explicit definition of *trustworthiness* that is suitable for service-oriented network environments. We have also given detailed definitions of the seven levels of *trustworthiness*, defined their *semantics* and discussed the corresponding postulates. We have related the levels to their corresponding representations in a star rating system and a QoS system.

In this chapter, we have provided a detailed definition of trustworthiness and have defined and described the seven Levels of the trustworthiness scale. The following are the important concepts discussed in this chapter:

- Trust (defined in Chapter 2) and trustworthiness are two distinct concepts.

- The important aspects of the definition of *trustworthiness* are as follows:
  - A *measure* means the ranking or rating system that quantifies the level of trust that the trusting agent has in the trusted agent.
  - The *level* implies either a numerical or a non-numerical representation of the strength of the trust relationship.
  - A *trust relationship* is an association between a trusting agent and a trusted agent.
  - A *given context* signifies that the rating reflects the strength of the trust relationship in a single scenario or context only.
  - A *given time slot* captures the dynamic nature of trust relationships.
  - *Quantify the trust values* refers to the determination of the trust value (numerical rating).
  - *Rating the trust* refers to qualifying the trust into levels (non-numerical rating).
  - *Trustworthiness scale* is a measuring system.
- We have defined seven levels of trustworthiness, provided semantics, postulates and corresponding trustworthiness values and a star rating system.
- We have distinguished trust, trust value, trustworthiness and trustworthiness value.

Additionally, we distinguish between trust, trustworthiness, trust value and trustworthiness value. Further, we introduce two more concepts; namely: trustworthiness measure and trustworthiness prediction

- Trust values represent the amount of the trust that the trusting agent has in the trusted agent. Trust values can represent a relationship in the past and a trust value in the past.
- Trustworthiness is measured to help determine a level of trust (the amount of trust).
- Trustworthiness value is a calculated trust value in a range that determines the level of trust in a trust relationship.
- A trustworthiness measure is a measure that helps determine the level of trust (the amount of trust).
- Trustworthiness prediction is the use of some historical data to predict future trust.
- The challenge of trustworthiness measures and prediction is that we are often unable to capture the internal factors of agents, in terms of their willingness and capability.

In summary, we conclude that

- *Trust* is the belief the trusting agent has in the trusted agent in a given context and a given time slot.
- *Trust values* represent the amount of trust the trusting agent has in the trusted agent.
- *Trustworthiness* is *a measure* that helps determine the level of trust (the amount of trust).
- *Trustworthiness value* is *a calculated trust value* in the range [0, 5] that determines the level of trust in a trust relationship. The method for calculation of this value will be discussed in chapter 6.

In the next chapter, we shall discuss other factors that have an impact on the determination of trust values.

References

[1] Wang Y. & Vassileva J. (2003) *Trust and Reputation Model in Peer-to-Peer*, Available: www.cs.usask.ca/grads/yaw181/publications/120_wang_y.pdf (15/10/2003).
[2] Wang Y. & Vassileva J. (2003) Bayesian Network Trust Model in Peer-to-Peer Networks, *Proceeding of Second International Workshop on Agent and Peer-to-Peer Computing*, Melbourne, pp. 23–24.

[3] Aberer K. & Despotovic Z. (2003) *Managing Trust in a Peer-2-Peer Information System*, Available: http://citeseer.nj.nec.com/aberer01managing.html (11/9/2003).

[4] Xiong L. & Liu L. (2003) *A Reputation-Based Trust Model for Peer-to-Peer eCommerce Communities*, Available: http://citeseer.nj.nec.com/xiong03reputationbased.html (9/10/2003).

[5] Kamvar S.D., Schlosser M.T. & Garcia-Molina H. (2003) *The EigenRep Algorithm for Reputation Management in P2P Networks*, Available: http://citeseer.nj.nec.com/kamvar03eigentrust.html (11/9/2003).

[6] Yu B., Singh M.P (2003) "Incentive mechanisms for Peer-to-Peer systems", *Proceeding of Second International Workshop on Agent and Peer-to-Peer Computing,* Melbourne, pp. 77–88.

[7] Chen R. & Yeager W. (2003) *Poblano: A distributed Trust Model for Peer-to-Peer Networks*, Available: http://www.jxta.org/docs/trust.pdf (20/9/2003).

[8] Rahman A.A. & Hailes S. (2003) *A distributed Trust Model*, Available: http://citeseer.nj.nec.com/cache/papers/cs/882/http:zSzzSzwww-dept.cs.ucl.ac.ukzSzcgi-binzSzstaffzSzF.AbdulRahmanzSzpapers.plzQznspw97.pdf/abdul-rahman97distributed.pdf (5/09/2003).

[9] Cornelli F., Damiani E., Vimercati S., Vimercati S.D.C., Paraboschi S. & Samarati P. (2003) *Choosing Reputable Servents in a P2P Network*, Available: http://citeseer.nj.nec.com/cache/papers/cs/26951/http:zSzzSzseclab.crema.unimi.itzSzPaperszSzwww02.pdf/choosing-reputable-servents-in.pdf (20/9/2003).

[10] Zadeh L.A. (1975) *"The Concept of a Linguistic Variable and its Application to Approximate Reasoning Part I – Information and Sciences*, vol. 8, pp. 199–249; *Part II – Information and Sciences*, vol. 8, pp. 301–357; *Part III – Information and Sciences*, vol. 9, pp. 43–80.

[11] Jang J.-S.R. & Gulley N. (1995) *'The Fuzzy Logic Toolbox for use with MATLAB'*, The Mathworks Inc, Natick, Massachusetts.

[12] Takagi T. & Sugeuo M. (1985) 'Fuzzy identification of systems and its applications to modelling and control', *IEEE Transactions on Systems man & Cybernetics*, vol. **15**, no. 1, pp. 116–131.

4

Trust Ontology
for Service-Oriented Environment

4.1 Introduction

In this chapter, we extend the trust concept defined in Chapter 2 to suit service-oriented environments. There are two extensions: (a) the 'trust' in a service-oriented environment means to trust the 'quality' of something, such as a product sold or a service provided; (b) the 'belief' about the *quality* of a product or a service or an agent in a service-oriented environment is a result of 'quality assessment'. This chapter defines trust ontologies that are to be used in service-oriented environments.

In a service-oriented environment, business entities or agents can carry out the buying and selling of services and goods, requesting or inquiring of information, bidding or offering of contracts, publishing or advertising of products, and so on (refer to Figure 1.6 in Chapter 1). As communication can be anonymous or pseudo-anonymous, and non anonymous, the quality of a service, or the quality of a product or the quality of a communicating agent is one of the biggest considerations for all online consumers. It is one of the major barriers to e-commerce development. Therefore, the study of trust and the development of trust technologies for the service-oriented environment has a strong impact on consumers, businesses and the economic environment. If the quality of e-services, e-products, and so on, via the Internet is guaranteed, many more people will use it, and it will bring about a revolution in e-commerce and push the networked economy to a new level.

In the service-oriented environment, a trust ontology can represent at least one of three domains:

- Agent trust ontology
- Service trust ontology
- Product trust ontology.

This chapter will give a detailed explanation of each of the above trust ontologies and its application to specific trust ontologies.

It is important to note the difference between *agent trust* and the *service trust*. *Agent trust* is measured against the obligations of informal agreements or implicit specifications, whereas *service trust* is measured against formal contracts or service standards or service level agreements, and so on. The measure is normally against explicit criteria.

Trust and Reputation for Service-Oriented Environments   Elizabeth Chang, Tharam Dillon and Farookh Hussain
© 2006 John Wiley & Sons, Ltd

4.2 Ontology

4.2.1 What is an Ontology

Definition: Ontology can be viewed as 'a shared conceptualization' [1] of a domain that is commonly agreed to by all parties. It is defined as 'a specification of a conceptualization' [1]. 'Conceptualization' refers to the understanding of the concepts and relationships between the concepts that *can* exist or *do* exist in a specific domain or a community. A representation of the shared knowledge in a specific domain that has been commonly agreed to refers to the 'specification' of a conceptualization.

From this definition, the key constructs are 'shared' and 'commonly agreed' knowledge. There is a common misunderstanding about ontologies. Some people may think *ontology* is only a set of words, definitions or concepts, while others think ontology is similar to an object-oriented technology or simply a knowledge base.

However, this is not true and these are not ontologies because the ideas contained in the signifiers 'shared' and 'commonly agreed' must be considered when talking about ontologies. It is important to note that:

- If some concepts or knowledge are shared but the meaning is not commonly agreed or vice versa, they are not ontologies.
- If a set of knowledge is only referred to as *ontology*, but nobody uses it or shares it, then it is also not an ontology.
- If a set of concepts is used to describe objects, it can be shared knowledge within a single application or a single object-oriented system; however, it is not an ontology, because an ontology is for a specific domain and not for a specific application.

Ontology is the agreed understanding of the 'being' of knowledge [2]. In other words, there is consensus regarding the interpretation of the concepts and the conceptual understanding of a domain. Every domain has a commonly used set of concepts or words, and a set of possible closely related meanings to select from. If these concepts are attributed the same meaning by the community, business parties or domain of service operators, then computers can easily be used to establish communication and carry out tasks automatically. If, within a domain, different parties were to use their own terms and definitions, and there are no translators, then tasks could become very time consuming. It would become virtually impossible for computers to do the tasks automatically using the Internet, especially in the service-oriented network or virtual collaborative environment.

A very simple example of an ontological viewpoint of a concept is that a 'Published Article' should include author name(s), article title, publisher of journal or conference name, or newspaper name, place of publication, date of publication and ISBN number, or volume number or page number. This single concept can be used cross-culturally and internationally, from the public sector to the private, from groups to individuals.

4.2.2 Generic Ontologies and Specific Ontologies

A domain of interest can be represented by *generic ontologies* and *specific ontologies*, which are also known as *upper ontologies* and *sub-ontologies (lower ontologies)*. The *specific ontologies* or sub-ontologies are also known as '*ontology commitments*' [3]. They commit to use all the upper ontology concepts and specifications. As an example, a generic concept may be called *human relationship* and a specific 'human relationship' concept can be 'boy and girl relationship' or 'business and customer relationship' or 'teacher and student relationship', and so on. These are specific relationship concepts and they are all *committed* to the *specification* of the generic concept 'human relationship'. The *specification* of the generic concept 'human relationship' can be defined as 'having at least two parties involved. If one party is missing, the relationship would not exist

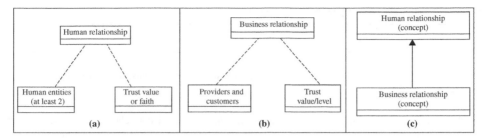

Figure 4.1

and the trust value has no meaning'. Therefore, all the specific ontologies or sub-ontologies should commit to this *specification*.

- Figure 4.1a shows the generic ontological representation of the *concept* of 'human relationship'.
- Figure 4.1b shows the specific ontological representation of the *concept* of 'business relationship'.
- Figure 4.1c shows the ontological view of generic and specific ontologies or the upper and sub-ontology hierarchy.

In the rest of this chapter, we shall focus on the ontological representation of trust in a service-oriented environment.

4.2.3 Notation System used for Ontology Representation

In this section, we present a notation system for *ontology representation*. This is because in the literature so far, there is no standard for ontology representation. A well-known notation such as UML, even its most recent version UML2, has limited power to represent the ontology and causes confusion between object technology and ontologies. Web Ontology Language (OWL) contains a powerful representation of ontologies. However, it is not readily understandable by domain experts. Hence, we define a notation system for an ontology using a graphical representation. We have chosen simplicity for ease of understanding and give a clear semantic for each of the constructs used.

There are five key notations or constructs, namely

- Ontology concept/ontology classes
- Ontology instance
- Ontology association relation
- Ontology generic-specific relation
- Ontology include/part-of relation.

Notation 1 – Ontology Concept
An ontology concept is defined as an abstraction of agreed terms, definitions and vocabularies (Table 4.1). A concept is an important element in modelling and representing reality. A concept is an abstraction that captures the properties of a group of individual instances that have commonalities. A concept is defined by its intention or extension. The intention of a concept is a specification that gives a definition of its properties and the axioms related to it. This gives a definition of its structure and behaviours. The extension of the concept is the enumeration of all its instances and relations.

In many proposed ontology modelling languages, the notion of concept is used interchangeably with the notion of class. However, in ontology modelling, it is important to recognize that a concept has to be shared across the domain and is commonly agreed upon between the participants in the domain. This is in contrast to the case of class/object in object-oriented modelling or knowledge modelling in knowledge-based systems where the system modeller dictates the definition of the class/object.

Table 4.1   Notation for representing ontology concept

Ontology notation	Semantics of the notation
	Double-field box represents the *ontological concept*

Note:
- Each concept has its own specification, including hierarchical or association relations with other concepts, attributes, and so on. In this chapter, many examples are given (Figure 4.1(a) and (b) and Sections 4.6–4.8 of this chapter).
- 'Specification of the concept' is specifying the common agreed definitions or axioms that are satisfied by every ontology instance of the concept. This is the intention of the concept. It can also be specified by enumerating the instances of the concept or extension of the concept.

Table 4.2   Notation for representing ontology instance

Ontology notation	Semantics of the notation
	Circle–line represents the instance of ontological concept.

Note:
- In ontology, it does not matter whether it is in the upper- or sub-class concepts, as they can all have 'instances'.

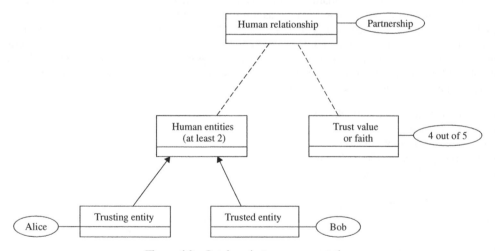

Figure 4.2   Ontology instance representation

Concept can be organized in generalization/specialization hierarchies or part of hierarchies and have association relationships with other concepts.

Notation 2 – Ontology Instance
An ontology instance is defined as a specific individual ontology concept or an example of the concept (Figure 4.2 and Table 4.2).

Table 4.3 Notation for representing ontology association relation

Ontology notation	Semantics of the notation
– – – – – – – – – –	A dotted line represents the ontology association relation, which represents that a concept is closely related to another concept. The cardinality and characteristics of the concept association can be added on top of the dotted line.

Note:
- The association relation between the concepts is an important part of the ontology representation. Often, one concept is related to another concept. It is important to comprehend this and the concepts themselves to understand its full semantics within a context.
- Here, the focus is on concept 'dependency' rather than 'relationship' between the concepts. Therefore, the cardinality is not central. However, for implementation constraints, if necessary, the notation '1:0, 1:1, 1:M or M:M' to express cardinality can be added.
- Here, the association relationship can be further classified by its relationship characteristics, such as one of the following:

- Explicit association (Chapter 2)
- Implicit association (Chapter 5.3.1)
- Asymmetry (Chapter 5.3.2)
- Transitivity (Chapter 5.3.3)
- Antonymy (or Inverse)(Chapter 5.3.4)
- Asynchrony (Chapter 5.3.5)
- Gravity (Chapter 5.3.6).

The notation to express the characteristics of the relationship is to add ≪Character≫ on top of the dotted line, as shown below.

| Explicit | Transitive | Asymmetry | etc |

In OWL, four association relationships are distinguished, and they are named as *Functional, Symmetric, Transitive* and *Inverse*.

Notation 3 – Ontology Association Relation

Association relation is defined as *a relation that represents one concept is closely related to another concept or concepts. It expresses a dependency between concepts*. The association relation is sometimes annotated with a name giving it specific semantics (Table 4.3).

Notation 4 – Generic- and Specific Ontology Relation

Generic–specific ontology relation expresses the generalization/specialization relation between the upper ontology concept, which is a generalization and a sub-ontology concept, which is a specialization (Table 4.4). The sub or lower ontology concept or concepts must commit to all the specifications of the upper ontology concept. The specific ontology involves additional constraints that stipulate the specialization.

Notation 5 – Composition/Part-of Ontology Relation

The ontology composition/part-of relation is defined as *a composition of lower level concepts to form an upper level composite* concept. The lower level concepts are in a 'part of' relation to the upper level concept (Table 4.5 and Figures 4.3(a–c)).

Table 4.4   Notation for representing generic–specific ontology relation

Ontology notation	Semantics of the notation
⟶	Line with solid arrows represents *the generalization and specialization relation*, which is a relation between upper–lower *generic and specific* concepts.

Note:
- The generalization and specialization relation is defined as *the relation between a super-class concept and a sub-class concept.* The sub-class concept inherits the properties of the super-class concept.

Table 4.5   Notation for representing include/part-of ontology relation

Ontology notation	Semantics of the notation
⟹	Line with open arrow represents composition and aggregation or part-of relationship between upper ontology concept and lower ontology concept.

Note:
- In an ontology presentation, we can distinguish part-of relation from *generalization and specialization* hierarchical relation and part-of relationship is a kind of aggregation relationship that may be mutually inclusive, exclusive or only part-of.
- The upper ontology concept may be a composite with a number of parts, and the lower ontology concepts are 'part of' the upper ontology concept.
- The composition/part-of relations can form a hierarchy.

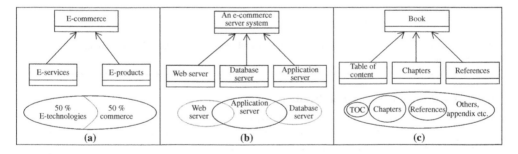

Figure 4.3   (a) Exclusive aggregation (b) Inclusive aggregation (c) Partial include or part-of
- In ontology, once a concept hierarchy is created, there is a consideration of whether it is making sense from the reverse point of view. For example, if a hierarchy if built from left to right, the reverse view is from right to left. In object/class relationships, the reverse view is very important and it must always be the case.

4.2.4 Summary of Ontology Notation

In Table 4.6, we give a summary of the five most important notations that are frequently used in ontology representations.

In the next section, we introduce key trust concepts and their concept hierarchy, for the ease of explanation of trust ontologies in the rest of the chapter.

Table 4.6 Ontology notation summary

Ontology notation	Semantics of the notation
	Double-field box represents the *ontological concept*.
	Circle–line represents the instance of ontological concept.
	A dotted line represents ontology concept association relation, which represents that a concept is closely related to another concept.
	Line with solid arrows represents *the generalization and specialization relation*, which is a relation between upper–lower, *generic and specific* concepts.
	Line with open arrow represents composition and aggregation or part-of relationship between upper ontology concept and lower ontology concept.

4.3 Hierarchy of Trust Concepts

This section provides the foundation for trust ontologies.

4.3.1 Trust Concept

Trust is realized by the concept of a trust relationship and trust value. Without a relationship, trust has no meaning. Each relationship denotes a trust value from the trusting party to the trusted party and it is for a particular context and time slot. A *context* can be decomposed into several aspects that can be used to derive the criteria for quality evaluation or measurement.

4.3.2 Hierarchy of Trust Concepts

In a service-oriented environment, trust can be validated at least in three domains, such as, agent, service and product. Each domain has many sub-domains, for example, for the agent, there could be service provider, or software agent, and so on. Figure 4.4 represents the Ontological view of trust and therefore, presents a hierarchy of upper and sub-trust domains. It has three levels:

At the top level of the diagram, we have the concept 'trust'.

At the middle-upper level of the diagram, there is a list of a number of generic concepts, namely, agent trust, service trust and product trust. These concepts shall commit (or confirm or inherit) the entire top layer ontology and its property specification. The detailed ontology for each of these domains is presented in the rest of this chapter.

At the lower level of the above diagram, we see a list of specific concepts, and the ontological view of each of these concepts is described as a specific ontology. Note that a specific ontology (sub-ontologies) shall commit to the specification of the upper ontology (or inherit properties from their generic ontology). These specific ontologies are detailed in the rest of this chapter.

Ontological representations of each of the above concept ('boxes') are shown in sections 4.6–4.8.

4.4 Hierarchy of Agents, Service and Product Concepts

This section gives a broad overview of agents, services and products that are available in a service-oriented environment. In order to understand the trust ontologies presented in this chapter, we need first of all to understand the hierarchy of these domain concepts.

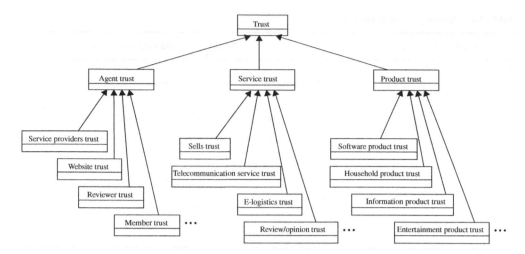

Figure 4.4   Trust concept and its domain hierarchy

4.4.1 Agent Concept Hierarchy

The trusting agent is an entity that has some level of faith or belief in the trusted entity. The trusted agent is an entity in which faith or belief has been placed by a trusting entity. Both trusting agent and trusted agent concepts (see definition in Chapter 2) have the same sub-class of concepts, such as business entity, human agent and software agent (Figure 4.5).

For details of the ontological concept presentation of each of the above in a trust context, please refer to Section 4.4.5.

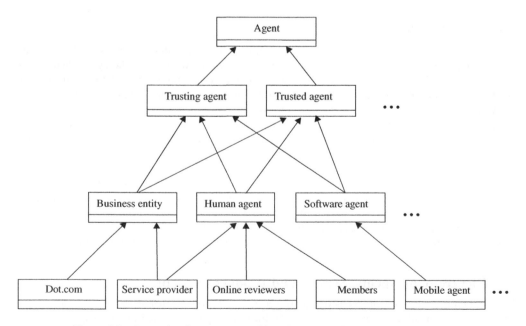

Figure 4.5   A sample of *agent concept* hierarchy in a service-oriented environment

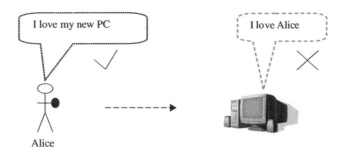

Figure 4.6   A trusting agent must have a human-like intelligence, whereas the trusted agent or entity may not

A trusting agent must be an intelligent agent, whereas a trusted agent can also be a product or a service that may not contain human-like intelligence. For example, Alice (human agent) can love her new PC (product); however, we would not normally say that a PC loves Alice (Figure 4.6).

4.4.2 Service Concept Hierarchy

In a service-oriented environment, there are many services, such as sales, supplies, marketing, advertising, middleman services, website hosting, travel agents services, producing products, information services, education services, IT&T services, customer opinions, security services, e-assessment, e-logistics, e-warehouses, and so on. They can be organized into groups or classified into several categories and can be organized in a hierarchical fashion (Figure 4.7).

For details of the ontological concept presentation of each of the above in the trust context, please refer to Section of 4.7 of this chapter.

4.4.3 Product Concept Hierarchy

In a service-oriented environment, there are many products (Figure 4.8).

For details of the ontological concept presentation of each of the above in the trust context, please refer to Section 4.8 of this chapter.

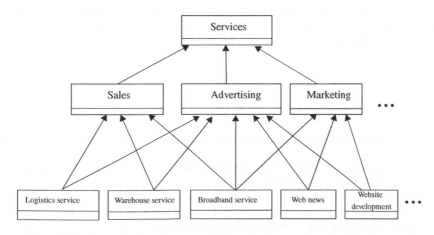

Figure 4.7   A sample of *service concept* hierarchy in a service-oriented environment

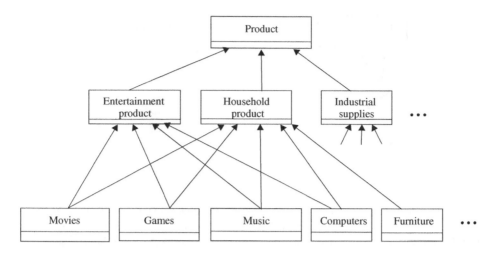

Figure 4.8   A sample of *product concept* hierarchy in a service-oriented environment

4.5 Hierarchy of Context and Association with Quality Assessment Criteria

Context, quality aspects and Quality Assessment Criteria are key concepts for the trust ontology in the service-oriented environment.

In the computer and IT fields, these concepts are widely used; however, they are not well defined and they have been presented in many ways and interpreted with many meanings. In this section, we formally define these concepts, especially for the service-oriented environment and their use in trust and reputation.

4.5.1 Example of the Context

Each context (a concept) has a number of quality aspects or conditions. For example, Alice trusts Bob in the *context* of 'delivery of a new TV'. There are several quality aspects of 'TV delivery', such as 'pick-up from specified shop on time', 'check overall new condition', 'shift TV without damage', 'deliver to the right address', and 'unload TV'. These *quality aspects* of a context 'TV Delivery' are the source for deriving the *Quality Assessment criteria* that can be used for the service quality measure (or measuring the trust). For example, 'pick up the right TV with the right product ID, with no mark, scratches or dent on the case when lifting and delivering, and unload TV and help set up. If all of these are achieved, the quality of service QoS is very good and the trust Alice has in Bob is high.

4.5.2 Hierarchy of Context, Quality Aspects and Quality Assessment criteria

Figure 4.9 depicts the formal representation of a context hierarchy in the service-oriented environment.

In Figure 4.9, the box on the far left represents context, such as 'The quality of Joe's dot.com company'.

In the middle section of Figure 4.9, the context is decomposed into a set of 'quality aspects' such as 'commercial value of the product' and 'percentage of discount', and so on.

In the boxes on the far right, we see a set of 'Quality Assessment Criteria' that are mapped to each of the quality aspects defined in the context. They are not part of the context; therefore, they are connected by the dotted line to the quality aspects. However, they are crucial to the context and quality aspects. Without criteria, one cannot judge the 'quality'.

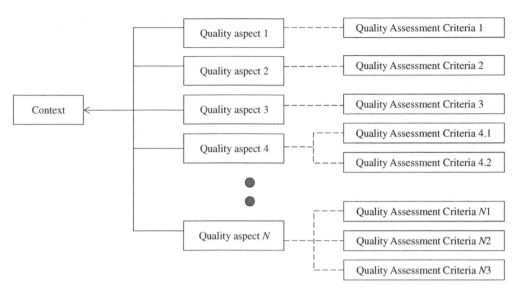

Figure 4.9   Hierarchy of a *context* with *quality aspects* and its relation to *Quality Assessment Criteria*

Examples of such criteria could include: 'commercial value of a product', 'number of customers that bought it', 'demonstration that many people are using it', 'always meets specification', and 'always provides 10% discount', and so on.

4.5.3 Definition of the Context

Context defines the nature of the agent and agent functions, service and service functions or product and product functions, and each *context* has a name, a type and a functional specification.

Example 1: Agent (context name), software agent (context type), mobile (context function).
Example 2: Camera (context name), product (context type), digital (context function), and so on.
Example 3: Local delivery (context name), transport (context type), pickup and distribution (context function).

A *context* can be decomposed into several quality aspects. With each quality aspect, there are always a set of Quality Assessment Criteria that can be used to measure the *quality aspect* under a particular *context*.

4.5.4 Definition of the Quality Aspects

Quality aspects decompose the *context* into several dimensions for the purpose of quality assessment or measurement. Quality aspects are sometimes interpreted as *conditions* for the *context*, for example, 'agent and its obligations 1, 2, 3', 'camera and its shutter speed and weight' or 'local delivery' and its 'Just-in-Time (JIT) service'.

4.5.5 Definition of the Quality Assessment Criteria

Quality Assessment Criteria define the quality metric for each *quality aspect* of the *context* for the purpose of measuring the *delivered quality* against the *defined quality*.

Quality Assessment Criteria must be explicitly defined against each *quality aspect*, so that the quality measure can take place and the results make sense to the public. Quality Assessment

Criteria help assessors or reviewers generate numerical values that quantify the quality aspects and the overall quality for a service or a product or an agent.

4.5.6 Examples Context, Quality Aspects and Quality Assessment Criteria
Example 1
If *context* is about quality of employee (agent trust), *quality aspect* may be 'attend work on time' (define each employee's obligation), *Quality Assessment Criteria* could be as follows: If 'always attends work on time', assign *excellent;* If '80 % of time attends work on time', assign *good* and if 'never attends work on time', assign *very bad*, and so on.

Example 2
If *context* is about quality of an academic (agent trust), *quality aspect* may be 'research', 'teaching' and 'administration' and *Quality Assessment Criteria* for 'research' could be 'publish at least two papers a year'; for 'teaching', the criteria could be 'the average positive feedback from students is above 70 %'; and for 'administration', the criteria could be 'served as a graduate coordinator' at least one year over the last five years, and so on.

Example 3
If *context* is about the quality of research (service trust), *quality aspect* may be 'publications', or 'external funding' or define quality aspects from service-level agreement or service standard, and so on. *Quality Assessment Criteria* for publications could be 'number of books, top journal publications, refereed conference papers', and if total publication is over five per year, assign 5 star; if total publication is three per year, assign 3 star and so on, and if no books, no journals and only two non-refereed conference papers are to the credit, assign 1 star and so on.

Example 4
If *context* is digital camera (product trust), *quality aspect* may be 'battery life' or 'zoom capacity' or define quality aspects from quality standard, and so on. *Quality Assessment Criteria* for battery life could be, 'up to 4 hours or 200 pictures', assign *Best*, and so on, and 'less than 20 minutes and can only take up to 20 pictures' assign 'poor', and so on.

4.6 Agent Trust Ontology
As explained earlier, trust in the service-oriented environment can represent three domains, namely, agent trust, service trust and product trust. In this section, we describe the agent trust ontology, including the generic and specific agent trust ontologies. The generic agent trust ontology is also known as *an upper ontology* and the specific agent trust ontology is also known as the *sub-ontology*.

4.6.1 Challenges in Agent Trust
An agent in the service-oriented network environment (as described in Chapter 1) is an intelligent autonomous entity. It can be a buyer, a seller, a supplier, a merchant, a website, a service provider or an interacting peer, a software component or an intelligent server running behind the scenes on the network.

Agent trust has the following common characteristics:

(1) Agent trust is private or individual. This implies that this trust does not apply to other agents. Consider the following example: Alice trusts Bob with a trust value of 5 out of 5. However, Sarah trusts Bob with a trust value of 2 out of 5. This means that the trust value is confined to each individual's trust.

(2) Agent trust is direct trust. Trust is established through direct interaction (see initiation of the relationship in Chapter 2). This is real trust because this trust is determined by the trusting agent, and not by others' recommendations. This is different from trust that is determined by the example of 'He is a friend of my friend, therefore I trust him as much as I trust my friend' which is called *transitive trust* – it is not a direct agent trust.

(3) Agent trust can be shared. An agent may be willing to share his or her trust, belief or opinion with others. This is commonly regarded as a good quality of the agent. If an agent likes to share this with other agents, the other agents receive great benefit from this valuable information. In sharing this trust with others, the agent may also receive, in return, other agents' beliefs, opinions or experience with their trusted agent or entities. This facilitates the formation of an open trusted community in the service-oriented environment. This in turn, benefits every agent in the community or networked economy. However, an agent should not take without giving, which is commonly referred to as *selfish behaviour*, as sooner or later the agent will be isolated and left out of the trusted community.

4.6.2 Choice of Agents

In the next few sections, we shall introduce the agent trust ontology as generic ontology and different agent types such as sellers, service providers, websites, brokers and reviewers as specific agent trust ontologies. Agents can be of many types including human agents and computer agents (see Chapter 1). We have chosen several agent types for discussing agent Trust. The reason these agents are chosen for discussion is because they represent the key players in the current generation of e-business and e-service operations.

4.6.3 Generic Agent Trust Ontology

Generic Agent Trust Ontology: In a service-oriented network environment, the generic agent trust ontology is defined as the conceptualization of the agent trust that the *trusting agent* believes the *trusted agent's ability to fulfil its obligation* in a given *context* and a given *time slot*.

The graphical view of generic agent trust ontology is shown in Figure 4.10 through the use of the ontology notation.

In the ontology diagram (Figure 4.10), boxes represent the ontological concept, next to each box could be examples or instances of the concept (such as Liz for trusting agent, and Bob as trusted agent, and 5/5 as trustworthiness value, and so on; here we will not present this visually). An arrow line represents the upper class and lower class of concepts. A dotted line represents navigation to an association concept. Association classes are used for associations that themselves participate in an association with another class.

We represent the *generic agent trust* ontology as the combination of the ontology name and a tuple where the elements of the tuple can be complex elements as defined below.

Agent trust (trusting agent, trusted agent, context, quality aspects, Quality Assessment Criteria, time slot, and trustworthiness value), where the following definitions hold good:

- 'Trusting agent' is one who determines and has the trustworthiness value of the trusted agent.
- 'Trusted agent' is one whose trust is being considered by other agents.
- 'Context' refers to agent functions or responsibility (such as a job title) for which the trust is being considered.
- 'Quality aspects' define the obligations of the agent.
- 'Quality Assessment Criteria' are metrics that are used to assess the agent's fulfilment of its obligations or trustworthiness.

Figure 4.10   Ontology representation of *agent trust concept* and its relation to other concepts

- 'Time slot' is a time or a time frame for which the trust value holds. During the time slot, the trust value remains constant.
- 'Trustworthiness value' is a measure of the trust against a trustworthiness scale.

Note that several elements in the tuple are complex elements; there can be complex data structures, for example, sets, tuples, tables, lists, and so on. *For example, Quality Assessment Criteria' is another tuple.*

There are three main features in the above ontology:

(a) The use of the words 'trustworthiness value' instead of 'trust value'. This is because the trust value can represent past trust relationships and past trust values, while the trustworthiness value represents the present and future only.
(b) The use of the word tuple. A tuple contains details of the trust relationship. The tuples are stored in the trusting agent's trust database. A trust database will contain tables that store the trust tuples. The tuples in the table form a matrix. Therefore, a table is a matrix. Each tuple is stored as one row in a table. When determining the trustworthiness of a trusted agent, the trusting agent checks the tuples against the context, time slot and trustworthiness value. By doing this, the trusted agent can determine the appropriate trust value or trustworthiness value of the trusted agent.
(c) The separate representation of agent trust from service trust and product trust.

A specific agent trust ontology could be for sellers, suppliers, brokers, websites, travel agents, manufacturers, information providers, education providers, IT service providers, security providers and the like. These agent functions and obligations are all different. The specific agent trust should commit to all of the generic agent trust ontology (concepts and property specification) explained in the preceding text. In the next few sections, we shall detail these specific ontologies or sub-ontologies.

4.6.4 Specific Agent Trust Ontology – Service Provider Trust

Service providers can be sellers, suppliers, brokers, websites, travel agents, manufacturers, information providers, and so on. Note that, we focus on a provider, and not on the service, although the quality of the service provider can be represented by the QoS they provide; in agent trust, there is no material data for the measure, such as there is no concrete contract but obligations between agents, whereas in the service trust, one can measure the QoS delivery based on the formal contract or service-level agreement or standards, and so on. Agent trust can also be viewed as an overall figure of merit of the context specified.

Service Provider Trust Ontology: In service-oriented network environments, the service provider trust ontology is defined as the conceptualization of the *service provider trust* that the *service requester trusts* the *service provider's ability to fulfil its obligations* in a given *responsibility context* and a given *time slot*.

The graphical view of service provider trust ontology is shown in Figure 4.11 through the use of the ontology notation.

We represent a *service provider trust* ontology as the combination of the ontology name and a tuple where the elements of the tuple can be complex elements as defined below.

Figure 4.11 Ontology representation of *service provider trust concept* and its relation to other concepts

Service provider trust (service requester, service provider, context, quality aspects, Quality Assessment Criteria, time slot and trustworthiness value), where the following definitions hold good:

- 'Service requester' is a trusting agent who has the trustworthiness value of the service provider.
- 'Service provider' is a trusted agent whose trust is being considered by the service requester.
- 'Context' refers to elements such as 'commitment' from the trusted agent or the service provider.
- 'Quality aspects' define the quality of the service provider from the customer's perspective.
- 'Quality Assessment Criteria' are metrics that are used for the assessment of a service provider's trustworthiness.
- 'Time slot' is the time frame for which a trust value holds, that is, during this period the trust value remains constant.
- 'Trustworthiness value' is a measure of the trust against a trustworthiness scale.

Consider the example of Third party logistics (3PL) provider trust:

Requester: GF manufacturer
Provider: Perth 3PL
Context: Local delivery
Quality aspects (obligations): On-time pick-up and delivery and allow track and trace
Quality criteria: If Perth 3PL always provides JIT pickup and delivery, assign 'highly reliable'; If JIT is not guaranteed and able to track and trace the 3PL, assign 'reliable', and so on.

In a similar manner, we can define the specific trust ontology for website, reviewers, feedback agents, virtual community members, and so on.

4.6.5 Specific Agent Trust Ontology – Websites Trust

A website is a special agent that acts like a front-end proxy for suppliers, manufacturers, service providers, brokers, retailers, shopping malls, or information resources. When it is the direct selling outlet for manufacturers, it is a seller, when it sells the goods or services for many business providers, it is a broker. A website that is evaluated here is one that can carry out business interactions and is able to deliver some services for customers or online users, including information repository websites.

As a website is manipulated by human intelligence, it is not just passive computer code providing information. It often contains intelligent software agents and is able to carry out tasks autonomously. Sometimes, human intervention is behind the functions or operations.

Website Trust Ontology: In service-oriented network environments, the website trust ontology is defined as the conceptualization of a website trust that an *online user or a customer trusts* a *website* and its *honesty in presenting true information (verifiable)* in a given *context* and a given *time slot.*

The graphical view of the website trust ontology is shown in Figure 4.12 through the use of the ontology notation.

We represent *a website trust* ontology as the combination of the ontology name and a tuple where the elements of the tuple can be complex elements as defined below.

Website trust (user, website, context, quality aspects, Quality Assessment Criteria, time slot and trustworthiness value) where the following definitions hold good:

- 'Customer or user' is a trusting agent who holds a trustworthiness value of a website.
- 'Website' is a trusted agent whose trust is being considered by the customer or users.

Figure 4.12   Ontology representation of *website trust concept* and its relation to other concepts

- 'Context' may be a site function or web content for which the trust is defined.
- 'Quality aspects' define a website's quality (from a customer's perspective).
- 'Quality Assessment Criteria' are metrics that are used for the assessment of the website's trustworthiness.
- 'Time slot' is the time frame for which the trust value holds, that is, during this period, the trust value remains constant.
- 'Trustworthiness value' is a measure of the trust against a trustworthiness scale.

Consider the example of an air travel broker site:

Trusting agent: Discount ticket buyer
Trusted agent: Hot deal centre
Context: Lowest international airfare
Quality aspect: Able to provide at least three low airfares (including the lowest) and for any airlines
Quality Assessment Criteria: Able to find three airlines and provide at least 10 % discount on the direct sales price, then assign 'very good'; Unable to find some airlines, and give the longest travel route (worst travel route) with minor discount, then assign 'poor', and so on; and never able to provide the right information and low airfares, then assign 'totally useless', and so on.

4.6.6 Specific Agent Trust Ontology – Reviewer Trust

From a commercial point of view, getting feedback or opinions from customers or buyers is an important business intelligence strategy. Alternatively, getting third party opinions or comments is also a way to justify the quality of providers or quality of agents. Recently, in the e-commerce environment, we have noticed several large websites asking users to give opinions and feedback (see Chapter 7). Feedback agents are also known as *reviewers*. They may be customers, end-users, buyers, online shoppers or professionally employed reviewers, and so on. In an anonymous, open, distributed network environment, a provider may not able to request physical customers to provide their opinion or give feedback. Therefore, one of the strategies could be to ask reviewers to provide feedback. Because the reviewer can be anonymous or pseudo-anonymous, it is important to carry out an automated trustworthiness validation for the reviewers (feedback agents) in order to make sure that the reviewer is honest and genuine.

Reviewer Trust Ontology: In a service-oriented network environment, the reviewer trust ontology is defined as the conceptualization of the reviewer trust that the *feedback receiver* has trust in the *reviewer's ability to provide quality feedback on a given topic and in* a given *time slot.*

The graphical view of the reviewer trust ontology is shown in Figure 4.13 through the use of the ontology notation.

We represent *a reviewer trust* ontology as the combination of the ontology name and a tuple where the elements of the tuple can be complex elements as defined below.

*Reviewer trust (feedback receiver, reviewer, topic, quality aspects, time slot and trustworthiness value), w*here the following definitions hold good:

- The feedback receiver is a trusting agent who holds the trustworthiness value of the reviewer.
- The reviewer is a trusted agent whose trust is being considered by the feedback receiver.
- 'Context' refers to elements such as a 'Topic' from the reviewer's feedback or performance of the reviewer for which they are areas of interest for which the trust is evaluated.
- Quality aspects represent the reviewer qualities along several dimensions that can be used to define the quality of the reviewer.
- Quality Assessment Criteria measure each quality aspect.

Figure 4.13   Ontology representation of *reviewer trust concept* and its relation to other concepts

- Time slot is the time for which the feedback is given.
- 'Trustworthiness' is a measure of the trust against a trustworthiness scale.

Consider, for example, a paper review:

Trusting agent: IE council
Trusted agent: Paper referee Dr Rob
Context: Journal papers reviewer
Quality aspect: Reviewer gives fair opinion and marks
Quality measure: If all the marks provided are in a normal distribution, assign 'trustworthy'. If all the marks provided are below 40 %, assign 'untrustworthy', and so on.

4.6.7 Specific Agent Trust Ontology – Member Trust

A member is part of a community. If the community is a virtual community, since agents may not be able to meet or know each other face to face, building member trust is important to every community member. The member may be part of an alliance, or in partnership, or in a virtual collaborative team. Such trust helps community productivity, efficiency and security.

Member Trust Ontology: In a service-oriented network environment, a member trust ontology is defined as *the conceptualization of the member trust that the trusting member has trust in the trusted member to fulfil its obligation in a given context and a given time slot'*.

The graphical view of the member trust ontology is shown in Figure 4.14 through the use of the ontology notation.

We represent *a member trust* ontology as the combination of the ontology name and a tuple where the elements of the tuple can be complex elements as defined below.

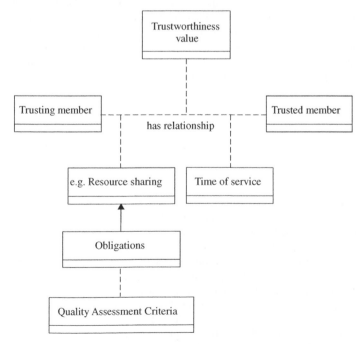

Figure 4.14   Ontology representation of *member trust concept* and its relation to other concepts

Member trust (trusting member, trusted member, context, Quality Assessment Criteria, time slot, trustworthiness value), where the following definitions hold good:

- The trusting member is a trusting agent who has the trustworthiness value for the trusted member.
- The trusted member is a trusted agent whose trust is being considered by the trusting member.
- 'Context' may be 'File Sharing' or 'Computer Power Sharing'.
- 'Quality aspects' could be conditions in the member contract or rules as set out in the virtual community agreement for which the trust relates.
- 'Quality Assessment Criteria' are metrics that are used to assess the member's fulfilment of its obligations or trustworthiness.
- Time slot is the time frame for which the trust value holds. During this period, the trustworthiness value remains constant.
- 'Trustworthiness' is a measure of the trust against a trustworthiness scale.

In this section, we have introduced the trust ontology for the trustworthiness of agents. The trustworthiness value is based on the fulfilment of its obligation as an intelligent agent. All agents in society can rate each other's trustworthiness. It is impractical to list every type of agent's trust (specific agent trust). While the generic ontological concept for agent trust has been provided, this can be easily adopted by another type of agent trust, for example, a seller trust, or a procurer trust, and so on.

4.7 Service Trust Ontology

As explained earlier, trust in service-oriented environments can represent three domains, namely, agent trust, service trust and product trust. In this section, we describe the service trust ontology, such as the generic and specific service trust ontology.

Service providers provide services to other agents or the public. The service may be free or may require a fee. Service providers in an integrated business may offer services such as sales of consumer goods, transportation, retail, broker service, mobile phone networks, voice over IP, e-education, e-books, information resource sharing, grid service, file sharing over P2P networks, or entertainment, and so on. The distinction made in this section is between the ontology of trust for the 'QoS' that the agent carries out compared to the last section where the ontology of *'agents'* was considered.

4.7.1 Key Issues in Service Trust

The key challenge of a QoS assessment (service trust) is to define a set of criteria for the measurement of QoS.

There are three issues:

(a) The criteria must be domain specific and domain knowledge rich.

Domain knowledge rich is specific expert knowledge from a specific domain, such as knowledge of logistics or knowledge of telecommunications. One cannot set criteria that do not reflect the specific service. The criteria must be domain knowledge rich.

Without long-term experience in the domain, it is not possible to develop a good set of criteria that can justify the QoS. An agent who is not a domain expert is unqualified to define the criteria for the measurement of domain-specific services. This must be done in conjunction with domain experts and through long-term evaluation validation and testing of services.

For example, one criteria to measure a 'Good Business' could be 'the business results in economic benefits'. Such criteria can be interpreted in many ways, and if a subset of these criteria is not defined, the assessment will be difficult and may result in unfair review. The challenge in the measurement of QoS is to define domain knowledge rich criteria that are relevant and specific to a domain; most importantly, direct materialized measurement and criteria are very useful. For example 'if a small business results in sales income at $80 K per year, assign the rating 'Good Business'.

(b) The criteria must be narrowly defined, clear and specific.

If the criteria are too broad or at too high a level, not clear, ambiguous or too implicit, then it may be impossible to accurately reflect or measure the QoS that is provided to the trusting party by the trusted party. For example, if we are asked to measure 'WestOil company's QoS in oil, gas and petroleum', the trusting party (customer) may not be able to fully understand what is meant by 'quality' and may ask 'Is oil and gas the same as petroleum, and so on' and they cannot measure the QoS. In this case, the trusted party (Oil and Gas Company) can abrogate their responsibilities to the trusting party (customer). Thus, another challenge in the measurement of QoS that is to make the criteria clear so that they can be understood by customers or quality assessors.

(c) The weight of each criterion is different.

Many service providers normally enter into service agreements that span a longer period, such as in telecommunication services, banking, insurance, and so on. They may contain a lot of the quality aspects of service and criteria for measurement. We must decide which criteria are important and which are not, that is, which criteria have the most impact on the services. The weight of each criterion should be quantified differently according to the importance of each criterion. For example, a power and electricity company may think JIT service is the most important criteria for the QoS; however, the customer may think that "to ensure there is no 'blackout'" is the most important criteria for power supply. There is a discrepancy between the importance of criteria according to the customer and the supplier (power and energy company), and this has to be worked out and agreed upon by both parties before the assessment can be carried out.

Once well-defined criteria have been established, the measurement of QoS is not a difficult task. In Chapter 6, we address the challenges in the measurement of QoS and provide examples. In Chapter 7, further examples are given.

Measuring service trust has the following characteristics:

(1) QoS is measured against two sources:
- Use of a QoS standard that is commonly understood by service buyers and suppliers; and
- Use of a service-level agreement or contract that is signed and understood by service buyers and suppliers.

2) QoS is judged by the fulfilment of the quality service aspects that are explicitly set out in the service standard or service agreement.

3) QoS is seen from two perspectives. The nature of standards or agreement about the services implies that the judgement of the quality has to consider both parties' perspectives. This is not like making a comment about an agent or product, where only one perspective is generally given.

4) QoS is measured by considering the customer's input as well as input from the service provider. This is often necessary because both parties' voices can be heard, thus, providing an open and fair assessment process.

4.7.2 Choice of Services

Note here that one should not confuse the service provider (the agent) with the service (provided by the agent). In order to explain the services, we normally refer to services such as sales, supplies, marketing, advertising, middleman services, websites hosting, travel agents services, producing products, information services, education services, IT&T services, customer opinions, security services, e-assessment, e-logistics, e-warehouses, and so on. They represent the most active service types that exist in the e-business and e-service network. We will give ontology representation examples for description of the service trust ontology.

4.7.3 Generic Service Trust Ontology

The measurement of QoS is done against a set of criteria that are domain rich and specific. Domain rich and specific refers to a substantial depth of knowledge represented in a particular domain. The measurement of QoS or trustworthiness of services shares the same ontology, the 'service trust ontology'.

Note here that we focus on *service* and not on the *provider*. Each service has a unique set of criteria or conditions that are used for the measurement of QoS.

Generic Service Trust Ontology: In service-oriented network environments, the service trust ontology is defined as the conceptualization of the service trust that the *customer* has trust in the service from the *service provider to have the required QoS* in the given service context and time slot.

Note that the *criteria* are the conditions that are usually found in the service standard or service agreement, which can be used to measure the QoS. They should be commonly understood by both parties.

Note that service here means a service context. A context defines the nature of a service.

The graphical view of service trust ontology is shown in Figure 4.15 through the use of the ontology notation.

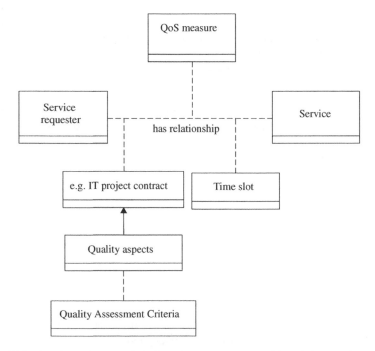

Figure 4.15 Ontology representation of *service trust concept* and its relation to other concepts

We represent the *generic service trust ontology* as the combination of the ontology name and a tuple where the elements of the tuple can be complex elements as defined below.

Service trust (service requester, service provider, service context, quality aspects, QoS Criteria, service time slot and trustworthiness value], where the following definitions hold good:

- 'Service requester' is a trusting agent who has the trustworthiness value of the QoS of the service from the service provider.
- 'Service' is the trusted entity whose QoS is being considered by the service requester.
- 'Context' refers to nature of service such as 'project management' or 'a service' or 'a service function'.
- 'Quality aspects' define the QoS.
- 'QoS Criteria' are metrics that are used for quality assessment of the trustworthiness of the services.
- 'Time slot' is the time frame for which the trust value holds, that is, during this period, the trust value remains constant.
- 'Trustworthiness' is a measure of the trust against a trustworthiness scale.

Specific service trust ontologies for sales, supplies, marketing, advertising, and so on, should commit to all of the ontology concepts and its properties as in the generic service trust ontology defined in this section.

4.7.4 Specific Service Trust Ontology – Sales Trust

Sales services are popular both online and off-line. All agents in society can offer sales services. However, not all can be successful. Buying and selling affect each individual and every business. Not everyone always has happy experiences with either method.

Sales Service Trust Ontology: In service-oriented network environments, the sales service trust ontology is defined as the conceptualization of the trust that the *buyer* has in the *seller's sevice* to fulfil sales standard (quality aspects) in the given sales *context* and a given *time slot*.

Note that the *criteria* are the conditions that are usually used in the service standard. They should be commonly understood by both buyers and sellers.

The graphical view of the sales trust ontology is shown in Figure 4.16 through the use of the ontology notation.

We represent the *sales service trust ontology* as the combination of the ontology name and a tuple where the elements of the tuple can be complex elements as defined below.

Sales trust (buyer, seller's service, selling context, quality aspects, QoS Criteria, time slot and trustworthiness value) where the following definitions hold good:

- 'Buyer' is a trusting agent who has a trustworthiness value of the QoS of the seller.
- 'Seller's service' is a trusted entity whose QoS is being considered by the buyer.
- 'Context' such as about 'sales' of any goods or services.
- 'Quality aspects' define the quality of the sales service from the buyer's perspective.
- 'QoS Criteria' are a set of criteria used to measure the quality of sales of the service provider.
- 'Time slot' is the time frame for which the trust value holds, that is, during this period, the trust Value remains the same.
- 'Trustworthiness' is a measure of the trust against a trustworthiness scale.

Examples of e-sale criteria could be the following:

- Price of sales
- Delivery options
- On-time delivery

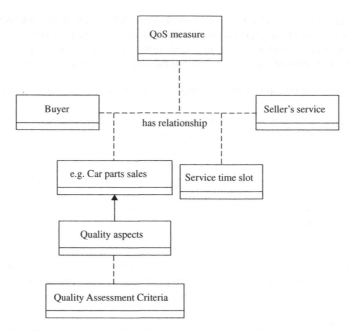

Figure 4.16 Ontology representation of *sales trust concept* and its relation to other concepts

Table 4.7 Trustworthiness levels for e-Sales

Trustworthiness scale (ordinal scale)	Semantics for sales service (linguistic definitions)	*Alternate* semantics for sales service
Level – 1	Unknown	Unknown
Level 0	Very untrustworthy	Poor
Level 1	Untrustworthy	Not good
Level 2	Partially trustworthy	Average
Level 3	Largely trustworthy	Fair
Level 4	Trustworthy	Good
Level 5	Very trustworthy	Excellent

- Ease of purchase
- Customer support.

The trustworthiness level for each of these criteria can be defined as given in Table 4.7.

In a similar manner, we can define specific ontologies for telecommunications, e-logistics, e-education, anonymous review, and so on.

4.7.5 Specific Service Trust Ontology – Telecommunication Service Trust

Telecommunication service is a lucrative service for generating revenue and providing service to the mass market. It includes telephones, computer communications, Internet, mobile service, VoIP (Voice over IP), teleconferencing, broadband, Asymmetric Digital Subscriber Line (ADSL) and cable TV services. It is a popular service in every home and business. However, you will not be surprised to learn that not all customers are satisfied with the services they pay for.

Telecommunication Service Trust Ontology: In service-oriented network environments, the telecommunication service trust ontology is defined as the conceptualization of the trust that the *customer* has in the *telecommunication service provided by the service provider to have the QoS* (quality aspects defined in the service standard or service agreement) in a given *context* (such as ADSL service) and a given *time slot*.

Note that the *criteria* are the conditions that are usually used in the service standard or service agreement. They should be commonly understood by both customers and providers.

The graphical view of telecommunication service trust ontology is shown in Figure 4.17 through the use of the ontology notation.

We represent the *telecommunication service trust ontology* as the combination of the ontology name and a tuple where the elements of the tuple can be complex elements as defined below.

Telecommunication service trust (service customer, telecommunication service, communication service context, quality aspects, QoS Criteria, time slot and trustworthiness value), where the following definitions hold good:

- 'Service customer' is a trusting agent who has a trustworthiness value of the QoS of the telecommunication service provider.
- 'Telecommunication service' is a trusted entity whose QoS is being considered by the service customer.
- 'ADSL service' is an example of context of the telecommunication service, another example of a 'context' could be 'mobile phone service'.
- 'Quality aspects' define the quality of the service for the purpose of measurement and sometimes can be used to compare the quality of other similar services.
- 'QoS Criteria' are a set of criteria used to measure the quality of the telecommunication services that the service provider provides.

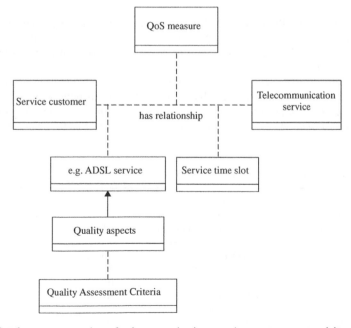

Figure 4.17 Ontology representation of *telecommunications service trust concept* and its relation to other concepts

- 'Time slot' is the time frame for which the trust value holds, that is, during this period, the trust value remains constant.
- 'Trustworthiness' is a measure of the trust against a trustworthiness scale.

Examples of telecommunication service criteria could be the following:

- Cost per call
- Monthly service charges
- Hidden charges
- Flexibility of the plan
- Cancellation options
- Free call period
- Account traceability
- Customer support

The trustworthiness level for each of these criteria can be defined as shown in Table 4.8.

4.7.6 Specific Service Trust Ontology – E-Logistics Trust

E-logistics provides online logistics bookings and orders for transportation (road, rail, sea and air), shipping requirements, the pick-up and delivery of goods, import and export document exchange, and so on. Logistics is a crucial part of any type of business, both online and off-line. However, inexpensive and good quality services can be difficult to find.

Logistics Service Trust Ontology: In a service-oriented network environment, the logistics service trust ontology is defined as the conceptualization of the trust that *the logistics customer* has in the *logistics provider's* capability of fulfilling its commitment set out in the service agreement and a given *time slot'*.

Note that the criteria are the conditions that are set out in the service agreement. They should be commonly understood and agreed to by both logistics customers and providers.

The graphical view of the logistics service trust ontology is shown in Figure 4.18 through the use of the ontology notation.

We represent the *e-logistics service trust ontology* as the combination of the ontology name and a tuple where the elements of the tuple can be complex elements as defined below.

Logistics service trust (logistics customer, logistics service provided, logistics service context, QoS Criteria, service time slot and trustworthiness value), where the following definitions hold good:

Table 4.8 Trustworthiness levels for telecommunication services

Trustworthiness scale (ordinal scale)	Semantics for *telecommunication service* (linguistic definitions)	*Alternate* semantics for *telecommunication service*
Level – 1	Unknown	Unknown
Level 0	Very untrustworthy	Terrible
Level 1	Untrustworthy	Bad
Level 2	Partially trustworthy	Average
Level 3	Largely trustworthy	Fair
Level 4	Trustworthy	Good
Level 5	Very trustworthy	Excellent

Figure 4.18 Ontology representation of *logistics service trust concept* and its relation to other concepts

- The logistics customer is a trusting agent who has the trustworthiness value of the QoS of the logistics service delivered by the provider.
- The logistics service is a trusted entity whose QoS is being considered by the logistics customer.
- Delivery is the context of the service, such as the transportation delivery services, to which the trust relates.
- 'Quality aspects' define the quality of the service for the purpose of measurement and sometimes can be used to compare the quality of other similar services.
- QoS Criteria is a set of criteria or conditions used to measure the quality of the logistics service of the logistics provider.
- Service time slot is a time or a time frame for which the trust value holds. During the service time slot period the trust value remains constant.
- 'Trustworthiness value' is a measure of the trust against a trustworthiness scale.

Examples of e-logistics' criteria could be the following:

- Agreed cost (no change on delivery)
- Intact delivery
- On-time pick-up
- On-time delivery
- Goods handling
- Track and trace capability
- Money back guarantee.

The trustworthiness level for each of these criteria can be defined as shown in Table 4.9.

4.7.7 Specific Service Trust Ontology – E-Warehouse Trust

E-Warehouse provides online warehouse space bookings ranging from frozen, to chilled goods or goods that are stored at normal temperature and also goods that require climate control such as

Table 4.9 Trustworthiness levels for logistics services

Trustworthiness scale (ordinal scale)	Semantics for logistics service (linguistic definitions)	Alternate semantics for logistics service
Level – 1	Unknown	Unknown
Level 0	Very untrustworthy	Awful
Level 1	Untrustworthy	Poor
Level 2	Partially trustworthy	Average
Level 3	Largely trustworthy	Fair
Level 4	Trustworthy	Good
Level 5	Very trustworthy	Excellent

humidity control. E-Warehouse handles goods movement online requests, such as 'inwards' (into the warehouse), 'outwards' (from the warehouse) or assignment of ownership (transfer the ownership of the goods when goods are sold). Warehouse services are one of the longest surviving businesses in the service industry. However, not all warehousing providers can provide a leading edge QoS.

Warehouse Service Trust Ontology: In a service-oriented network environment, the warehouse service trust ontology is defined as the conceptualization that represents the trust that *the warehouse customer has in the warehouse service given by the provider its fulfilment of commitments set out in the service agreement in a given context and a given time slot'*.

Note that the criteria are the conditions that are set out in the service agreement. They should be commonly understood and agreed to by both warehouse customers and providers.

The graphical view of the e-warehouse service trust ontology is shown in the Figure 4.19 through the use of the ontology notation.

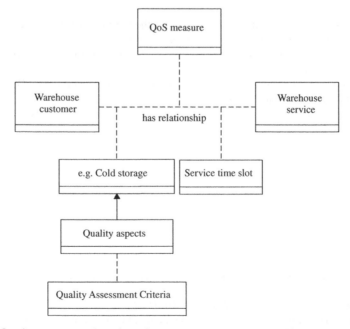

Figure 4.19   Ontology representation of warehouse *service trust concept* and its relation to other concepts

We represent the *e-warehouse service trust ontology* as the combination of the ontology name and a tuple where the elements of the tuple can be complex elements as defined below.

Warehouse service trust (warehouse customer, warehouse service, storage, QoS aspects, QoS Criteria, time slot and trustworthiness value), where the following definitions hold good:

- 'The warehouse customer' is a trusting agent who has the trustworthiness value of the QoS of the warehouse service given by the provider.
- 'The warehouse service' is a trusted entity whose QoS is being considered by the warehouse customer.
- 'Storage' is the context of the service, such as cold storage services, to which the trust relates.
- 'Quality aspects' define the quality of the service for the purpose of measurement and sometimes can be used to compare the quality of other similar services.
- 'QoS Criteria' are a set of criteria or conditions used to measure the quality of the warehouse service of the warehouse provider.
- 'Time slot' is a time or a time frame for which the trust value holds. During the service time slot period the trust value remains constant.
- 'Trustworthiness value' is a measure of the trust against a trustworthiness scale.

Examples of e-warehouse criteria could be the following:

- Cost of storage (rental)
- Order fulfilment
- Material handling (goods handling)
- Automation of goods check-in and check-out
- Track and trace capability
- Customer service.

The trustworthiness level for each of these criteria can be the same as for the logistics services.

4.7.8 Specific Service Trust Ontology – E-Education Trust

E-Education is now provided by most countries, many universities and even some schools as they move more and more towards commercialization. It provides a social and economic benefit to learners and education providers. However, education services in both the online and off-line realms, have quality issues that affect both the learners and the education providers. It is important to learn the trustworthiness of education providers and their service as well as to examine the quality of learning from the learners or graduates.

Teaching Service Trust Ontology: In a service-oriented network environment, the education service trust ontology is defined as the conceptualization of the trust that the learner has in the education service given by the provider to deliver the quality of teaching (context) set out in the service standard and *a given time slot'*.

The graphical view of the education service trust ontology is shown in Figure 4.20 through the use of the ontology notation.

We represent the *education service trust ontology* as the combination of the ontology name and a tuple where the elements of the tuple can be complex elements as defined below.

Education service trust (learner, education service, teaching, QoS Criteria, time slot and trustworthiness value), where the following definitions hold good:

- 'The Learner' is a trusting agent who has the trustworthiness value of the QoS of the education service given by the provider.

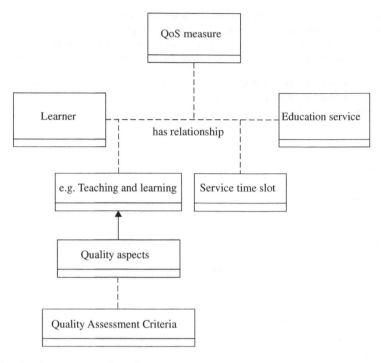

Figure 4.20   Ontology representation of *education service trust concept* and its relation to other concepts

- 'The education service' is a trusted entity whose QoS is being considered by the Learner.
- 'Teaching' is the context of the service, such as teaching an MBA degree course, to which the trust relates.
- 'Quality aspects' define the quality of the service for the purpose of measurement and sometimes can be used to compare the quality of other similar services.
- 'QoS Criteria' is a set of criteria or conditions used to measure the quality of the teaching service of the education provider.
- 'Time slot' is a time or a time frame for which the trust value holds. During the service time slot period the trust value remains constant.
- 'Trustworthiness value' is a measure of the trust against a trustworthiness scale.

Examples of education service criteria could be as follows:

- Teacher's qualification
- Teacher's knowledge
- Teaching content
- Course materials
- Mode of delivery
- Assessment
- Tutorial and consultation support
- Approaches of lecturers.

The trustworthiness level for each of these criteria can be defined as shown in Table 4.10.

Table 4.10 Trustworthiness levels for teaching services

Trustworthiness scale (ordinal scale)	Semantics for *teaching service* (linguistic definitions)	Alternate semantics for *teaching service*
Level – 1	Unknown	Unknown
Level 0	Very untrustworthy	Worst
Level 1	Untrustworthy	Bad
Level 2	Partially trustworthy	Average
Level 3	Largely trustworthy	Fair
Level 4	Trustworthy	Good
Level 5	Very trustworthy	Very good

4.7.9 Specific Service Trust Ontology – Review or Opinion Trust

Collecting customers' opinions or obtaining third party reviewer's perspectives are services that are becoming more and more important for both online and off-line businesses, governments and individuals. If managed properly, this service could provide unbiased assessment, fair judgement and true evaluation that is beneficial to all parties involved and for the society as a whole. It provides a validation of the service provider's quality, service and performance, and so on. The review or opinion could be in the form of surveys, questionnaires, form completion, open-ended comments or feedback, criteria measures and ticks, or yes or no options. It can involve both qualitative and quantitative information.

Here we treat an opinion as a simple version of a review. Reviews can be much more complex and lengthier than opinions. Reviews normally have a formal process, whereas opinions can be more freestyle in nature. Reviews can be driven by rules and regulations, and may have legal and social responsibilities, whereas opinions are free speech with relatively little legal and social responsibility. We shall discuss the review trust ontology, which will also adequately cover the opinion trust ontology.

Review Trust Ontology: In a service-oriented network environment, the review trust ontology is defined as '*the conceptualization of the trust that the receiver has in the review given by a reviewer with respect to the reviewers capability to provide unbiased reviews or feedback (context)* according to each of the Quality Assessment Criteria set out in the assessment standard in *a given time slot*'.

Note that each organization may have its own assessment standards specific to a domain of assessment. Not all the standards may be fully developed to ensure fair judgement.

The graphical view of the review trust ontology is shown in Figure 4.21 through the use of the ontology notation.

We represent the *review trust ontology* as the combination of the ontology name and a tuple, where the elements of the tuple can be complex elements as defined below.

Review trust (receiver, review, review topic or feedback, Quality Assessment Criteria, time slot and trustworthiness of each assessment criterion), where the following conditions hold good:

- The receiver is a trusting agent who has the trustworthiness value of an unbiased review (quality of review) of the reviewer.
- The reviewer is a trusted agent whose QoS is considered by the receiver.
- Review topic or feedback is the context of the service, such as the review of a mobile service provider, to which the trust relates.

Figure 4.21 Ontology representation of *review service trust concept* and its relation to other concepts

- Quality Assessment Criteria are a set of criteria or conditions used to measure the quality of the reviewer's service.
- Time slot is a time or a time frame for which the trust value holds. During the service time slot period the trust value remains constant.
- 'Trustworthiness value' is a measure of the trust against a trustworthiness scale.

Examples of review Quality Assessment Criteria could be as follows:

- A track record of positive or negative reviews
- Domain experience or expertise in reviewing
- Specific achievements in reviewing capacities
- Reviews completed successfully in time frames
- Active review membership in a given time context/slot
- Adherence to open transparency principles in reviewing

The trustworthiness level for each of these criteria can be defined as shown in Table 4.11.

4.8 Product Trust Ontology

As explained earlier, trust in a service-oriented environment can represent three domains, namely, agent trust, service trust and product trust. In this section, we describe the product trust ontology, including the generic and specific product trust ontologies.

The ontology for product trust or the measurement of the quality of a product is not as complex as either the agent trust ontology or the service trust ontology. Agent trust is dynamic and implicit

Table 4.11 Trustworthiness levels for review services

Trustworthiness scale (ordinal scale)	Semantics for *review service* (linguistic definitions)	Alternate semantics for *review service*
Level – 1	Unknown	Unknown
Level 0	Very untrustworthy	Useless
Level 1	Untrustworthy	Not helpful
Level 2	Partially trustworthy	Somewhat helpful
Level 3	Largely trustworthy	Largely helpful
Level 4	Trustworthy	Helpful
Level 5	Very trustworthy	Very helpful

because of the internal factors and the intellectual capacity of agents. Service trust involves service contracts, service agreements or service standards in which both service requesters and Service providers have to have a common understanding and mutually agree before transactions take place. Therefore, the measurement of service (MoS) has to take into consideration both parties' opinions. However, when discussing product trust, it is the trust of an agent's belief in a particular product, not the converse. There is no dynamism or internal factors in the product nor is there any mutual understanding that occurs before the transaction takes place.

4.8.1 Evaluation of the Product Quality

We often evaluate products we buy as to whether they are worth the expenditure, or whether they are of good or bad quality. Positive evaluations attract us to buy more or make recommendations to acquaintances to buy more. When we evaluate products, it is based purely on our opinion, as a 'product' is unable to respond if we give unfair opinions. This is different from providing opinions about agents or service providers, because the judgement has to take both parties' opinion into consideration. This sometimes involves witnesses or third parties in order to make fair judgements.

Products tend to be independent from product producers or sellers. This is unlike services that tend to be directly influenced by service providers. Therefore, the measurement of a quality of product (QoP) is simpler than the measurement of the QoS.

Some products that even have the same brand may be produced by many manufacturers and sold by many shops and selling agents, such as sports shoes. They might be produced in China or Malaysia; however, other products may only come from one source in the world, such as special information resources or specialized medical or health databases.

4.8.2 Choice of Products

There are many products that exist in our day-to-day lives, such as food products, household products, industrial supplies, entertainment and information products, and so on. At the end of the chapter, we provide two example products, namely: entertainment products and information products and consider their specific ontologies. The ontological representation of product trust can be applied to any type of products.

4.8.3 Generic Product Trust Ontology

The measurement of QoP is based on a set of criteria that are domain rich and specific. 'Domain rich and specific' refers to substantial deep knowledge of a particular product domain. The measurement of QoP or trustworthiness of product shares the same ontology as the 'product trust ontology'.

Note also that we focus on the *product* not the *seller or the producer*. Each *category of product* has a unique set of criteria or conditions that are used for the measurement of the QoP; The QoP represents the product aspect analogous to the QoS providers.

Product Trust Ontology: In service-oriented network environments, the product trust ontology is defined as the conceptualization of the trust that the customer has in a *product* and its quality aspects defined in the product specification and satisfactory fulfilment of all the assessment criteria in a given *time slot*.

The graphical view of the product trust ontology is shown in Figure 4.22 through the use of the ontology notation.

We represent the *product trust ontology* as the combination of the ontology name and a tuple where the elements of the tuple can be complex elements as defined below.

Product trust (buyer, product, context, quality aspects, Quality Assessment Criteria, time slot and trustworthiness value), where the following definitions hold good:

'*Buyer*' is a trusting agent who has a trustworthiness value of a particular product.

- '*Product*' is a trusted entity or object and its quality is considered by the buyers.
- '*Context*' defines the sub-domain or domain functions, such as a digital camera.
- 'Quality aspects' defines the quality of a product, it might be established by the product producer or end customers or buyers for the purpose of measurement or comparison of the quality of the product
- 'Quality Assessment Criteria' are a set of criteria specified by the producer or end customers or buyers to help measure or compare the quality of the product.
- 'Time slot' is the time frame for which the trust value holds, that is, during this period, the trust value remains the same. Time slot may be relevant here as a product may change or deteriorate with time.

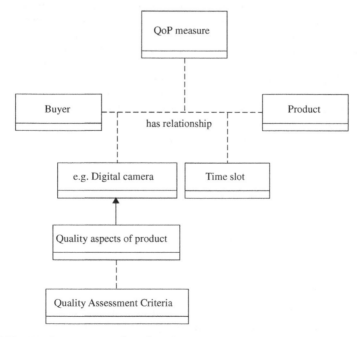

Figure 4.22 Ontology representation of *product trust concept* and its relation to other concepts

- 'Trustworthiness value' is a measure of the trust against a trustworthiness scale.

Specific trust ontology for the QoP should commit to all the ontology concepts and their properties from the generic product trust ontology defined in this section. In the next two sections, we consider two specific product trust ontologies.

4.8.4 Specific Product Trust Ontology – Entertainment Product Trust

Most families around the world enjoy buying entertainment products. However, not every purchaser is happy (or entirely happy) with the product purchased and not all the product producers are satisfied with their sales. Therefore, to improve product quality, it is important to learn what the customer wants. It is equally important for customers to know which products are best suited to satisfy their needs.

Entertainment Product Trust Ontology: In service-oriented network environments, the entertainment product trust ontology is defined as the conceptualization that represents the trust that the buyer has in the *entertainment product* in a given *context* and criteria set out by the buyers in a given *time slot*. This definition is the same as that used for the product trust ontology. Note that the *criteria* are the conditions that set out in the standard for the measurement of a range of products that have similar features.

The graphical view of entertainment product trust ontology is shown in Figure 4.23 through the use of the ontology notation.

We represent the *entertainment product trust ontology* as the combination of the ontology name and a tuple where the elements of the tuple can be complex elements as defined below.

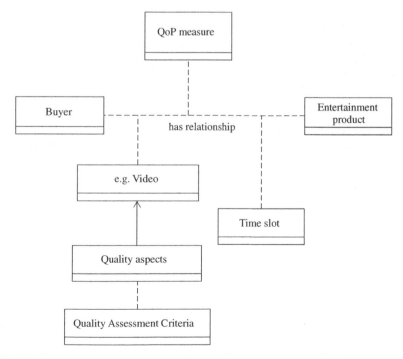

Figure 4.23 Ontology representation of *entertainment product trust concept* and its relation to other concepts

Entertainment product trust [buyer, entertainment product, context, quality aspects, Quality Assessment Criteria, time slot and trustworthiness Value), where the following definitions hold good:

- *'Buyer'* is a trusting agent who gives a trustworthiness value to a particular entertainment product.
- *'Entertainment Product'* is a trusted entity or object where its quality is considered by the buyers.
- 'Context' could include videos or games, and so on.
- 'Quality aspects' define the quality of the product, it might be established by the product producer or end customers or buyers for the purpose of measurement or comparison of the quality of the product
- 'Quality Assessment Criteria' is a set of criteria to be used to qualify and quantify the quality of the product.
- 'Time slot' is the time frame for which the trust value holds, that is, during this period, the trust value remains the same.
- 'Trustworthiness value' is a measure of the trust against a trustworthiness scale.

Examples of criteria for an entertainment product such as a video camera may be the following:

- Use of operation
- Photo quality
- Durability
- Battery life
- Shutter time lag
- Customer support.

Note that different products should have different criteria. The trustworthiness level for each of these criteria can be defined as shown in Table 4.12.

4.8.5 Specific Product Trust Ontology – Information Product Trust

Information products are products like articles, e-book, information, data, documents, text, experimental results, pictures, images, dictionaries, ontologies, research outcome, software, and so on, that are shared, downloadable and, mostly, free of charge. However, in the open anonymous network, how do we ensure the information comes from secure sources and is accurate or scientifically correct?

Information Product Trust Ontology: In service-oriented network environments, the information product trust ontology is defined as the conceptualization of the trust that the online users have in the *information* in a given *context* and criteria set out by the users in the given *time slot*.

Table 4.12 Trustworthiness levels for product quality

Trustworthiness scale (ordinal scale)	Semantics for *product quality* (linguistic definitions)	*Alternate* semantics for *product quality*
Level – 1	Unknown	Unknown
Level 0	Very untrustworthy	Terrible
Level 1	Untrustworthy	Poor
Level 2	Partially trustworthy	Average
Level 3	Largely trustworthy	Fair
Level 4	Trustworthy	Good/Very good
Level 5	Very trustworthy	Perfect

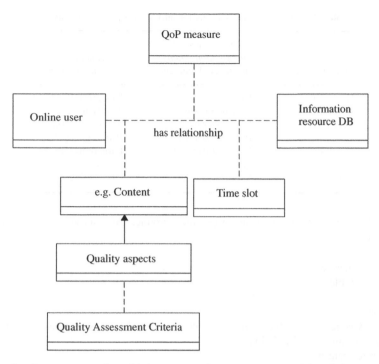

Figure 4.24 Ontology representation of *information product trust concept* and its relation to other concepts

This definition 'commits' to the product trust ontology.

The graphical view of information product trust ontology is shown in Figure 4.24 through the use of the ontology notation.

We represent the *information product trust ontology* as the combination of the ontology name and a tuple where the elements of the tuple can be complex elements as defined below.

Information product trust (online user, information description, information context, quality of information, Quality Assessment Criteria, time slot and trustworthiness value), where the following definitions hold good:

- '*Online user*' is a trusting agent who has a trustworthiness value of a particular product.
- '*Information description*' is a trusted entity or object and its quality is considered by the *online user*.
- 'Information context' is a general name for all the customized information criteria.
- 'Quality aspects' define the quality of the information product, it might be established by the information provider or end customers for the purpose of measurement or comparison of the quality of the product
- 'Criteria' is a set of criteria defined by the online users to measure the quality of the information product;
- 'Time slot' is the time frame for which the trust value holds, that is, during this period, the trust value remains the same. For example, if an information provider's web server is up and down all the time, it creates difficulties for online users to access information, or the information may be valid only for some time period and needs to be updated.
- 'Trustworthiness value' is a measure of the trust against a trustworthiness scale.

Table 4.13   Trustworthiness levels for information product quality

Trustworthiness scale (ordinal scale)	Semantics for *information quality* (linguistic definitions)	*Alternate* semantics for *information quality*
Level – 1	Unknown	Unknown
Level 0	Very untrustworthy	Totally useless
Level 1	Untrustworthy	Not very useful
Level 2	Partially trustworthy	Partially useful
Level 3	Largely trustworthy	Largely useful
Level 4	Trustworthy	Useful
Level 5	Very trustworthy	Outstanding

Examples of Assessment Criteria for an information product for diseases where the information criteria could be:

- Information owner details
- Organization profile
- Abstract
- Introduction
- Body of knowledge (disease type, virus type, environmental factors, symptoms, causes, etc.)
- Experimental results and simulations
- Information references
- Online customer support.

Note that different products should have different criteria. The trustworthiness level for each of these criteria can be defined as shown in Table 4.13.

4.9 Trust Databases

4.9.1 Agent Trust Database

In this section, we give a preliminary introduction to trust databases. The design of a trust database should be no different from the design of any information system. For every trust relationship that the trusting agent has, it produces a trust tuple, which contains details of the trust relationship. These tuples can be and should be stored in the trusting agent's trust database. How large will a trust database be? Generally, a database for customers or account inventory can be very large. This is not the case with a trust database. A trust database will only contain a few tables that store the trust tuples. Each tuple is stored as a row in a table. This is best demonstrated by the example in Table 4.14.

Table 4.14   An example of a third party agent's trust database for Alice

Trusting agent	Trusted agent	Context	Time slot	Assessment #	Trustworthiness value
Alex	Bob	Local delivery	2003	123	4
Alex	Bob	Local delivery	2004	124	3
Peter	Liz	Sweet corn export	2004	224	5
Sarah	Liz	Sweet corn expert	2004	126	4
Sarah	Jo	Cold storage	2003	332	4
Tom	Jo	Cold storage	2003	121	4

A trust database can be divided into three categories:

(a) Third party agent trust database
(b) Private agent trust database
(c) Federated agent trust database (includes both the third party agent database as well as the private trust database).

The trust database serves three purposes for business intelligence:

- Recording the trustworthiness value of the trusted agents. It can come from two sources:
  - The trust value is assigned directly by the trusting agent through direct interaction with the trusted agent (private trust database).
  - The trust value is obtained from other agents if it is valuable for future reference (third party trust database).
- Keeping a history of the trusted agent's quality and performance for future benefit. For example, if a retailer had ten interactions with a particular supplier, and the supplier delivered 70 % of the commitment on average, it is important that the trusting agent can check it on its own trust database to determine whether to go ahead again with the supplier or find a better one who can deliver 90 % of the commitment on average.
- Sharing information with other agents. This is especially useful when a supplier cheats a retailer. It is important to share this information online with others to prevent vulnerability of other customers and end-users.

In an agent trust database, there are several trust tuples we need to consider:

(1) The trusted agent database (trusted agent, context, time slot, assessment #, trustworthiness value)
(2) The third party agent trust database (trusting agent, trusted agent, context, time slot, assessment #, trustworthiness)
(3) Agent quality aspects and criteria database (agent quality aspects #, context #, Criteria1, Criteria 2 ... Criteria N)

4.9.1.1 Trust Database Example 1 – Alice's Third Party Agent Trust Database

Note that the trust database stores the trustworthiness value of the trusting agent who assigned it to the trusted agent. The table stores the trustworthiness values of the trusted agent that are assigned by a third party agent (trusting agent, not Alice). Alice may think that this data might be important or useful for her in the future. Note that Alice can have several third party trust databases. Each database is categorized in a particular business or service domain, such as entertainment, auctions or travel, and so on. She can have as many as she wants, all of which can be organized in a tree structure for ease of search. If Alice, the trust database owner, decided only to record her own direct experiences without considering other agent's opinions, the trust database would appear as in Table 4.15

Table 4.15 An example of a private trust database for Alice

Trusted agent	Context	Time slot	Assessment #	Trustworthiness value
Bob	Local delivery	2003	123	4
Bob	Local delivery	2004	124	3
Liz	Sweet corn export	2004	224	5
Jo	Cold storage	2003	126	4

Table 4.16   A sample of a federated trust database for Alice

Trusting agent	Trusted agent	Context	Time slot	Assessment #	Trustworthiness value
Alex	Bob	Local delivery	2003	112	4
Alex	Bob	Local delivery	2004	113	3
Peter	Liz	Sweet corn export	2004	124	5
Sarah	Liz	Sweet corn export	2004	133	4
Sarah	Jo	Cold storage	2003	132	4
Tom	Jo	Cold storage	2003	234	4
Alice	Bob	Local delivery	2003	244	4
Alice	Bob	Local delivery	2004	222	3
Alice	Liz	Sweet corn export	2004	323	5
Sarah	Liz	Sweet corn export	2004	324	4
Sarah	Jo	Cold storage	2003	344	4
Alice	Jo	Cold storage	2003	333	4

4.9.1.2   Trust Database Example 2 – Alice's Private Trust Database

In comparing Tables 4.14 and 4.15, we notice that as Table 4.15 is a private database, a trust database does not have a column for trusting agents, because it is about Alice and no one else.

If Alice decided to keep both a third party agent trust database as well as a private database, she can do this by organizing the databases into groups or sub-groups without the categorization of each group. This trust database is known as a *federated database*.

A federated trust database is defined as a group and sub-group categorized trust domain database that includes both third party agent trust data as well as private trust data.

4.9.1.3   Trust Database Example 3 – A simple version of a Federation Agent Trust Database

Compare Table 4.16 to Table 4.14. The difference is that the Federated trust database includes the owner of the business' opinion, such as Alice as trusting agent, whereas third party trust database (Table 4.14) does not include Alice herself, but only third parties.

The above example may be sufficient for an organization to keep track of the trustworthiness of its business partners or interaction agents or users. However, if an organization is a multi-business, multinational organization, then a more sophisticated federated trust database is needed. The conceptual view of such a federated database is shown in Figure 4.25.

4.9.2   Trust Database for Quality of Service

There are two kinds of trust databases for the trustworthiness of service. One is the QoS measure based on the service standard. Therefore, the standard is not specifically tied to a service provider. However, the other service trust database is where the service quality is measured against the service agreement. In this case, the database is much more complex and each service criteria will be indexed and categorized against each service provider.

In the above type of service trust database, there are several trust tables (databases) to consider.

(1) The customer or service requester database (Customer ID, CName ... QoS measure #):
    Trusting agents' (feedback customer) databases of those trusting agents who share their direct experiences or trust values with the other agents
(2) The service provider database (Provider ID, PName, ... QoS Index #):
    Trusted agents databases – in this case, we use the service database as the trusted entity.
(3) QoS Context from service agreement (provider ID, customer ID, service context #, QoS agreement #): This is suitable for the criteria that are derived from the service agreement.

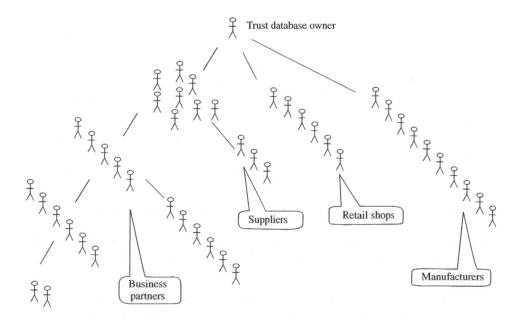

The federated trust database can be categorized by each specific business domain of interest or virtual communities units; it can include both private and third-party trust information or opinions.

Figure 4.25 A conceptual view of a federated trust database

(4) QoS context from standard (service context #, QoS Aspects #, QoS Aspects):
QoS context database has two types: one is standard (ISO standard, for example) and the other is customer made (from their service-level agreement). These databases are look-up databases (which are predefined tables, also called *look-up tables* in database terms, where the data is normally persistent, i.e. no edit or delete operations etc. are possible).

(5) QoS aspect and criteria database (QoS Aspect #, Context #, Criteria1, Criteria 2 . . . Criteria N):
The QoS Criteria database stores criteria, and this is also a look-up database. The criteria database is a weak entity and must associate with one of the QoS context databases.

(6) QoS measure (customer ID, QoS Assessment #, QoS Index #, Provider ID, Context #, trustworthiness value):
QoS measure database contains each customer's feedback (trust values and opinions). Note that each customer's feedback is represented by a trusting agent's trust value and opinion in the trustworthiness system.

In this section, we have given a high-level introduction of a service trust database. However, a complete tutorial containing a detailed description of trust databases for trustworthiness of services and a prototype system can be downloaded from this book's website.

4.9.3 Trust Database for QoP

There are two main trust databases for QoP: one is a quality standard that sets out the criteria for the measurement of a product. The other one is the product catalogue. Other databases include customers and feedback.

In a product trust database, there are several trust tuples we need to consider.

(1) The customer database (Customer ID, CName ... QoS Measure #):
     This is a trusting agents (feedback customer) database of those trusting agents who share their
     direct experience or trust value with other agents. In the service-oriented environment, they can
     submit this to the trustworthiness systems, sometimes called *customer feedback system.*
(2) The products database (Product catalogue #, Product ID, Product Type ... QoS Index #):
     There could be millions of products and thousands of categories of products existing in product
     databases, and it is quite normal that a trustworthiness system keeps updating and maintaining
     the details of the products.
(3) Product aspects and criteria database (Product Aspects #, Context #, Criteria1, Criteria 2 ...
     Criteria *N*):
     This database is a look-up database (the predefined tables, which are also called *look-up tables*
     in database terms, where the data is normally persistent, i.e., no edit or delete operations, etc.
     are possible).
(4) QoP (Customer ID, QoP Assessment #, QoP Index #, Provider ID, Context #, trustworthiness
     value):
     QoP measure database contains each customer's feedback (trust values and opinions). Note
     that each customer's feedback is represented by his or her trust value and opinion in the
     trustworthiness system.

In this section, we have given a high-level introduction of a product trust database. However, a
complete tutorial with a detailed description of trust databases for trustworthiness of products and
a prototype system can be downloaded from this book's website.

4.10 Summary

In the previous two chapters, we introduced the concepts of trust and trustworthiness. In this chapter,
we have introduced trust ontologies. An ontology represents a set of concepts that are commonly
shared and agreed to by all parties in a particular domain.
    We have introduced generic trust ontologies and specific trust ontologies:

* Agent trust ontology. This ontology helps the e-service community to share the understanding
   of agent trust, such as seller trust, service provider trust, website trust (software agent), broker
   trust, supplier trust, buyer trust or reviewer trust.
* Services trust ontology. This ontology assists in the shared understanding of the QoS that agents
   provide in the service-oriented environment. Example of services are sales, orders, warehousing,
   logistics, education, governance, advertising, entertainment, trading, online databases, virtual
   community services, security, information services, opinions and e-reviews.
* Goods or products trust ontology. This ontology is helpful in the shared understanding of the
   quality of the products such as commercial products, information products, entertainment products
   or second-hand products.

In addition, we have given a brief introduction to trust databases for agents, services and products.
    Establishing trust, building trust and maintaining the trust are great challenges to both customers
and businesses.
    Building trust in a service-oriented environment adds value to a business entity and boosts
consumer confidence in open, distributed networked environments. In particular, it assists in helping to provide business intelligence by understanding customers' needs, market preferences and
user requirements. It promotes continuing effort in customer service improvement through the
strengthening of consumer and business trust relationships.
    One of the primary methods of building trust is through mining the feedback from customers or
other users, including collecting end-user opinions, or partners' suggestions, as these help business

operators ascertain what the market needs or what customers want. This analysis enables detection of competitor strategies, all of which can be inferred from consumer behaviour. The output of this process provides input that can reshape strategic business directions, improve customer services and assist the organization in determining its productivity and performance, and so on.

References

[1] Gruber T.R. (1993) 'A translation approach to portable ontologies', *Knowledge Acquisition*, Vol. **5**(2) pp. 199–220.
[2] Hazdic M. (2005) "Ontology based information system modelling and application to human disease ontology development" PhD Thesis, School of Information Systems, Curtin University of Technology.
[3] Spyns P., Meersman R. & Jarrar M. (2002) 'Data modelling versus ontology engineering' *SIGMOD*, Vol. **31**(4): 12–17.

5

The Fuzzy and Dynamic Nature of Trust

5.1 Introduction

Trustworthiness measurement and prediction are both complex and limited. This is because the nature of trust is *fuzzy, dynamic and complex*. The following three concepts are important:

- The *fuzzy nature* of trust means it is indefinite or imprecise, and sometimes it results in vagueness when we express trust or try to explain a trust level.
- The *dynamic nature* of trust refers to trust not being constant or stable but always changing as time passes.
- The *complex nature* of trust arises from the fact that there are multiplicity of ways for determining trust and a variety of views about trust.

We note that when something cannot be explicitly defined and is not stable and always changes, coupled with a variety of views and opinions, it is always difficult to manage and to predict its future values.

For example, tradesmen often have random service times. They often agree to do things on a date and within a certain time frame. However, it happens quite often that most service deliveries are not just-in-time (JIT). This sometimes causes difficulties to customers in managing their duties or tasks [1].

Another example of this is best illustrated by the following situation. At the beginning of a school semester, schoolteachers ordered some textbooks online for students. There were three possible options:

- Delivery within 3 days, cost US$30
- Delivery in around 10 days, cost US$18
- Delivery in approximately 3 weeks, cost US$8.

The teachers and students selected the third option, because they thought that they could wait for three weeks and the saving was important. However, the textbooks arrived three months after ordering, which was at the end of the semester. However, when they checked with the online bookstore, a 'delivered' message status had been posted, and they could not do anything about the delay or seek information as to where the books were and when they would come. In fact, when the book carton finally arrived, no one bothered to open the box, because the course had finished, and the books were not needed.

Trust and Reputation for Service-Oriented Environments   Elizabeth Chang, Tharam Dillon and Farookh Hussain
© 2006 John Wiley & Sons, Ltd

In this example for time delivery, trust is quite explicit and simple. However, the *trust* they had in the online bookstore changed over the time period. It was a situation where they *predicted* that they would be able to manage their study because they could get the books within a certain time frame, but unfortunately they could not predict before the due date that the shipment would be delayed. Therefore, their level of trust in the integrity of the bookstore was reduced, as their expectations were not met.

In this chapter, we will discuss the fuzzy, dynamic and complex nature of trust. Throughout this discussion, we will come to realize the challenges that are faced in the measurement and prediction of trustworthiness.

5.2 Existing Literature

After reviewing the extant literature on trust, it is evident that there have been few studies of fuzziness, dynamism and complexity of trust and their impact on trust, trustworthiness measurement and prediction. The lack is particularly apparent in the world of e-business and service-oriented network environments. Some studies by Egger (including those of Shelat [2] & Egger [3–6] and Egger [6]) consider how factors like the usability of websites (a website may represent a service provider), the way content is organized and how security and privacy issues are addressed to communicate trust to their human users. Factors considered by Egger are applicable for B2C (Business to Customer) e-commerce, where the customer (usually the client) interacts with the service providers through websites. Kim and Moon [7] investigated how graphic design elements in a website can communicate trust to human users. However, the studies mentioned above do not investigate how the usability of a website can assist in communicating and establishing trust between providers and customers and in trustworthiness measurement and prediction. Other works only provide reference to a single trust value and a single context for trust management. They have not considered other factors such as context dependence, time slots for frames, or internal factors of interacting parties or agents, and so on, nor have they examined all possible fuzzy and dynamic characteristics of trust [8].

5.3 Fuzzy and Dynamic Characteristics of Trust

In this section, we shall explain six important fuzzy and dynamic characteristics of trust:

- Implicitness in trust
- Asymmetry in trust
- Transitivity in trust
- Antonymy in context
- Asynchrony in time space
- Gravity in relationship.

These factors create big challenges for the trustworthiness measurement and prediction. They are important in understanding the complexity of trust and how trust can be measured or predicted. This will help us to appreciate the conceptual definition of trust and methodologies for the measurement of trustworthiness and prediction in the service-oriented network environment.

Note that we use the term fuzzy in this chapter *not* in the sense of the precise definitions given in the fuzzy systems literature but to indicate a certain vagueness, complexity or ill-definition and qualitative characterization rather than quantitative representation.

5.3.1 Implicitness

Trust is implicit. This means a trusting agent may not be able to explicitly articulate and specify the belief, willingness, capability, *context* and *time dependency* of trust.

Figure 5.1   The implicitness of trust

The reason for this is that there are many elements that go into making up trust in a relationship. Some of these are clearly discernable, while others are not so easily differentiated, as they may be built up slowly and incrementally through the experience of the relationship. This has some similarities with deep expertise involving judgement. In such situations, trust can only be an estimate at best. While human agents are coping with the implicitness of trust, software or intelligent agents too behave like humans in determining trust. Figure 5.1 shows the word 'trust' is used in many situations of everyday life. It shows vagueness and leaves many aspects unspoken or undefined.

In Figure 5.1, in a trust context, Alice says 'I trust myself', Bob says 'I do not trust anybody', Charlie says I trust my Boss' and Doug says 'I trust the bank'. The fuzziness of trust is clearly illustrated.

The above examples demonstrate there are many aspects of trust that are implicit, because a trusting Agent cannot explicitly specify the level of 'belief', 'willingness', and 'capability' or cannot specify the context and time frame in relation to trust. Therefore, there is an implicit assumption. This is because belief', 'willingness', and 'capability' are not easily observable or explicitly defined. The fuzzy value of trust is also context and time dependent. For example, the opinion 'I trust myself' may change because the context may be evolving or understanding of oneself changes. As time passes, the belief in oneself also may evolve.

A trust relationship can involve oneself only (i.e., 'I trust myself'), another party or agent (i.e.,'I trust my boss'), a group, an organization (i.e., 'I trust the bank'), and so on. We can most often define the context and time frame relating to a trust relationship, but we cannot count on (explicitly state) the *willingness* and *capability* of individuals or others involved in the trust relationship. The understanding of the context may change, and as time passes belief also changes. The understanding of trust by parties in relation to *fuzziness* is implicit rather than explicit, and it is also *dynamic*.

Example 1:

Alice says 'I trust myself and I trust myself in everything'. This sounds explicit, but when she says 'I do not know what I am doing', or 'I am sorry ... ', it means she does not trust herself all of the time. We know that Alice's trust of herself is an implicit trust, which changes as time passes, and that fuzziness and dynamism are characteristics of trust.

Example 2:

Bob says 'I do not trust anybody'. However, does that include everything and everyone and every situation? We must question 'What is this about? Why is this so?' in the context. This trust is always implicit to a person such as Bob who does not have confidence in anybody else. It may also be due to the fact that Bob himself is not a confident person. But Bob is growing and changes his views over time. Therefore, the trust or belief is dynamic. This also illustrates the fuzzy and dynamic characteristics of trust.

Example 3:

Charlie says 'I trust my boss.' People may naturally assume that this implies the trust is in the work situation. However, there are so many contexts at work, and outside work Charlie may mean something different. The qualification may be that Charlie should say 'I trust my boss sometimes'. Charlie may also mean something else when he refers to trust in his boss, such as the boss's intelligence, rather than any explicit trust in the relationship he has with his boss. He may mean that the trust he has in his boss varies, but he cannot explicitly list every situation or context where he does or does not trust his boss. In this situation, the trust is implicit and may change over the time period. This also demonstrates the fuzzy and dynamic characteristics of trust.

Example 4:

Doug says 'I trust the bank'. This trust may come from somebody else's recommendation that banks are safe. Doug trusts the bank without explicit details or personal experience. Doug's trust in the bank in this situation is dependent on exogenous (external) factors, rather than personal experience, and his views are therefore subject to change in the context of his own experiences. There may be an explicit level of trust, and this level of trust may change over time. This further indicates the fuzzy and dynamic characteristics of trust.

In the above examples, we see that individuals may be able to define the 'context' and 'time slot' relating to trust but they *cannot* give explicit definitions of the 'willingness' and 'capability' with regard to an individual or others about his or her trust. Therefore, trust is implicit. The only thing we can do is to give an estimate of 'willingness' and 'capability' through behaviour monitoring, evaluation and correlations between an individual's or an organization's behaviour. For example, no one would really trust a bank all the time. People generally monitor their accounts to make sure that the expected and actual monetary transactions are correct or approximately correct. Therefore, sometimes the level of trust can be explicit; however, trust will never be absolutely explicit.

The challenge in trustworthiness measurement or trustworthiness prediction is the degree of the implicitness of the trust, which is the explicit measure of 'belief', "willingness' and 'capability' in the trust dynamics. The only way to provide *an estimate* is through a well-known scientific method, namely, the correlation or regression of behaviour or a correlation between what people say and what people do.

In a business situation, we can correlate committed services with actual delivered service to validate the trust level.

5.3.2 Asymmetry

Trust is asymmetric. This means that a Trusting Agent 'A' has a certain belief in the Trusted Agent 'B' in a particular context. It does not imply that the Trusted Agent 'B' should have the same belief in the Trusting Agent 'A' in the same context. Hence, owing to the non-mutual reciprocal nature of the trust relationship, trust is asymmetric.

For example, the trust relationship involving an exchange from Alice to Bob is not the same as an exchange from Bob to Alice. The characteristics of the trust relationship are also influenced by the agents' internal factors (characteristics). The context of the situation becomes very important in relation to the agents' internal factors, as well as the symmetry (or lack of it) between the trusting agent and the trusted agent.

In the example in Figure 5.2, we assume the highest trust value is 5 and the lowest value is 0. Alice assigns a trustworthiness value of 5 to Bob in the context of borrowing a credit card, but Bob assigns a trustworthiness value of 2 to Alice in the same context. The value differences are influenced by internal characteristics of an individual. There is no explicit understanding of the value of the trust in the relationship between the two parties unless it is, in a human context,

Context: Sharing research

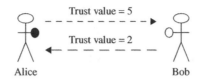

Figure 5.2   The asymmetric nature of trust

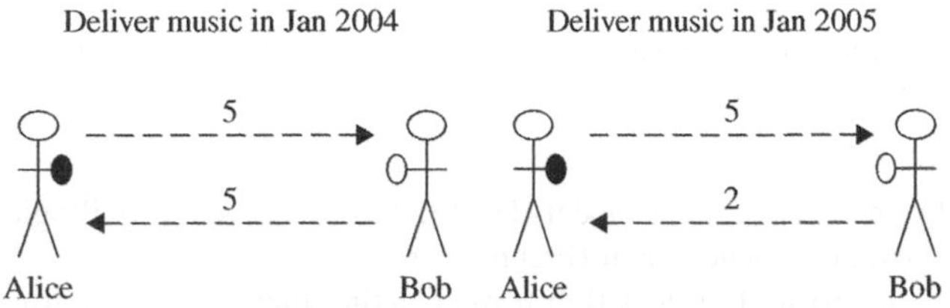

Deliver music in Jan 2004   Deliver music in Jan 2005

Alice   Bob   Alice   Bob

Figure 5.3   From symmetric trust to asymmetric trust as time passes

verbalized. In general terms, agents do not explicitly verbalize a numeric trust value; they generally verbalize a level of trust. Fuzziness, therefore, is evident.

Trust can move from being symmetric to being asymmetric. Let us assume that Alice and Bob trust each other to exchange or deliver high-quality music to each other in 2004. With the passage of time in 2005, Bob's capability or willingness to deliver high-quality music to Alice decreases. As a result of this, the trust that Alice has in Bob in the context of delivering high-quality music decreases or becomes null. Hence, we see that trust, which was initially *symmetric* (equal) between two agents, has become *asymmetric* with the passing of time as in Figure 5.3. This is also related to the *dynamic* nature of trust with time.

In Chapter 2, we learnt that trust is unidirectional (goes in one direction). This means that if we assume the trusting agent to be Alice and the trusted agent to be Bob, the trustworthiness measure or estimation is only from Alice to Bob, but not both. However, Bob can also be a trusting agent and Alice his trusted agent. We consider that this is a different trust relationship, because it has a different trust value. The trust level and the trust value go in one direction from the trusting agent to the trusted agent only.

It needs to be clearly understood that the trust measure or prediction is asymmetric, regardless of whether it is in the physical world or the virtual world.

The challenge of a trustworthiness measure and prediction is conditioned on the asymmetric character of trust. Therefore, one trust value does not represent both parties in a trust relationship. This is often implicitly assumed in a static social world, which is conceptually loose, as trust in the social world can also imply a dynamic exchange between individuals that is sometimes multidirectional. A trustworthiness measure in a service-oriented network environment must be *unidirectional* and only from a trusting agent to the trusted agent. It is only meaningful to the trusting agent and for use by the trusting agent. The trust value can move from being symmetric to asymmetric or vice versa. Fuzziness and dynamism is therefore again apparent in the situation.

5.3.3 Transitiveness

Trust is transitive. This means that if Agent A trusts Agent B then Agent A may trust any agent that Agent B trusts in the same context or in a pre-defined different context. This character of trust

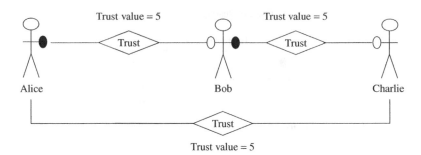

Figure 5.4   A transitive trust situation indicating constancy in trust values (if Alice trusts Bob and Bob trusts Charlie, then Alice trusts Charlie)

can be evidenced by the various trust models [8–13, 19]. However, it is illogical to assume that transitive trust is an explicit phenomenon (Figure 5.4).

For example, if Alice trusts Bob and Bob trusts Charlie, then Alice may trust Charlie as well, even though Alice does not know Charlie at all. However, if you ask why Alice trusts Charlie, she would say it is because she trusts Bob.

The transitivity of trust, also known as a *derived* trust, means that trust is derived from an existing trust between agents. Note that the derived trust and the trust from which it is derived should be considered within the same context. It is important to understand that this derived trust may be explicit, but generally it is very hard to quantify it accurately. We assume then some level of implicitness (*or vagueness*).

The level of trust through a transitive introduction may hold and is dependent on the strength of the original Agent's trust relationship. For example, Alice trusts the authenticity of Bob, and Bob trusts the authenticity of Charlie. Owing to the transitive nature of trust, Alice should trust Charlie. However, the transitive nature of trust may not hold in situations when the trust from Alice to Bob is not strong. This is illustrated in Figure 5.5. Also, Alice could trust Bob but she may be uncertain about Bob's ability to judge another person even if it was in the same context. In this case, she may reduce the trust value to some extent for Charlie.

Transitive trust is a very important concept in service-oriented network environments where anonymous users or agents often want to identify quality service through a transitive introduction, also known as 'recommendation' or 'reputation'. The recommendation or reputation is fuzzy in the sense that a transitive introduction is context and time dependent, and the dependencies are

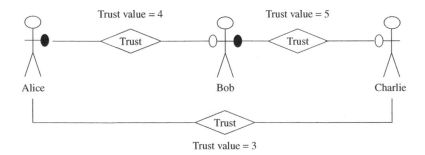

Figure 5.5   A transitive trust situation indicating inconstancy in trust values (the trust level may vary and depends on the strength of the trust relationship between the trusting agent and the trusted agent)

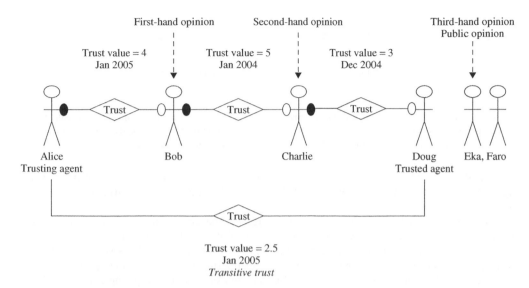

Figure 5.6 Transitive trust depending on the opinions of trust of others

not always explicit as there may be an inability to have the same view or understanding about the context and exact time frame in which the trust value or level was assigned.

Transitive trust is also time dependent, which means that it depends on when the trust value was assigned and when this trust value was recommended. Trustworthiness prediction has to take account of aggregated time frames or slots in order to determine the trust value. Note that this value can change when time passes. This is a dynamic characteristic of trust. For example, each time Alice asks for references about Sonya, the recommendation or trust value changes as time passes, because the trust value refers to different time spots.

Transitive trust is affected by the first-, second- and third-hand opinions. Figure 5.6 gives an illustration of this transitive trust and shows that the trust values in relationships often depend on whether the opinions are first hand, second hand or third hand. The figure also shows the time frames (time slots) when the trust values were valid. Chapters 8–10 will provide an introduction on how to calculate or predict trustworthiness in this context.

The *challenge* of a trustworthiness measure and prediction is the method of using the *transitive trust value,* also known as the *recommendation value* or *reputation value*. Often, we get different trust values from different agents about 'a particular agent' or 'a service'. They relate to the different time frames (time slots) in which we often have to deal with first-hand, second-hand and third-hand opinions. This involves the *fuzzy and dynamic* characteristics of trust. We will introduce these methods in Chapters 8–10.

5.3.4 Antonymy

The *antonymous* nature of trust is related to 'context'; therefore, the context may be understood differently by the two Agents, A and B, involved in a trust relationship. Therefore, what may be clear to one agent may not be clear to another.

The context, as seen from the perspective of Agent A, may be the opposite of or different from that seen from the perspective of Agent B. This context is used in a different way and is often implicitly recognized by either party or agent.

For example, Alice trusts Bob in the 'context' of selling a book. From Alice to Bob, the context is 'sell' and from Bob to Alice it is 'buy', as in Figure 5.7(a).

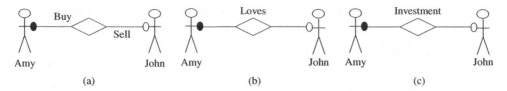

Figure 5.7 (a) An antonymous perspective of trust (both the agents understand the 'context' in opposite ways) (b) An antonymous perspective of trust in context (both the agents understand the context' in their own way) (c) Antonymous perspective of trust in the future context (both the agents understand the 'context' differently)

In another example, Agent A trusts Agent B in the context of 'learning', but from Agent B to Agent A, the context may be of 'teaching'. This is due to the opposing perspectives (context) each agent takes in an interaction.

In the example in Figure 5.7(a), each party perceives the context to be the opposite of that perceived by the other party. Amy sees the context of the trust relationship as one of 'buy', whereas John sees the context of the trust relationship as one of 'sell'. Fuzziness is evident because of the antonymous nature of the perspective of the relationship between both the agents.

In the example in Figure 5.7(b), Amy loves John and thinks 'Love means giving'. However, from John's perspective, 'Love means taking'. By extrapolating this example further, Amy believes that if people love each other, they should be married; however, John believes you can love someone but do not necessarily have to be married to her.

In the example in Figure 5.7(c), Amy carries out an investment with John, which she believes is for the family's future, or for children. However, John carries out an investment with Amy that he believes is for money; he does not think of children or the family's future.

The *challenge* of the trustworthiness measure and trustworthiness prediction is to define the context *clearly*. This is difficult to do even in the real world. For example, A thinks the context is very clear, but this is not the case for B or others involved in the situation. If the context is not clear, trustworthiness has *no value*, and worse, it affects all parties involved in the relationship. In other words, the context influences the ability to evaluate trust. Therefore, trust is not only dynamic, but its fuzziness is triggered either positively or negatively with respect to the context in which the trust relationship exists.

In the example in Figure 5.8, Sonya, a company director gives a job to Mark, a manager. Sonya thinks the job description is very clear, but Mark does not think so. At the end of the task Sonya gives Mark 1 out of 5 as a trustworthiness value (assuming 5 is the highest trust ranking). If Sonya is right, the trustworthiness value is useful for future planning or job distribution especially for her organization. However, if Sonya is wrong, the trustworthiness value has no meaning. Moreover, it is not good for Sonya's organization and it is also not good for Mark. It damages everyone involved in the trust relationship. Conversely, Mark may not care what Sonya thinks of him as he could find a job elsewhere. However, it is important in Sonya's organization to validate an individual's performance or the performance of business partners as adequately and accurately as possible.

The methodology of how to determine the *clearness of context* is a big challenge in both the physical world and the virtual world. This will be introduced in the next chapter as part of the concept of trustworthiness prediction.

5.3.5 Asynchrony

The asynchronous nature of trust refers to asynchrony in a 'time slot', that is, the time slot of the trustworthiness may be understood or defined differently between trusting agents and trusted agents. *Fuzziness* is inherent in any situation that becomes unclear to either party or agent in the trust relationship.

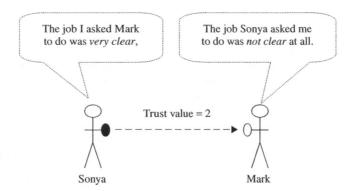

Figure 5.8 A trust context (The 'context' may or may not have been communicated *clearly*, resulting possibly in the wrong trustworthiness value prediction in a given situation)

For example, Agent A may think that the relationship with Agent B spanned a period of 5 years (time slot), but Agent B may think that the Agent B to Agent A relationship has never had any meaning and, therefore, does not acknowledge the relationship for that period, as they may have just known each other without meaningful interaction. Agent B may also think that the relationship from B to A only lasted for a shorter period of time, 1 year, when they have had meaningful interaction.

For example, Alice loves Charlie. She thinks they were in love for 5 years, but from Charlie's perspective, he may believe that the relationship was for a maximum of 1 year, illustrating asynchrony in the time slots for which the relationship holds. Moreover, Charlie may not agree there was a relationship at all (Figure 5.9).

The Agents may understand the time slot differently for a given context. The time slots between the Agents may be the same, completely different or partially overlapping.

As a result of the asynchronous nature of the time slots, trustworthiness prediction cannot be straightforward. The *challenge* of the trust measure and trustworthiness prediction is that we have to deal with different time slots in time/space. We need both to *aggregate the time slot and also average the trust value* over the aggregated time slots. This is especially important when recommendations of trust and reputation take place.

For example, we may want to know about Sonya's job references from the previous year. However, three referees may give a reference for 1 year each, spread over the last eight years. In this case, we cannot simply just take all these values and use them as a *trustworthiness value* for Sonya, partly because her capability and/or attitude towards the job may have changed.

Chapters 8 and 9 will introduce the notion of a *trustworthiness value* in detail.

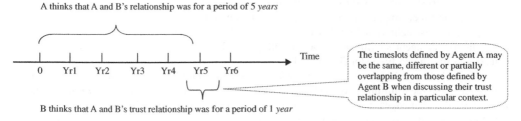

Figure 5.9 Asynchrony in the trust time slot (Agents A and B have their own definitions of the time slot with regard to the same *context*)

5.3.6 Gravity

The *Gravity of trust* refers to the *gravity* of the relationship, which means the seriousness of the relationship from the perspective of each party or its *influence* on each party. This means that for the two agents, A and B, the relationship from A to B may be important but that from B to A may not necessarily be important. The agents have their own views on whether the relationship means much to them, and the *influence* it could have on their business or lives. As stated in all the previous examples, the fuzziness and dynamism of trust is also inherent in this characteristic of trust.

For example, Angel may think the relationship with Trish is not important to Angel, but Trish thinks that relationship with Angel is very important to Trish. This indicates that each agent has a right to determine the importance of the trust relationship.

Regardless of who is the trusting agent and who is the trusted agent, from Agent A's point of view, the Agent A to Agent B relationship within a particular context may not be important to Agent A at all. However, from Agent B's point of view, the relationship may be very important.

For example, consider a situation that illustrates a relationship between a business provider (known as the 'ABC Bookstore') and a customer Bob (known as a Book Buyer). Bob thinks ABC is important to him but ABC *does not* think whether Bob trusts them or does not trust them is important.

We assume that 'ABC bookstore' thinks that the trust relationship from ABC to Bob is not important to ABC, as he is only one of many customers. However, customer Bob, the buyer, thinks the trust value of ABC is very important. For instance, customer Bob dislikes ABC. ABC feels that this does not matter very much to them. However, from customer Bob's point of view, the trust relationship with him and ABC is very important. For instance, if Bob does not trust ABC, he will never buy anything from them. It will also affect the recommendation and reputation value of ABC when he is asked for reference.

As another example, consider Liz as the boss who manages employee Maja. Boss Liz trusts employee Maja and employee Maja trusts boss Liz. They both believe the relationship is important to both parties.

From boss Liz's point of view, the trust relationships from Liz to Maja and from Maja to Liz are both important. From employee Maja's point of view, the trust relationships from Maja to Liz and from Liz to Maja are both important. Therefore, the relationships in both directions are important to both parties.

The *challenge* of the trust measure and the trustworthiness prediction is that of the measure of *gravity*, using a method known as *'influence'*. The method of trustworthiness measurement and *influence* or gravity is introduced in the next chapter.

5.4 Endogenous and Exogenous Characteristics of Agents

5.4.1 Internal Factors of Trusted Agents

To study why trust is fuzzy and dynamic, we now look at the agents who are involved in the trust relationship and their impact on the fuzziness and dynamism of trust.

As per our trust definition from Chapter 2, trust is defined as the belief that the trusting agent has in the trusted agent's willingness and capability to deliver a mutually agreed service in a given context and in a given time slot (Figure 5.10).

The key challenge is how to measure w*illingness* and *capability* so that the trust value can closely represent the truthfulness or quality of the trusted agents. *Willingness and capability* (internal factors) are the challenges in the measurement and prediction of trustworthiness.

Willingness symbolizes the trusted agent's volition to act or be in a state of readiness to act honestly, truthfully, reliably and sincerely in delivering a mutually agreed behaviour (Chapter 2). As this factor is internal to Agents, it is very hard to estimate even with scientific measurement or

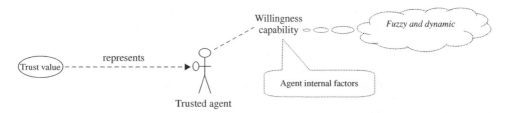

Figure 5.10 The trust value represents the *willingness and capability* of the trusted agent. It signifies the level of trust that the trusting agent has in the trusted agent

research methods. The willingness of a person or an agent could change as time passes, as it could be dependent on the mood or situation of a person. Therefore, it makes the trust model dynamic.

Capability is defined as 'the talent, competence, aptitude and ability of the trusted agent in delivering on the mutually agreed services' (Chapter 2)'. *Capability* signifies the agent's intelligence, skills, knowledge and experience. It is internal to an Agent. A person or an agent's capability changes with time owing to internal or external influences. Examples of external influences could be further training or study or broader and varied experiences. These changes could happen in any given time slot or over many time slots.

These two factors are internal factors of agents. As they are internal, we cannot have direct measures because we cannot obtain them by direct observation. Therefore, when we derive a trust value, it is only an estimate or an approximate value. Fuzziness and dynamism are doubled when taking the views of both parties together.

The *willingness* and *capability* of an agent is *dynamic*. If we assume the *context* is *clearly* defined and clearly understood by all the parties involved, with the passage of time, the *trust* or *belief* that the trusting agent has in the trusted agent may still change owing to the following factors:

(1) With further dealings, it is possible for the trusting entity to get a better idea of the 'capability' and 'willingness' of the trusted entity to act in the way the trusting entity wants it to act in a given context.
(2) The 'capability' or 'willingness' of the trusted entity to act in a given context as desired by the Trusting entity may vary with time, depending on the trusted entity's circumstances.
(3) Upon getting more recommendations from other entities, the trusting entity could get a better idea about the 'capability' and 'willingness' of the trusted entity to act in a way that it wants it to act, in a given context.

5.4.2 Psychological Factors of the Trusting Agent

When we consider an agent who is a human being, internal factors, namely, willingness and capability, are influenced by the agent's psychological predisposition. However, if we consider that an agent can also be a machine or an object, psychological factors do not apply. However, in the service-oriented network environment, agents are usually service providers, customers and merchants as well as websites. Some of these agents have strong human and intellectual involvement and inputs. If we assume the agents (whether the trusting agent or the trusted agent) are human beings in the service-oriented network, the psychological nature of the trusting agent is a very important factor that influences an agent's decision to trust the 'trusted' agent. We note the following here:

In psychological terms, according to Myers [14] and Mallach [15], it is reasonable to make the following assumptions:

• People with a *'sensing preference'* will not trust any person with whom they did not have any previous interaction. Both Myers and Mallach indicate that people with a 'sensing' preference have a tendency to rely on facts and experience [14, 15].

- People with an *'intuition preference'* may trust a person with whom they have not had any previous interactions. The preference of the trusting agent will influence its decision to trust a given trusted agent, with or without detailed information on the trustworthiness of the trusted agent. Myers and Mallach contend that persons with an 'intuition' preference have a tendency to rely more on possibilities and taking risks [14, 15].
- People with a *'thinking preference'* have a tendency to analyse things in an objective and logical fashion with little or no regard for personal values before they reach or take a decision [14, 15]. We could also believe that if the trusting agent has a *thinking preference*, he/she will pay little or no attention to the personal values of the trusted agent or to personal feelings about the trusted agent and make an objective and logical decision regarding whether to trust the trusted agent.
- People with a *'feeling preference'* place primary importance on personal values, before reaching a decision [14, 15]. We could also believe that the trusting agents who give preference to *feeling* will place greater importance on his/her personal feelings about the trusted agent and values of the trusted agent while they decide whether to trust the trusted agent.

Table 5.1 shows a summary of psychological factors and their impact on trust decision making.

Whether a trusting agent gives preference to 'thinking' or 'feeling' will determine whether he/she makes the decision on the basis of facts or personal values of the trusted agent. We believe that the psychological nature of the trusting agent is a very important factor in influencing the trusting agent's decision to trust the trusted agent.

The attitude towards centralized web services or Peer-to-Peer (P2P) e-commerce is yet another important factor that will influence the trusting agent's decision to trust the trusted agent. Many people are reluctant to use the Internet or web services as a means of carrying out business or transactions owing to the inherent risks involved in electronic business. Many people regard it as unsafe as they are not totally convinced about how the other entity behaves in tasks like possessing credit card details, handling privacy issues, and so on. Although technologies like cryptography, digital certificates and some legislation have been introduced to mitigate the risk involved in carrying out online transactions, some sections of the community are still not convinced that the Internet is a safe place to carry out transactions. An example of a defensive measure used to reduce the risk of carrying out online transactions is the verification of the *identity* of the website with the help of digital certificates before carrying out an electronic transaction.

In distributed environments, such as Peer-to-Peer (P2P), Grid and mobile networks, the problem is graver than compared to client–server communication. In client–server communication, many of the security measures taken to ensure that client–server–based e-commerce is a safe approach to carry out transactions that rely on trusted certification authorities. However, in the other communication environments mentioned, there may not be a centralized authority that oversees security because of the decentralized nature of the environments.

The outcome of previous transactions between the trusting agent and the trusted agent will have a major bearing on the decision of whether to trust the trusted agent again. Depending on the outcome

Table 5.1 Four different types of psychological factors affecting the intellectual decisions of trusting agents

Psychological nature of the trusting agent	Impact on the trust decisions to another agent
Sensing	Rely on facts and experience
Intuition	Rely more on possibilities and taking risks
Thinking	Analyse things in an objective and logical fashion with little or no regard for personal values
Feeling	Place primary importance on personal values, before reaching a decision

of a previous transaction, the trusting agent may or may not be more confident in deciding whether to trust the trusted agent. If the outcomes of the previous transactions are positive, then trust in the trusted agent will grow. The trusting agent is most likely to trust the trusted agent in future transactions. On the other hand, if the outcome of the previous transaction is negative, this will have a negative impact on the perceived trustworthiness of the trusted agent by the trusting agent.

5.4.3 Endogenous Factors of Agent

Endogenous factors of agents refer to their internal factors such as psychological considerations, personal characteristics, knowledge or skills.

For example, while considering the previous section (Section 5.3.1, Figure 5.1) *trust* (as illustrated again in Figure 5.15) in the case of Alice and Bob may be too optimistic; they have the tendency to believe that only good things can happen to them in this world (optimists). Charlie may be pessimistic and has a pessimistic attitude towards life, tending to underrate his capabilities; however, Doug may be a very simple person and easily trusts everybody and everything. The endogenous factors, including psychological factors, are factors internal to the trusting parties. These factors can never be captured explicitly and they change over time.

Endogenous characteristics influence trust value and prediction. These characteristics cannot be captured directly. An agent's 'willingness' and 'capability' are considered under the heading endogenous factors (Section 2.4 of Chapter 2 and Section 3.7 of Chapter 3). When predicting trust in a relationship, the factors that influence the trust decision, which cannot be explicitly managed, are the endogenous factors. For example, if a person's thinking is changed; no one could know or capture this immediately. *Endogenous factors* cause the changes in the trust relationship (Figure 5.11). However, endogenous factors cannot be apprehended directly by observation, and thus the measure and prediction of trustworthiness of an agent is only an estimate or an approximation.

The challenge of the trustworthiness measure and prediction is that we are unable to explicitly capture the endogenous characters of agents. Therefore, we have to develop methods for trustworthiness measurement and prediction that can, through recognition of specified external behaviours, enable us to estimate the trustworthiness of the agent.

5.4.4 Exogenous Factors of Agent

Exogenous factors are known as external factors of trust, such as external activities, that is, behavioural changes such as making a commitment to deliver a service or evaluating an actual service delivery. These external activities can be identified and predicted. *Exogenous activity* influences the trust value and prediction. It may be caused by the environment where a business interaction is carried out or where a service provider is unable to fulfil the commitments.

In Chapter 1, we described the service-oriented environment as being heterogeneous and consisting of anonymous, pseudo-anonymous, and non-anonymous users or machines communicating with each other for services.

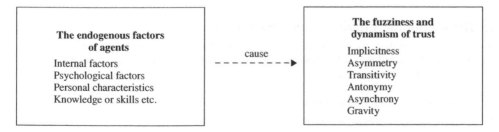

Figure 5.11 The fuzzy and dynamic nature of trust is caused by the e*ndogenous factors* of the agents

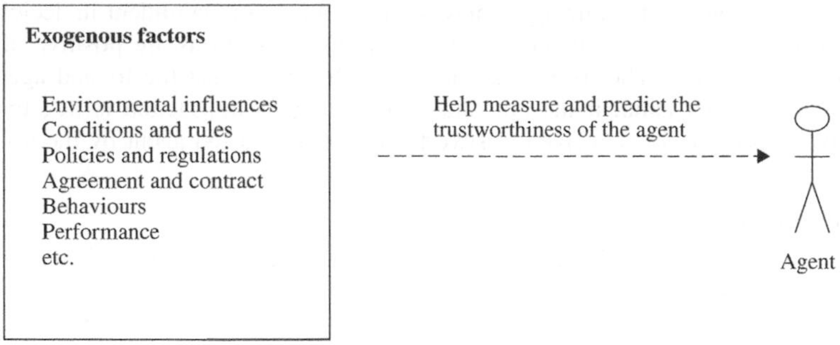

Figure 5.12   Exogenous factors are the source for trustworthiness measurement and prediction

In Peer-to-Peer (P2P) service-oriented networks, file sharing applications such as Gnutella [16] and Napster [17] enable users to share files among themselves. Free Net [http://freenet.courceforge. net] is a P2P-oriented service network for anonymous storage. SETI@HOME (http://pwp.netcabo.pt/ knology/SETI_ENG.htm) is an example of a pseudo-anonymous P2P application for distributed computing. In non-anonymous service-oriented environments, such as Logistic networks [1, 18], Agents make use of each other's resources. These resources can be either physical resources (like warehouse space or transport capabilities) or digital resources (like each other's track-and-trace applications). However, the exogenous factors or external activity can be captured, analysed, measured and calculated to determine a level of trustworthiness.

For example, Jon said '"PriceLine" asked me to give my credit card details before giving me a hotel's name. I do not want to do that'. Liz said 'I have used 'PriceLine' before. It is very reliable. Just give your credit card details and see'. Mark said '"PriceLine" is a US-based business. I would not use it if I am not living in the United States'. This example implies that an environment (anonymous, pseudo-anonymous and non-anonymous) is an *exogenous factor* that affects the trust value and prediction.

Consider another example where a service provider, John, said he would deliver the furniture by Monday 24 Dec 2004, and so Liz paid the money in full on the spot because Liz trusts John. In Jan 2005, Liz still had not received her furniture. She went back to the shop and asked John about it. John signed another delivery note to commit that he would definitely deliver the table within 3 weeks. Liz's *trust value* for John dropped to 3 out of 5. By February 2005, Liz still had not received any furniture, so her trust in John has dropped to zero. She has no choice but to take John to court and try to get her money back.

The *challenge* of the trust measure and trustworthiness prediction is to develop an estimation method that can *handle heterogeneous environments and anonymous, pseudo-anonymous, and non-anonymous users and service providers* and service interactions for predicting the trustworthiness value. This is illustrated in Figure 5.12. The proposed method is known as *the correlation of behaviour method*, which will be introduced in the next chapter.

5.5 Reasoning the Fuzziness and Dynamism

In the previous section, we illustrated the six characteristics of the *fuzziness and dynamism* of trust. We also illustrated the endogenous and exogenous factors of the agents. Now we would like to show how these are related to each other.

5.5.1 Fuzzy and Dynamic Characteristics in the Trust Model

The six fuzzy and dynamic characteristics of trust are as follows:

1. Implicitness in trust

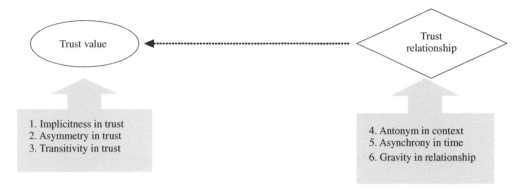

Figure 5.13   The alignment of the *six* characteristics of trust to the preliminary trust model

2. Asymmetry in trust
3. Transitivity in trust
4. Antonym in context
5. Asynchrony in time
6. Gravity in relationship.

Out of the fuzzy and dynamic characteristics of trust, characteristics 1–3 are related to trust values and characteristics 4–6 are related to trust relationships.

From Figure 5.13, we observe the following:

- 'Implicitness of trust', 'asymmetry in trust' and 'Transitivity of trust' are related to the *trust value*, because an agent here is making a *decision* on the trust value only; therefore, 'implicitness', 'asymmetry' and 'transitivity' are relevant only to trust values. For example, if Alice says she trusts Bob in lending him her car, the implicitness is in the trust value and not in the relationship. Because the relationship in this context is explicit, the trust value, however, may be implicit.
- 'Antonymy in context', 'asynchrony in time space' and the 'gravity of the relationship' are related to the *trust relationship*. The trust relationship is context and time dependent. An agent is constructing/experiencing a perception of the context, time or gravity of the relationship. This is the agents' own view or opinion of what they see or believe in a trust relationship.

5.5.2 Endogenous and Exogenous Characteristics in the Trust Model

In the trust model, the relationship involves agents. Each agent has endogenous and exogenous factors that impact on the trust decision making, and this in turn affects the trust relationship.

In this trust model illustrated in Figure 5.14, we see that both *endogenous* factors and *exogenous* factors are related to agents in the trust model.

5.5.3 Reasons for Fuzziness and Dynamism

The six fuzzy and dynamic characteristics of trust are triggered by agents. The agent's internal factors are hard to capture and predict; however, they have a strong impact on an agent's own development and decision making. External factors, however, can be captured and therefore used to help estimate an agent's trustworthiness.

Figure 5.15 illustrates this point.

The fuzzy and dynamic characteristics of trust are triggered by endogenous and exogenous factors. However, only exogenous factors can be captured and used for the measurement of trustworthiness. In view of the trust model, we see that both endogenous factors and exogenous factors are related to peers in trust relationships.

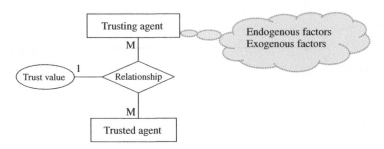

Figure 5.14   Alignment of the agent's endogenous and exogenous factors in the trust model

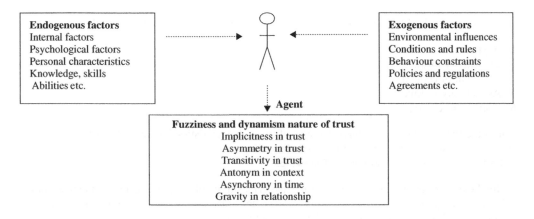

Figure 5.15   Exogenous and endogenous 'triggers' of trust

The dynamism of trust is influenced by the factors associated with trust and trust relationships. We now analyse the eight characteristics of trust aligned with the definition of trust and the trust model, in order to reason about the factors that determine the trust dynamics.

Table 5.2 depicts the characteristics of trust and an agent's endogenous and exogenous characteristics related to the definition of trust and the trust model.

In general, *fuzziness* and *dynamism* mean '*unclear*' and '*changing* all the time'. While some changes (such as external behaviour) can be predicted, because they can be explicitly defined, others cannot be predicted (such as the internal factors of agents), because they cannot be explicitly defined. We can only give a measure or an estimate of the dynamism of trust in the trust relationship.

In Table 5.2, the following may be noted:

(1) The first column describes the concepts of trust.
(2) The second column describes the trust model and all the related concepts in the trust model. Column 2 is a pictorial representation of column 1.
(3) The third column aligns the fuzzy and dynamic characteristics to the definition of trust and trust model.
(4) The fourth column aligns the endogenous and exogenous characteristics of an agent to the definition of trust and the trust model.
(5) The fifth column shows that trustworthiness measure or prediction and which concepts can be explicitly defined and which concepts cannot be explicitly defined.

Table 5.2 The factors determining the nature of trust

Trust definition contains concepts	Trust model contains concepts	Trust characteristics	Agent characteristics	Trustworthiness
Belief	Trust value	Implicitness of trust Asymmetry in trust Transitivity of trust		Fuzzy result
Trusting agent and Trusted agent	Relationship	Gravity of relationship		Explicit result
Willingness capability	Agents (conceptual behaviour)		Endogenous	Fuzzy result
Delivery of the mutually agreed services	Agents (external behaviour)		Exogenous	Explicit result
A given context	Context	Antonym in context		Explicit result
A given time slot	Time slot	Asynchrony in time space		Explicit result

Table 5.3 Aspects of the definition of trust and their relationship to an agent's internal and external factors and their impact on the dynamism of trust

Aspects of the trust	Relation to agents	Dynamism analysis
Belief willingness capability	Agent's internal activity	Internal factors are very hard to capture. They could be changed by both internal influence and external influence. For example, more education or an accident could change the capability of a person, or psychological advice could affect the level of willingness. However, no one can predict how much they will change with time. Therefore, *internal factors cause the dynamic* nature in the trust model.
Trusting agent and trusted agent	Agent's external activity	Identifying a trusting agent or a trusted agent is an external activity. They can be explicitly defined.
Deliver the mutually agreed services	Agent's external activity	External factors can be captured though correlation of expected delivery of the service compared to the actual service that is provided.
A given context, A given time slot	Agent's external activity	These are external factors and can be captured; therefore they are not the cause of the dynamism in the trust model.

In Table 5.3, we further explain what can be measured and what cannot be measured. We note that change can be caused by *external factors* as well as internal factors.

In real life, we note that both factors could cause the change. However, the internal factors are hard to capture and predict, even after intense empirical studies of a person's psychology. This is unlike external factors, where one can directly observe them, predict them and try to manage them.

Therefore, the *internal factors* cause the dynamism or changes and they are the factors that humans or machines cannot manage. Humans or machines can manage the external factors that cause the dynamism or changes. Since they can be captured, they are considered to be stable in the trust model.

5.6 Managing the Fuzziness of Trust

In most Internet trustworthiness systems, there are features for measuring service providers, merchants or online shops. However, if a provider offers a very good service on books but very bad service on delivery, then the trustworthiness value for the merchant should be separate and distinguishable from that of the service that is provided.

5.6.1 Measuring the Service Quality

When talking about measuring the quality of service, one needs to consider the input from both the service requester and the service provider. Measuring service is one of the most difficult tasks with respect to Internet service provision in the service-oriented environment. It is different from measuring a product, such as a book. When we express an opinion as to whether a book is good or bad, the book itself will not answer or argue back. However, when we are measuring a service, and not the service provider, the customer and the provider/s of the service/s all have their say. However, the calculation of trusted services is not going to be straightforward (as elaborated in the next chapter). However, we develop a provider service rating, or the Trustworthiness of service in the service-oriented network environment.

Each service provider may have a number of services. The objective here is to measure the service of the provider, rather than the provider itself, because you are valuing service quality from the perspectives of two agents. To have a fair process, we need to consider inputs of both parties as both parties have their own opinions. This is also different from rating merchants or online shops, as we are rating the service, not the provider or merchant or shop.

5.6.2 Measuring the Product Quality

Various forms of measurements have been used in most Internet trustworthiness systems. The measure is much simpler than measuring the service where there is human intelligence behind the service. They have the prerogative to provide input on the quality of service. In measuring a product or website, we do not require input.

If we say a camera is good or bad, we do not get the 'camera's' opinion about comments that the customer made. The features often include, however, in addition to the product, measuring service providers, merchants or online shops.

Measuring the product is a much simpler process than measuring service, and it has similarities to measuring websites or online shops.

Trust in the transaction is only dependent on one party. The valuation of the service quality is therefore unidirectional and dependent on the *capacity* of the online provider to provide the service. A valuation is therefore much simpler and more straightforward.

When we value an item or product or an online shop, it is much simpler in the sense that the item or product does not have an opinion. We (as a human agent) would want to argue back if the customer's opinion is believed to be wrong.

5.7 Managing the Dynamism of Trust

In the existing literature, the methods of managing trust focus on assigning the trust value with the assumption that there is only one context and the trust value is assigned only for that context. This is due to the fact that many e-service providers only provide a single service (single context). However, this assumption becomes less relevant as the concept of e-services has expanded to multiple services over the last few years. Additionally, the methods of managing trust only consider one trust value that does not change. These methods do not consider the dynamic nature of trust and the change in trust values with time.

5.7.1 Dynamism of Trust in Time Space

The trust value remains stable over a particular time slot or time period. When the trust value changes, a new time slot is created. Trustworthiness prediction is defined as the process of determining the future trust value known as trustworthiness value of the trusted entity or agent, given its past repute values or historical trust value or direct interaction from the given time spot, slot and space.

The repute value is defined as a trust value for an entity, that is, its reputation in a given context and in a given time slot as recommended by a third-party agent or recommendation agent or witness agent.

5.7.2 Managing the Trust Dynamism

In the previous section, we described six characteristics of trust and two characteristics of agents. In the measurement of trust and trustworthiness prediction, we need to consider all of these characteristics.

To manage trust over the network as adequately as possible, one must consider the fuzziness and dynamism of the trust. To give an estimate of the trustworthiness value, we will carry out the correlation between the *expected service* and the *actual delivered service* to predict the trustworthiness value.

For the determination of the trustworthiness value for a trusted entity or agent, we choose to apply the technique used in the human world, that is, the correlation between the expected behaviour and the actual behaviour to determine the level of trust. We adopt this approach for the trustworthiness prediction in e-business or e-services to overcome the dynamism of trust.

The expected behaviour of the trusted agent is related to the quality aspects or the mutually anticipated conduct of the trusted agent prior to its interaction with the trusting agent. The correlation is the degree of difference between the expected delivery by the trusted agent and actual delivery by the trusted agent during the interaction.

5.7.3 Correlation Agent Quality Aspects of the Context

Correlation refers to how similar the following two factors are:

- The quality aspects that the trusting agent expects are offered by the trusted agent in a given context.
- The actual delivered quality of the service or product or fulfilment of the obligation from the trusted agent in that particular context in a given time slot

The greater the correlation between these two parts, the higher the trustworthiness value assigned to the trusted agent by the trusting agent and vice versa. Strong correlations between the above two parts indicate that the trusted agent met the expectation held by the trusting agent, in that context. On the other hand, a weak correlation indicates that the trusted agent failed to meet the expectation held by the trusting agent in that context.

5.7.4 Strategies for Trust Measurement and Prediction

In Chapter 2, we gave the formal definition of trust as follows:

'Trust is the belief the trusting agent has in the trusted agent's willingness and capability to behave in a given context and at a given time slot, as expected by the trusting agent'.

To determine the level of trust, we introduced the notion of trustworthiness.

The measurement of trustworthiness is an assessment of current trust values that depicts the level of the trust relationship that the trusting agent has with the trusted agent in a given context, in a given time slot and with a given type of initial association relationship. Trustworthiness is an

Table 5.4 The strategies for trustworthiness measurement and prediction to address the fuzzy and dynamic characteristics of trust and agents

Eight characteristics	Strategies for trust measurement and prediction
Implicitness of trust	Clearly define the quality aspects for each context, carefully develop the assessment criteria, and measure against a well-defined scale system.
Asymmetry in trust	Must define direction from trusting agent to the trusted agent
Transitivity of trust	Know the trustworthiness of recommendation agent, and also consider whether the opinion is first hand, second hand or third hand.
Antonym in context	Utilize the trust ontology, the agreed and shared conceptualization.
Asynchrony in time space	Aggregate the trust value over different time slots and use advanced algorithms, such as Markov model, and so on.
Gravity in relationship	Identify the weight (*Influence*) of each quality aspect of the context.
Endogenous characteristic of agent	Capture external behaviour for estimation.
Exogenous characteristic of agent	Correlate the fulfilment of the quality of the service/product, and so on.

assessment of the trust level for a given context and time and for the relevant type of initiation of relationship (as indicated in Chapter 2) for a trust relationship.

The assessment can only be done by correlating the actual behaviour with the expected behaviour, in a given context and in a particular time slot with respect to a given method of initiation. We shall learn how to do this in the next chapter. Prediction of trustworthiness is an assessment of future trust values that depicts the level of trustworthiness in future time slots. Prediction of trustworthiness also utilizes the value in the current time slot as well as values in previous time slots. Hence, it would also take into account trends in variation of trust, among other things. This is discussed in Chapter 12.

The constraining aspect in the measurement of trust and the prediction of trustworthiness lies in the inability to handle the 'internal factors' of agents, namely, 'willingness' and 'capability'.

In Table 5.4, it is proposed as to how the eight characteristics are to be treated or measured when carrying out trust measures and trustworthiness prediction.

5.8 Summary

The following important concepts related to the dynamic nature of trust have been discussed in this chapter:

- The limitations introduced in the measurement of trust and the prediction of trustworthiness lie in the inability to adequately quantify the internal factors of agents, namely, willingness and capability. Much of this is related to the characteristics of fuzziness and dynamism.
- Since capability and willingness are by and large not directly observable, we arrive at an estimation of these by utilizing the external factors (expected and actual behaviours) of agents within the context of the relationship.
- In addition to the endogenic factors of willingness and capability, the psychological factors of a trusting agent contribute to the dynamic nature of trust.
- The preference of the trusting agent for 'sensing' or 'intuition' will influence its decision to trust a given trusted agent, with or without detailed information on the trustworthiness of the trusted agent.

- Whether a trusting agent gives preference to 'thinking' or 'feeling', will determine whether he/she makes decisions related to trust based on facts or personal values of the trusted agent.
- The psychological nature of the trusting agent, as well as the attitude of the trusting agent towards P2P e-commerce and previous transactions with the trusted agent, affects the trusting agent's decision to trust the trusted agent before any interaction with the trusted agent takes place. Hence, we have named these factors as *pre-interaction factors*.
- Reputation factors are those factors pertaining to the reputation of the trusted agent and can influence the trusting agent in deciding whether to trust the other agent.
- Personal interaction factors are defined as those factors that help the trusting agent to associate a trustworthiness value to the trusted agent based on his/her personal interaction with the trusted agent.
- Trustworthiness prediction is defined as the process of determining the future trust value known as the trustworthiness value of the trusted entity or agent, given its past repute values or historical trust value or direct interaction from the given time spot, slot and space.
- The repute value is defined as a trust value for an entity, that is, its reputation in a given context and in a given time slot as recommended by a witness entity or witness agent.
- The time slot of a trustworthiness prediction is defined as the breadth or duration of time over which the Trust Value from the historical trust value or repute value is collected.
- The time space of a trustworthiness prediction is defined as the total duration of time over which the behaviour of the trusted entity will be analysed and the process of trustworthiness prediction carried out.
- The expected behaviour of the trusted agent is the mutually anticipated conduct of the trusted agent prior to its interaction with the trusting agent. The correlation is the degree of similarity between the expected behaviour of delivery of the trusted agent and actual behaviour of delivery of the trusted agent during the interactions.
- The endogenous characteristics of trust, specifically the implicitness in trust, asymmetry in trust and transitivity in trust are related and aligned to trust value in the trust model.
- The endogenous characteristics of antonymy in context and asynchrony in time and gravity in the relationships are related specifically to the trust relationship.
- There is a relationship between the endogenous factors, the agent and the exogenous factors in the form of 'triggers' relating to the fuzziness and dynamism of Trust.
- We can conclude the following from our analysis of the dynamic nature of trust:
  - The internal factors of agents determine the dynamic nature of the agents.
  - The dynamic nature of the agents leads to the dynamic nature of trust, trust relationships and trust values.
  - 'Context', 'time' and 'initiation of relationship' are not dynamic as these factors are defined by the agents and, once defined, they do not change.

Trust is one of the most *fuzzy, dynamic* and *complex* human and business *concepts* in our world. Measuring trust or carrying out trustworthiness predictions are not easy tasks [19]. The difficulty in measuring trust or predicting trustworthiness in service-oriented network environments poses many questions. These include issues such as how to measure the 'willingness' and 'capability' of individuals in the trust dynamic and how to provide for a *concrete* and *explicit* level of trust. It is also necessary to understand why trust is fuzzy and changes when time passes and how we make sure that a 'context' is *clear* when we say 'trust is context dependent'. We also need to know what human activities we can capture in the real world and use them for measuring and predicting trust.

In this chapter, we studied the *fuzzy, dynamic and complex* nature of trust. The dynamic nature of trust creates the biggest challenge in measuring trust and in the prediction of trustworthiness. We should now understand what can be measured and what cannot and what can be done and what cannot.

References

[1] Chang E., Dillon T.S., Gardner W., Talevski A., Rajugan R. & Kapnoullas T. (2003) A Virtual Logistics Network and an e-Hub as a Competitive Approach for Small to Medium Size Companies, *2nd International Human.Society@Internet Conference*, Seoul, Korea.

[2] Shelat B. & Egger F.N. (2002) *'What makes people Trust online gambling sites?'* Available: [http://www.ecommuse.com/research/publications/chi2002.pdf] (10/08/2003).

[3] Egger F.N. & Groot B. (2000) *'Developing a model of Trust for electronic commerce: An application to a permissive marketing website'*, Available: [http://www.ecommuse.com/research/publications/WWW9.htm] (20/6/2003). http://freenet.sourceforge.net/.

[4] Egger N.F. (2000a) *'Trust me, I'm an online vendor'*, *Towards a Model of Trust for E-Commerce System Design*, Available: [http://www.zurich.ibm.com/~ mrs/chi2000/contributions/egger.html] (10/09/2003).

[5] Egger N.F. (2000b) *'Towards a model of Trust for e-commerce system design'*, Available: [http://www.zurich.ibm.com/~ mrs/chi2000/contributions/egger.html] (29/05/2003).

[6] Egger N.F. (2003) *'Deceptive technologies: Cash, ethics and HCI'*, Available: [http://www.ecommuse.com/research/publications/sigchi_bulletin.htm] (23/05/2003).

[7] Kim J. & Moon J.Y. (1997) *'Emotional usability of customer interfaces'*, Available: [http://hci.yonsei.ac.kr/non/e02/97-CHI Emotional_Usability_of_Customer_Interface.pdf] (23/08/2003).

[8] Hussian F., Chang E. & Dillion T.S. (2004) 'Taxonomy of trust relationships in peer-to-peer (P2P) communication', *Proceedings of the Second International Workshop on Security in Information Systems*, Porto, Portugal, pp. 99–103.

[9] Smith J.H. (2002) *'The Architectures of Trust'*, M.Sc Thesis, The University of Copenhagen, Copenhagen, Denmark.

[10] Rahman A.A. & Hailes S. (2003a) *'Relying on Trust to find reliable information'*, Available: [http://www.cs.ucl.ac.uk/staff/F.AbdulRahman/docs/dwacos99.pdf] (7/08/2003).

[11] Rahman A.A. & Hailes S. (2003b) *'A distributed Trust Model'*, Available: [http://citeseer.nj.nec.com/cache/papers/cs/. . ./abdul-rahman97distributed.pdf] (5/09/2003).

[12] Wang Y. & Vassileva J. (2003) *'Trust and Reputation Models in Peer-to-Peer'*, Available: [www.cs.usask ca/grads/yaw181/publications/120_wang_y.pdf] (15/10/2003).

[13] Burton K.A. (2002) *'Design of the open privacy distributed reputation system'*, Available: [http://www.peerfear.org/papers/openprivacy-reputation.pdf] (10/11/2003).

[14] Myers S. (2003) *'Working out your Myers Briggs Type'*, Available: [http://www.teamtechnology.co.uk/tt/t-articl/mb-simpl.htm] (27/12/2003).

[15] Mallach E.G. (2000) *'Decision support and data warehouse systems'*, McGraw Hill Companies, Irwin.

[16] http://www.gnutella.com/, accessed 2005.

[17] http://www.napster.com/, accessed 2005.

[18] Chang E., Talevski A. & Dillon T. (2003) 'Web service integration in the extended logistics enterprise', *Proceedings of the IEEE Conference on Industrial Informatics, INDIN*, Banff, Canada.

[19] Rahman A.A. & Hailes S. (2003) *'Supporting Trust in Virtual Communities'*, Available: [http://citeseer.nj.nec.com/cache/papers/cs .. ../abdul-rahman00supporting.pdf] SETI@HOME, Available: [http://pwp.netcabo.pt/knology/SETI_ENG.htm] (28/09/2003).

6

Trustworthiness Measure with CCCI

6.1 Introduction

In Chapter 5, we saw that trust is dynamic and complex in nature. In Chapter 4, we have come to know the importance of the *context* to which trust refers. We note that trust is the belief that the expectation was met in a context (Chapter 2). We extended this concept to service-oriented environments, and state that trust in the service-oriented environment implies quality of service, quality of products or quality of agent over anonymous communications. In this chapter, we address a crucial component in studies of trust, that is, the measurement of trust. For the service-oriented environment, it implies the measurement of the 'quality' of agent, services or products for improving or realizing consumer confidence.

We note that in a measurement methodology, the *Quality Assessment Criteria* play key roles. We also need to have a means to define the criteria, and to take into account the *clarity* of each *criterion* conveyed to the trusted agent by the trusting agent before the interaction, the *influence* of each criteria that impacts on the overall quality measure, and the *correlation* between what was actually delivered and the quality of services or product against the *Quality Assessment Criteria* (also known as the *defined quality*).

In this chapter, we also give a detailed explanation of quality assessment through CCCI (Correlation of delivered quality against defined quality, Quality Commitment to each of the defined Quality Assessment Criteria, Clarity of each criterion from both parties' views and Influence of each criterion on the overall quality assessment) method. This method is based on determining the correlation between the actually delivered quality of service or product by the trusted agent and originally defined quality of services (the Quality Assessment Criteria) to provide a gauge of the trustworthiness of the trusted agent.

6.2 Trustworthiness Measure Methodology

Trustworthiness characterizes the *quality* of service or product. Trustworthiness of an agent, a service or a product in the service-oriented environment implies the *'quality'* of an agent, a service or a product. The *primary measure* of the quality of the service or product, and so on, in service-oriented environments is to correlate the *delivered quality* of service against a *defined quality* of service that is mutually agreed upon between the trusting agent and the trusted agent (a consumer and a provider, for example).

Trust and Reputation for Service-Oriented Environments   Elizabeth Chang, Tharam Dillon and Farookh Hussain
© 2006 John Wiley & Sons, Ltd

The purpose of the *trustworthiness measure* is to record the *quality* of the trusted agents (such as business partners or service providers and their quality of services and products, and so on.) in heterogeneous and sometimes anonymous networks for the trusting agents to have a future reference or for recommendations to others who might query about some service they do not know or have no experience with. This helps *trusted* business transactions and virtual collaboration and keeps the service-oriented environment safe and trustworthy, as well as helping to provide a transparent and harmonious nature to the distributed heterogeneous, anonymous, pseudo-anonymous, and non-anonymous e-service networks.

6.2.1 Conceptual Framework for Measurement of Quality and Trust

When we talk about the *quality* of something or the *trust* in someone, normally the 'context' is already defined. Many examples have been given in Chapter 4 in conjunction with the description of trust ontologies. However, 'context' alone does not provide enough information to justify the quality referred to. For example, a service provider (Peter) provides a service, 'Interstate delivery' (*context*). For each job, Peter and a customer sign a 'service agreement', or 'contract' or 'terms and conditions'. Peter then carries out the service. However, Peter's service is always very poor (the left column of Table 6.1).

From a quality assessor's (Tom) perspective (the middle column of Table 6.1), to assess the service quality of Peter, Tom does not make simple statements like 'Peter is not good', or 'Peter's interstate delivery is poor'. He decomposes the *service* known as '*Interstate delivery*' (context) into several quality dimensions (quality aspects). He also makes reference to the 'service-level agreement' or 'contract', which has some indication of quality delivery. Tom then develops the Quality Assessment Criteria for each of these quality dimensions (quality aspects).

For example, quality dimensions may be (a) just-in-time service, (b) handling the goods in the right manner or (c) maintaining a low cost. The Quality Assessment Criteria for each of the quality dimensions are as follows:

(a) Just-in-time pick up and delivery with a minimum delay of no more than four hours.
(b) There should be no damage to fragile goods.

Table 6.1 Mapping the service process, quality assessment process and methodology framework

Service process	Quality assessment process	Framework process
Service provider: **Peter**	Quality assessor: **Tom**	Quality measure framework
1 Provide service: interstate delivery.	Obtain domain knowledge.	Define the knowledge domain of the context and obtain expert knowledge.
2 Enter the 'service-level agreement' with customer.	Decompose the *context* into several quality dimensions, or abstract the quality aspects from the 'service-level agreement' or 'contract', and so on.	Identify the quality dimensions or quality aspects, or abstract the quality aspects from 'service-level agreement' or 'contract' or service 'terms and conditions', or 'standard', and so on.
3 Derive the service quality criteria from the Service Level Agreement.	Develop the criteria for the measure of each quality dimension/aspect.	For each quality aspect, develop a set of Quality Assessment Criteria.
4 Deliver the service.	Correlate the delivered quality of service against each of the Quality Assessment Criteria.	Measure the trust through CCCI metrics.

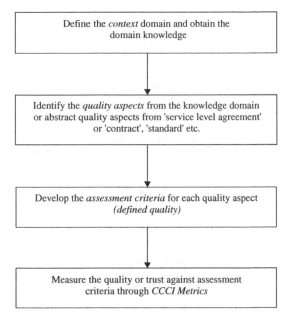

Figure 6.1 The conceptual framework of the trustworthiness measure methodology

(c) There should be no double charges when delivery is completed. To determine whether Peter's service is very poor, there is a detailed correlation of what Peter actually did against the Quality Assessment Criteria.

Note that the right-hand side of Table 6.1 gives an overall view of the *quality measure* or framework of *trustworthiness measure methodology*.

From Table 6.1, we see that a *quality measure* or framework, *trustworthiness measure methodology* contains four steps.

The framework of the trustworthiness measure methodology given in Figure 6.1 contains four major steps:

(1) Obtain context-associated domain knowledge.
(2) Identify the quality aspects from the knowledge domain.
(3) Develop Quality Assessment Criteria for each quality aspect.
(4) Measure the quality and trust against Quality Assessment Criteria through CCCI metrics.

- Step 1 of the methodology framework focuses on the context and its domain knowledge.
- Step 2 of the methodology framework focuses on the identification of the quality aspects. One can develop the quality aspects through the context and its domain knowledge. If this is difficult, an alternative approach is from the 'service-level agreement' or 'contract' or 'standards', and so on. Note that the 'contract' or 'service-level agreement' may not specifically outline 'quality aspects'. However, there may be some indication of quality aspects. One can therefore identify and develop a set of quality aspects from the 'service-level agreement' or 'contract' or 'service standards'.
- Step 3 of the methodology framework is to develop Quality Assessment Criteria for each of the quality aspects and this enables quantification of the quality into a numerical representation or a ranking system presentation. These defined criteria are defined quality that can determine the quality or trustworthiness of the agent, service or product.

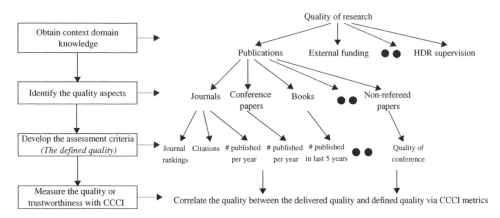

Figure 6.2 An example of a trustworthiness measure for the 'quality of research'

- Step 4 of the methodology framework includes the correlation of delivered quality (such as quality of service or product, etc.) against the Quality Assessment Criteria. Here, it is also important to state that the *influence* (the weight) of each criterion may be different and the impact on the overall quality will vary. The criteria may be developed by a professional body or an assessor. However, the criteria may not be clearly understood by the trusting agent or the trusted agent, and may evolve with time. Therefore, it is important to state the *clarity* of the criteria defined. For example, if an assessment criterion is defined as 'There is a significant national benefit', although it sounds a very appropriate criterion, it is not clear how it should be measured.

In Figure 6.2, we illustrate the methodology with a simple example, 'the measurement of the quality of publicly funded research'.

'Research' is a type of service that is part of a strategic plan of many large organizations and government bodies. There may be experimental research, pure research or applied research, and so on. However since research involves huge expenditure, either by the public and/or the government, the judgment of the 'quality' of the products of research is a significant part of a *measure of trustworthiness* so that there is some assurance that research institutions will not waste public money.

In a similar manner, we can assess the quality of government sponsorship (either fully or partially) for small-medium enterprises (SMEs), and their productivity. In the same approach, we can assess the quality or the trustworthiness of service of any business domain, service domain or product domain. We can also assess the quality of other domains, such as quality of teaching or learning, quality of medicare, transport and logistics, and so on. In the next few sections, we shall explain each of the four steps in detail.

6.2.2 Step 1 – Define the Context Domain and Obtain Expert Knowledge

In Chapter 4, we have given a detailed definition of context for the service-oriented environment. Since context is used in many disciplines and fields, the meaning of *context* changes; therefore, we must define what context means in the service-oriented environment.

A *context* defines a specific domain by designating the name of the domain and includes distinct function(s) for the given context.

The 'context' can be further decomposed into '*criteria*' that can be arranged in a hierarchy. The '*criteria*' like the '*context*' are domain specific and knowledge rich. A context may be a representation of knowledge domain of one of knowledge layers or a node in the layered knowledge

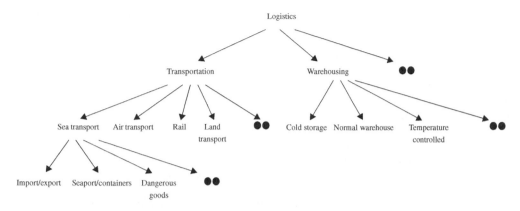

Figure 6.3   Example of super context and sub context

hierarchy, such as a business domain or medical domain. Each of these domains can be decomposed further and further. Thus, context can be decomposed downwards to further layers, known as the *sub context* and the *sub–sub context* (Figure 6.3).

In the service-oriented network environment, *context* represents the agent or agent functions, service or service functions, product or product functions. Each *context* has a name, a type and a functional specification. This has been defined in Chapter 4. A *context* has three parts, namely, name, type and function description. *Context name* is the name of the service; *context type* is the type of service and *context description* describes the detail of service. See Table 6.2 for an example. Table 6.2 shows a description of a *context*. This context can be super context or sub context.

A *context* can represent a physical domain and a conceptual domain. In a physical domain, a context can represent an agent, an object, a service, a product, and so on. In the conceptual domain, a context can represent abstractions, such as 'e-business', 'virtual team', 'management', and so on.

In order to measure the quality of an agent, service or product *(context)*, we need to decompose a *context* into several quality dimensions. Each of these dimensions is used to carry out the measure for a quality of the context. Each of these dimensions is also known as the *quality aspect* of a *context*. This is introduced in the next section.

6.2.3 Step 2 – Identify the Quality Aspects

Context usually only gives us the name of a domain, such as 'sales', 'logistics service', or 'management'. To assess whether the quality of 'sales', or 'logistics service' or 'management' is good or bad, we need to decompose the context into several quality dimensions (quality aspects) in order to provide a reason 'why' it is good or bad.

A *quality aspect* is defined as a quality dimension of a context for the purpose of quality assessment. *Quality aspects* define the quality of the *context*. For example, in 'logistics service', the quality aspects could be 'just-in-time service', 'handling goods in the right manner', 'allowing customers to track and trace their goods', and so on. If all the criteria are satisfied, we can then

Table 6.2   Example of a context

Context	Example
Name:	Cold storage management
Type:	Warehousing
Function description:	Store climate-controlled cargo

state whether they .are good or bad with some degree of explanation of 'why' they may be good or bad.

The strategies to identify the quality aspects of a context, such as the quality of an agent, quality of a service or quality of a product is from the following sources:

- From the context domain knowledge
- From the contracts
- From the agreement
- From terms and conditions
- From the service level agreement
- From the commitment promised by the provider
- From the quality standards defined by the professional bodies, and so on.

'Contract', 'agreement' between trusting agent and trusted agent, 'mutually agreed service', 'terms and conditions', 'commitment', 'quality standard' and so on, capture some *quality aspects* that the trusted agent has made to the trusting agent. It creates the trusting agent's *expectations* with respect to the trusted agent.

Consider a simple example in which Charlie bought a backpack over the Internet. He selected 'green' and 'sport' as the colour and style options respectively. Upon payment, the order was confirmed and the delivery date was set. However, the provider actually delivered a 'blue' 'school' backpack to Charlie, instead of the right colour and the right product. When Charlie contacted the provider for an explanation, he was informed that he would have to either wait for six months until they had a new one in stock or just accept the 'blue' 'school' backpack. Therefore in this scenario, the provider did not deliver the mutually agreed service or contract or the *original commitment*.

6.2.4 Step 3 – Develop the Quality Assessment Criteria

Quality Assessment Criteria are used to define the quality metric for each *quality aspect* for the purpose of measuring *quality*. The quality metric is defined as *a set of conditions or facts for which each delivered quality aspect is compared, correlated, quantified and judged with*. Quality Assessment Criteria set the quality standards. They also act as guidelines for the trusted agent to follow when conducting business.

In the service-oriented network environment, when measuring the trustworthiness or the quality of the trusted agent, we do not consider the trusting agent's expectations of the trusted agent. We also do not consider the obligations of the trusting agent, because the measure is only one way, from the trusting agent to the trusted agent. We only consider the original obligation or commitment of the trusted agent and the actually delivered service.

If *Quality Assessment Criteria* are misunderstood by either party, the measure of trustworthiness may fail. In this case, both parties have to share the responsibility. If the trusting agent carries out a measurement of trustworthiness that it intends to use for the future, it is the trusting agent's duty to make sure that the agreed quality of service and Quality Assessment Criteria are commonly understood by both parties. If it is misunderstood or unclear, it will create the possibility of fraudulent activities as well as errors in obtaining the trustworthiness measure.

6.2.5 Step 4 – Measure the Quality or Trust with CCCI Metrics

In a service-oriented network environment, to obtain a trustworthiness value of the trusted agent, we carry out a *comparison* or *correlation* between the defined quality of service and the service actually delivered (see Figure 6.4).

The measurement of the trustworthiness against Quality Assessment Criteria is carried out through CCCI metrics. This is described in the next section.

Measure the trustworthiness against the assessment criteria

Tom
Trusting peer

Peter
Trusted peer

Figure 6.4 Through the correlation of the *Quality Assessment Criteria* with the *actual delivered service, we* determine the trustworthiness of trusted agents

6.3 CCCI Metrics

The CCCI metrics is especially designed for measuring the quality of the agent and the quality of service (QoS) provided by the agent and product in the service-oriented environment. There are four *key metrics*, as given in Table 6.3.

6.3.1 Correlation of Defined Qualities

Correlation has been widely used in statistics. The primary function of correlation is to study the gap or differences between two comparable entities, objects, classes, or variables, for a period of time and to monitor the movement of gaps, differences and relationships for the purpose of validation, estimation or prediction.

When we carry out a correlation for the purpose of determining the trustworthiness measure or estimation, there are at least two comparable things: one that could be used as a criterion or variable and the other that is used to match or map to the criterion. For example, the original commitment from the service provider could be used as a criterion for measuring the actual service delivered.

In the CCCI metrics, the *criterion* would be the defined Quality Assessment Criteria. A correlation of defined qualities means to correlate quality delivery against all of the Quality Assessment Criteria.

6.3.2 Commitment to the Criterion

Commitment to the criterion measures *how much* the delivery quality aspect is committed to the defined assessment criterion. What is the difference between the two?

Table 6.3 Four key metrics of CCCI Methodology

$Corr_{qualities}$	The *correlation* of *defined quality against* the *actual delivered quality*. In other words, the correlation between the original committed service and the actual delivered service from the trusted agent.
$Commit_{criterion\,c}$	The *commitment* to the criterion depicts how much the delivered quality by the trusted agent committed to the defined Quality Assessment Criteria.
$Clear_{criterion\,c}$	The *clarity* of the criterion depicts the level of common understanding of the Quality Assessment Criteria by trusting agent and trusted agent.
$Inf_{criterion\,c}$	The *influence* of the criterion depicts that different Quality Assessment Criteria have different impact on the overall quality.
	These metrics formulate the *CCCI* trustworthiness measure methodology.

Figure 6.5   The correlation between the defined *quality of service* and the *actual delivered service* for the determination of the level of commitment of the trusted agent

Figure 6.5 is an example where Peter originally committed to deliver a 5-star service; however, he only delivered a 2-star service.

6.3.3 Clarity of the Criterion

It is important to make each criterion explicitly clear when determining the trustworthiness of the trusted agent. *Quality aspects and Quality Assessment Criteria* should be *clearly specified, commonly understood* and *mutually agreed to* (see Chapter 2 for definitions). They should be part of the *terms and conditions* that serve as a foundation for deriving the Quality Assessment Criteria for the trustworthiness measure. If the *criteria* are not explicitly defined, the trustworthiness measure may not be carried out properly and disputes may occur.

Figure 6.6 is an example where a customer (Charlie) and a provider (Peter) encounter a dispute in a service interaction.

Sometimes, a lack of industrial or commercial experience could cause the *quality aspects* or *Quality Assessment Criteria* to be ambiguously defined. However, in this case, both parties share the responsibility. In Figure 6.6, for example, Peter may dislike Charlie because Charlie says 'Peter is untrustworthy', and Charlie may dislike Peter because he lost money. Both parties in this case are at fault. Therefore, both parties should check the *clarity* of the agreement or criteria set out in the agreement before their actual interaction in a given context.

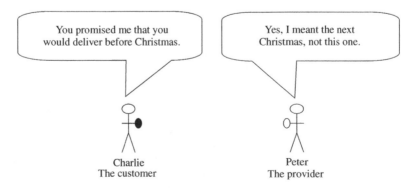

Figure 6.6   The quality assessment criteria for quality of service must be *clearly* defined to avoid dispute

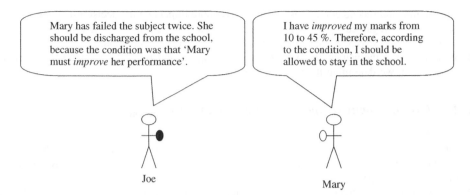

Figure 6.7 The *conditions* or *criterion* set out in the agreement was *not clear*

In Figure 6.7, Joe is advised not to expel Mary from school. If the condition was 'if she does not *pass* the subject again, she will be discharged from school', Joe would be able to remove Mary easily without any dispute.

If one party feels that the criteria set out in the agreement are clear and the other party feels they are not, then the service interaction should be delayed to achieve clarification to avoid future disputes.

Clarity is a fuzzy term just like 'trust', or 'friendliness', and so on. We have proposed a scale of *clarity* in the CCCI Methodology to help explain the degree of *clarity,* which is introduced in the next two sections.

6.3.4 Influence of a Criterion

It is important to *weigh the influence* of a criterion when determining trustworthiness because there may be multiple criteria existing. Some criteria may be more important than others.

The term *influence* signifies the *importance* of a criterion when deciding trustworthiness. For example, in Figure 6.8, an agreement states the criteria that are to be fulfilled.

If we give equal importance to all conditions or criteria, Mary may have to be expelled from school because she did not fulfil two of the four conditions. However, in reality, we give more

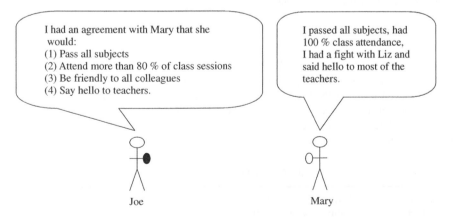

Figure 6.8 Some *conditions* or *criteria* set out in the agreement have more influence than others when carrying out measurement or making a decision

weight to the first and second conditions and they have more impact on the decision-making process. Therefore, Mary is regarded as 'having achieved a lot and fulfilled most of the criteria'.

In the next few sections, we define all the possible levels and values of $\text{Commit}_{\text{criterion}}, \text{Clear}_{\text{criterion}}$ and $\text{Inf}_{\text{criterion}}$. Using the values of $\text{Commit}_{\text{criterion}}, \text{Clear}_{\text{criterion}}$ and $\text{Inf}_{\text{criterion}}$ assigned to all criteria in the interaction, we determine the value of $\text{Corr}_{\text{qualities}}$.

6.4 The Commitment to the Criterion – $\text{Commit}_{\text{criterion}}$

6.4.1 Definition

The *fulfilment of each commitment* is defined as a measure of how much of the trusted agent's commitment to each criterion (i.e. the original commitment set out in the service agreement) has been fulfilled by the service delivered.

The level of the *commitment* of the quality criteria is represented by $\text{Commit}_{\text{criterion}\,c}$ and is set out in Table 6.4.

For example, if West Warehouse has committed $6000\,\text{m}^2$ space to East Logistics, the evaluation of the fulfilment of the commitment is given as

$$\text{Commit}_{\text{criterion}\,c} = \text{Commit}_{\text{space}}$$

6.4.2 Description

In a given interaction with a trusted agent, the trusting agent must consider the correlation between the defined QoS and the service actually delivered, corresponding to each individual *criterion* for the interaction in order to determine the correlation between the defined QoS and the actual delivered service for the entire interaction.

We define the correlation between the defined QoS and the actual delivered service for a given criterion in an interaction as *a numeric value that quantifies the extent of concurrence between the commitment made by the trusted agent for that criterion in that interaction and what was actually delivered*.

We denote this correlation as $\text{Commit}_{\text{criterion}}$.

6.4.3 Application of the Metric

If a service contains a number of criteria, we will need to evaluate the delivered service against the commitment for each criterion. That is represented as

$$\text{Commit}_{\text{criteria}} = \text{Commit}_{\text{criterion}\,1} + \text{Commit}_{\text{criterion}\,2} + \cdots + \text{Commit}_{\text{criterion}\,N}$$

6.4.4 Seven Levels of $\text{Commit}_{\text{criterion}}$ and Values

Here, we represent *seven levels of commitment* and their user-defined values. These values correspond to the seven levels of the trustworthiness measure explained in Chapter 3.

We define seven levels of $\text{Commit}_{\text{criterion}\,c}$. We give a semantic description of these levels for $\text{Commit}_{\text{criterion}\,c}$. Each of these seven levels corresponds to a different degree or extent to which the trusted agent fulfils its commitments.

Table 6.4 $\text{Commit}_{\text{criterion}\,c}$ of CCCI Methodology

Commit:	Metric for measurement of the level of commitment to the quality criteria
Criterion:	Quality Assessment Criteria
c:	A particular criteria

Table 6.5 Commit$_{criterion}$ levels and values

Seven-level scale	Semantics (deliver the service)	Description	Values of Commit$_{criterion\,c}$	Visual representation (star rating system)
−1	Ignore	No agreement was drawn up.	$x = -1$	Not displayed
0	Nothing is delivered	The provider (trusted agent) did not fulfil any of the commitments.	$x = 0$	Normally not displayed
1	Minimally delivered	The provider only delivered a little bit of what was committed.	$0 < x \leq 1$	From ⯪ to ★
2	Partially delivered	The provider only delivered half of what was committed.	$1 < x \leq 2$	From ★⯪ to ★★
3	Largely delivered	The provider delivered a large proportion of service that was committed.	$2 < x \leq 3$	From ★★⯪ to ★★★
4	Delivered	The provider's service delivery is satisfactory and all the commitments were honoured.	$3 < x \leq 4$	From ★★★⯪ to ★★★★
5	Fully delivered	The provider has fully delivered the commitment.	$4 < x \leq 5$	From ★★★★⯪ to ★★★★★

Commitment is represented by the *criterion* for assessment of the fulfilment of the commitment that the trusted agent has originally agreed to. These seven values lead to seven levels of Commit$_{criterion}$, which are explained in Table 6.5.

6.4.5 Maximum Value and Relative Value

This is similar to the trustworthiness measure. We have a seven-level scale of the trustworthiness measure and corresponding trustworthiness values. Here, the calculation is carried out by using either levels or values. If we use levels for calculation purposes, we utilize an ordinal scale. We can then convert the levels into values. If we use values for calculation, we then achieve trustworthiness values. If we want to know which value corresponds to which level, it can be mapped.

Again, as was the case with the definition of trustworthiness, this is always measured against the highest possible ratings with levels or values. For purposes of simplicity, we use *values* to carry out trustworthiness measures (Table 6.6).

The *relative value* is the calculated value that is the Commit$_{criterion\,c}$ actual value against MaxCommit$_{criterion\,c}$.

Table 6.6 The maximum value and relative value of the commitment to the criterion

The maximum value of Commit$_{criterion\,c}$	MaxCommit$_{criterion\,c}$	MaxCommit$_{criterion\,c} = 5$
The relative value is the Commit$_{criterion\,c}$	$Rela$Commit$_{criterion\,c}$	Commit$_{criterion\,c}$/MaxCommit$_{criterion\,c}$ or Commit$_{criterion\,c}$/5

6.4.6 Example

As mentioned previously in the example of the interaction between the two logistics companies (East Logistics and West Warehouse), we note that East Logistics, as the trusting agent, can assign a trustworthiness value to West Warehouse based on whether the following requirements are met:

- West Warehouse allocated a warehouse space of $6000 \, \text{m}^2$.
- The warehouse space was provided for the duration of 2 months.

In other words, East Logistics will assign a *trustworthiness value* to West Warehouse based on the above two criteria for the amount of space allocated and the duration of time for which it is allocated.

- We denote the fulfilment of warehouse space criterion measured as the actual space allocated by West Warehouse to East Logistics as against the initially promised space ($6000 \, \text{m}^2$) as Commit_{space}.
- We denote the fulfilment of the period criterion measured as the duration for which the warehouse space was actually provided by West Warehouse as against the duration that was initially promised (2 months) as Commit_{days}.

Therefore the total fulfilment of the commitment can be represented as

$$\text{Commit}_{warehouse-booking} = \text{Commit}_{space} + \text{Commit}_{days}$$

6.5 Clarity of a Criterion – $\text{Clear}_{criterion}$

6.5.1 Definition

The *clarity of a criterion* is defined as *a measure of how clearly that criterion is set out in the Service Agreement between both communicating parties* (Table 6.7).

The *clarity of each commitment* is represented by $\text{Clear}_{criterion\,c}$.

For example, if West Warehouse has committed $6000 \, \text{m}^2$ space to East Logistics Limited, the evaluation of clarity of this condition or criterion is written as follows:

$$\text{Clear}_{criterion\,c} = \text{Clear}_{space}$$

6.5.2 Description

The clarity of a criterion ($\text{Clear}_{criterion}$) expresses whether or not a criterion is clearly defined, expressed, understood and agreed to by both agents, especially, the trusted agent. This is because we need to consider the trustworthiness of the trusted agent and not that of the trusting agent. Each criterion has to be made explicitly clear prior to the signing of a service agreement.

This involves determining if the defined QOS is communicated and each criterion is explicitly and clearly defined, commonly understood and mutually agreed upon. Each assessment criterion is going to be used as part of the evaluation of the trustworthiness of the trusted agent and should be communicated to both parties explicitly and in unambiguous terms.

Table 6.7 $\text{Clear}_{criterion\,c}$ of CCCI Methodology

Clear:	Metric for clarity measure
Criterion:	Assessment criterion
c:	A particular criterion

Normally, when making a service deal, both the customer and the provider have to sign a service agreement. This service agreement is drawn up by the service provider and is tailored to the customer's requirements.

In many situations, the service agreement or contract, may either be treated as insignificant (the customer may not bother about it), or be too long and full of legal jargon (the customer may not quite understand it). Many customers also simply *trust* a service providers' high-level explanations and read only the key pages of their agreements. There may however be hidden costs, hidden conditions or hidden statements in some insignificant place in an agreement where a customer does not easily notice them. There are many such examples in both the virtual and the real worlds.

Some credit card providers send out forms informing us about a first year waiver of an annual fee, lowest annual interest rate, 55 days interest-free period and Frequent Flyer points for every dollar you spend, and so on. If you sign up the same day, you get a free travel bag. However, after signing up, you may get a bill in the first month without even having used the card followed by another bill in the second month, and so on. When you request an explanation from the credit card company, they might simply describe the amounts on the bills as bank and government charges over which they have no control.

In some restaurants, sometimes the words 'extra surcharge' appears in small print. Peanuts and butter offered on the table that you might think are free, may end up costing you an extra $5.00. You might only notice the charge when you pay the bill. The restaurateur could show you that this is indicated on the menu, but this might be on the last line of the last page in the smallest font possible.

In some hotels, sometimes you may think that bathroom items or drinks on the table are free. However, on your way out, you will find you were charged 5 times more than what they cost in shops. However, you may not have read any information anywhere in the room or in your hotel bookings indicating the excessive charges.

Nowadays, people relocate frequently as their jobs take them all over the country. It is quite common to have logistics service providers transport and deliver household items across the country. Very often, one may find that the actual charges are not the same as those stated in the quotation for the service. Sometimes, the final charges could be more than double the charges originally quoted. There are also cases of hidden costs. For example, a customer received a quotation of $4000 (including insurance) from a logistics provider for shifting 10 computers from the Eastern States to the West Coast in Australia. However, when the computers arrived, the bill was almost $8000. The explanation given by the provider was that as the computers were delivered to the second floor of a building, therefore an extra $4000 was charged. They had the customer's credit card number when they paid the deposit so they went ahead and charged further against it.

The following example illustrates the experience that many people have with mobile phone service providers. A customer purchased a mobile phone on a fixed contract for two years at a total cost of $ 1000.00, at a call rate of approximately $0.30 per 30 seconds call with free calls up to $45.00 per month. However, there were many hidden costs that they did not know about upfront. For example, the phone and the service were purchased in the Eastern States of Australia, where the calls are charged at the rate of $0.30 per 30 seconds. However, in other States the cost is much higher, sometimes being as high as $1.20 for every 30 seconds. The reasoning given by the provider was that since the phone was purchased in the Eastern States where the provider was, if the phone was used in the Western State, the call had to be routed back to the provider's base in the East and therefore, the cost was higher. It would take customers a long time to discover this bit of reasoning on their own. Another hidden cost is discovered by customers when they tried to cancel the phone contract. After approximately 3 months of use, the cost of cancellation of service was about $1000 to be paid upfront. The customers decided against cancellation at that point in time because they had paid so much for the service. After one year, they again tried to cancel it. The cost of cancellation was still $1000. The explanation given by the Help Desk was that, as a formula had been applied, the calculated cancellation cost was $1000, despite the contract saying that the customers could cancel the service at anytime. The customers are not told in the beginning about the use of any formulae for calculating a cancellation cost. This example illustrates a lack of clarity given by the service provider with respect to the service user.

6.5.3 Application of the Metric

If a service contains a number of criteria, we will need to evaluate the clarity of each criterion. This is represented as:

$$\text{Commit}_{\text{criterion 1}} * \text{Clear}_{\text{criterion 1}} + \text{Commit}_{\text{criterion 2}} * \text{Clear}_{\text{criterion 2}}$$
$$+ \ldots + \text{Commit}_{\text{criterion } N} * \text{Clear}_{\text{criterion } N}$$

6.5.4 Seven Levels of Clear$_{criterion}$ and Values

Here, we represent seven levels of clarity corresponding to the seven levels of trustworthiness measures laid out in Chapter 3.

We propose seven different levels and values for Clear$_{criterion}$, to represent how clearly the criterion is defined. These seven levels lead to seven value ranges for Clear$_{criterion}$, which are explained in Table 6.8.

6.5.5 Maximum Value and Relative Value

This is the same as the trustworthiness definition and the Commit$_{criterion}$ definition. The evaluation has to be made against the highest possible ratings level or value.

For purposes of simplicity, we shall just use the *value* of Clear criterion to undertake the determination of the trustworthiness measure.

Table 6.8 Complex Clear$_{criterion}$ levels

Seven-level scale	Semantics	Description	Value of Clear$_{criterion\ c}$	Visual representation (star rating system)
−1	Ignore	No agreement was drawn up.	$x = -1$	Not displayed
0	Not clear	This criterion is not clearly defined.	$x = 0$	Normally not displayed
1	Some part unclear	Some criteria are not clear.	$0 < x \leq 1$	From ☆ to ★
2	Partially clear	Only 50 % of service conditions or criteria are clear.	$1 < x \leq 2$	From ★☆ to ★★
3	Largely clear	The criterion or service condition is basically clear.	$2 < x \leq 3$	From ★★☆ to ★★★
4	Clear	The criterion is generally clear.	$3 < x \leq 4$	From ★★★☆ to ★★★★
5	Very clear	This criterion is made explicitly clear.	$4 < x \leq 5$	From ★★★★☆ to ★★★★★

Table 6.9 The maximum value and relative value of the clearance of the criterion

The maximum value of Clear$_{\text{criterion}\,c}$	MaxClear$_{\text{criterion}\,c}$	MaxClear$_{\text{criterion}\,c} = 5$
The relative value of the Clear$_{\text{criterion}\,c}$	$Rela$Clear$_{\text{criterion}\,c}$	Clear$_{\text{criterion}\,c}$/MaxClear$_{\text{criterion}\,c}$ or
		Clear$_{\text{criterion}\,c}$/5

The *relative value* is the calculated value, that is, the actual value of Clear$_{\text{criterion}\,c}$ as a proportion of the MaxClear$_{\text{criterion}\,c}$, as shown in Table 6.9

6.5.6 Example

In order to further illustrate the importance of clarity in the *communication* of *criteria* from the trusting agent to the trusted agent prior to the interaction, let us reconsider the example of the interaction between East Logistics and West Warehouse. Let us assume that in the agreement the criterion 'space' is defined as an area for storing 100 tonnes of rubber without specifying the size of the space required. West Warehouse has assumed that $1000\,\text{m}^2$ would be adequate, but when the goods arrive, West Warehouse indicated that it was not an adequate calculation. In this case, if East Logistics takes into account the size of the warehouse space allocated to it by West Warehouse when deciding the *trustworthiness* of West Warehouse, then the trustworthiness value assigned to West Warehouse will not be correct. The terms or the conditions were not clearly defined and the criteria were not clear. In this case, the criterion is unambiguous if a warehouse space of $6000\,\text{m}^2$ is specified by the trusting agent and allocated by the trusted agent prior to the interaction. The trustworthiness measure could be useful and valid if

- we denote the clarity of the warehouse space specification, as Clear$_{\text{space}}$;
- we denote the clarity of the duration, as Clear$_{\text{days}}$.

Therefore, the total fulfilment of the commitment and the degree of clarity can be best represented as

$$\text{Commit}_{\text{space}} * \text{Clear}_{\text{space}} + \text{Commit}_{\text{days}} * \text{Clear}_{\text{days}}$$

6.6 Influence of a Criterion – Inf$_{\text{criterion}}$

6.6.1 Definition

The *influence of each commitment* is defined as *a measure of how important the criterion is when deciding the trustworthiness of the trusted agent* (Table 6.10).

The *influence of each commitment* is represented by Influence$_{\text{criterion}\,c}$.

6.6.2 Description

We denote the *influence of a criterion* as Inf$_{\text{criterion}}$. The importance of a criterion in an interaction is determined on the basis of a perception of the trusting agent. Therefore, the influence of each criterion in an interaction may vary depending on the following:

(1) From the perspective of the trusting agent, some criteria may be more important than others in determining the outcome of an interaction. In other words, the trusting agent might consider the

Table 6.10 Influence$_{\text{criterion}\,c}$ of CCCI Methodology

Influence:	Metric for influence measure
Criterion:	Criterion
c:	A particular criterion

interaction as unsuccessful if certain criteria are not fulfilled. On the other hand, the trusting agent might consider some other criteria to have minimal effects on determining the outcome of the interaction.

(2) One trusting agent may have a completely different perception, compared to that of another trusting agent, of the importance of each criterion in the same type of interaction with the trusted agent.

6.6.3 Application of the Metric

If a service contains a number of criteria, we will need to weight each criterion because each of them may have a different impact on the trustworthiness value. This is represented as

$$\text{Commit}_{\text{criteria 1}} * \text{Clear}_{\text{criterion 1}} * \text{Inf}_{\text{criterion 1}} + \text{Commit}_{\text{criterion 2}} * \text{Clear}_{\text{criterion 2}} * \text{Inf}_{\text{criterion 2}}$$

$$+ \ldots + \text{Commit}_{\text{criterion } N} * \text{Clear}_{\text{criterion } N} * \text{Inf}_{\text{criterion } N}$$

6.6.4 Values of Inf_criterion

Here, we consider two approaches, a simple approach and complex approach. In order to express the influence of each criterion in the service interaction, we propose seven levels that denote, in increasing order, the weight that can be assigned to a criterion in any given interaction. The seven levels of $\text{Inf}_{\text{criterion}}$ lead to the seven value ranges of $\text{Inf}_{\text{criterion}}$, which are explained in Table 6.11.

The influence of each criterion in determining the outcome of the interaction will be taken into account when determining the trustworthiness of the trusted agent. The influence of each criterion is different with regard to the overall quality assessment.

6.6.5 Maximum Value and Relative Value

The maximum value and the relative value are defined in the same manner as in the trustworthiness definition, and in the $\text{Commit}_{\text{criterion}}$ as well as $\text{Clear}_{\text{criterion}}$ definitions. The evaluation has to be against the highest possible ratings with level or value (Table 6.12).

For the purposes of simplicity, we shall just use *value* to carry out the determination of the trustworthiness measure.

The relative value is the calculated value, that is, the actual value of $\text{Inf}_{\text{criterion } c}$ as a proportion of the $Max \text{Inf}_{\text{criterion } c}$.

6.6.6 Example

In the example of the interaction between East Logistics and West Warehouse in Section 6.5.6, we could have one of the following scenarios:

(i) West Warehouse allocates a warehouse space of *less than* 6000 m^2 but it is for a duration of *60 days*.
(ii) West Warehouse allocates a warehouse space of 6000 m^2 for the requested duration of *60 days*.
(iii) West Warehouse allocates a warehouse space of 6000 m^2 but it is for a duration of *less than 60 days*.

Let us assume that East Logistics considers the size of the warehouse space very important. It attaches less importance to the number of days for which the space is allocated.

Since East Logistics attaches more importance to the size of the warehouse space than for the duration for which it was allocated, we need a metric to take this into account. If we do not take into account the size of the warehouse space allocated to East Logistics, then West Warehouse might be assigned the same trustworthiness value for the first and third scenarios, assuming both criteria (for the size of space and duration) were communicated clearly by East Logistics to West Warehouse before the interaction.

Table 6.11   Complex $Commit_{criterion}$ Levels

Seven-level scale	Semantics	Description	Value of $Inf_{criterion\,c}$	Visual representation (star rating system)
−1	Ignore	No agreement was drawn up.	$x = -1$	Not displayed
0	Unimportant	If a criterion is assigned this weight, it means that this is some additional information requested by the trusting agent and, in case the trusted agent does not satisfy this criterion, it will not be assigned a low trustworthiness value.	$x = 0$	Normally not displayed
1	Very little importance	This weight indicates that the criterion has minimum importance.	$0 < x \leq 1$	From ☆ to ★
2	Partially important	This weight indicates that the criterion is of 50 % importance.	$1 < x \leq 2$	From ★☆ to ★★
3	Largely Important	This weight indicates that the criterion is important in general and its importance is more than 50 % but less than the next two levels.	$2 < x \leq 3$	From ★★☆ to ★★★
4	Important	This weight indicates that the criterion is important for the trusting agent and if the trusted agent does not satisfy this criterion, it may be assigned a low trustworthiness value.	$3 < x \leq 4$	From ★★★☆ to ★★★★
5	Very important	This weight indicates that the criterion is *crucial* for the trusting agent and if the trusted agent does not satisfy this criterion, it may be assigned a low trustworthiness value.	$4 < x \leq 5$	From ★★★★☆ to ★★★★★

Table 6.12   The maximum value and the relative value of the influence of the criterion

The maximum value of $Inf_{criterion\,c}$	$Max\,Inf_{criterion\,c}$	$Max\,Inf_{criterion\,c} = 5$
The relative value is the $Inf_{criterion\,c}$	$Rela\,Inf_{criterion\,c}$	$Inf_{criterion\,c}/Max\,Inf_{criterion\,c}$ or $Inf_{criterion\,c}/5$

Therefore, we need a mechanism by which West Warehouse can be assigned a high trustworthiness value only in the second and third scenarios. In the first scenario, when the size of the warehouse space is not as is desired by East Logistics, West Warehouse should be assigned a lower trustworthiness value compared to the other two scenarios.

Hence, we see that we need a way by which the trusting agent can express the significance that it attaches to each criterion in the interaction with the trusted agent, when determining the trustworthiness of the trusted agent. Additionally, the influence of each criterion should be communicated in clear terms to the trusted agent. Henceforth, we assume that the trusting agent communicates the influence of each criterion to the trusted agent, prior to the interaction between them. In the above example,

- we denote the *Influence* of the warehouse space specification as Inf_{space};
- we denote the *Influence* of the duration as Inf_{days}.

Therefore the total fulfilment of the commitment and the degree of clarity as well as the weight of each criterion can be represented as

$$Commit_{space} * Clear_{space} * Inf_{space} + Commit_{days} * Clear_{days} * Inf_{days}$$

The above situation could fit into one of the following scenarios:

(i)   West Warehouse allocates a warehouse space of *less than* $6000\,m^2$ but it is for the duration of *60 days*.
(ii) West Warehouse allocates a warehouse space of $6000\,m^2$ for the requested duration of *60 days*.
(iv) West Warehouse allocates a warehouse space of $6000\,m^2$ for duration of *less than 60 days*.

6.7 Correlation of Defined Quality – $Corr_{qualities}$

6.7.1 Definition

The correlation of the overall delivered quality is defined as a measure of how much the trusted agent delivered its quality against the defined quality (Table 6.13).

The *correlation of the overall delivered quality* is represented by $Corr_{qualites}$

6.7.2 Description

The *correlation between the defined QoS and the actual delivered service for the overall interaction* ($Corr_{qualities}$) can be determined by the following:

- the correlation between the defined quality and the actual delivered QoS corresponding to each criterion that the trusting agent is looking for in the service interaction;
- accumulating all the correlation values for the entire interaction.

6.7.3 The Correlation Metric

$Corr_{qualities}$ can be expressed as the sum of the correlation values corresponding to all the *criteria* in the service interaction. Here, we assume that there are N criteria in an interaction and $Commit_{criterion\,c}$ denotes the fulfilment of the '*cth*' criterion.

The contribution of the '*cth*' criterion to the overall value of $Corr_{qualities}$ can be represented as

$$f(Commit_{criterion\,c}, Clear_{criterion\,c}, Inf_{criterion\,c}) = Commit_{criterion\,c} * Clear_{criterion\,c} * Inf_{criterion\,c}$$

Table 6.13 $Corr_{qualities}$ of CCCI Methodology

Corr:	Metric for clarity measure
Qualities:	Delivered qualities by the trusted agent, that is, the entire Quality Assessment Criteria is to be used to measure the trustworthiness value
c:	A particular criteria

Therefore we can express Corr$_{qualities}$ as

$$Corr_{qualities} = \sum_{C=1}^{N} f(Commit_{criterion\,c}, Clear_{criterion\,c}, Inf_{criterion\,c})$$

$$= \sum_{C=1}^{N} Commit_{criterion\,c} * Clear_{criterion\,c} * Inf_{criterion\,c}$$

Equation 6.1: Corr$_{qualities}$ for a given service as a sum of $f(Commit_{criterion\,c}, Clear_{criterion\,c}, Inf_{criterion\,c})$

6.7.4 Values of Corr$_{qualities}$

As indicated earlier, in order to determine the correlation value for an interaction (Corr$_{qualities}$), the trusting agent has to consider the following:

- the criteria utilized to determine the trustworthiness of the trusted agent;
- the value of Commit$_{criterion}$ for each of these criteria;
- the sum of all the Commit$_{criterion}$ appropriately weighted by Clear$_{criterion}$ and Inf$_{criterion}$ in an interaction to get the Corr$_{qualities}$.

Additionally, we argue that each Commit$_{criterion}$ should be evaluated by

- whether the criterion as a basis of assigning trustworthiness was clearly communicated to the trusted agent (Clear$_{criterion}$);

and

- the influence of each criterion in the interaction (Inf$_{criterion}$) from the point of view of the trusting agent.

Corr$_{qualities}$ provides a framework as to how the correlation value of a service (Corr$_{qualities}$) can be obtained using the above-defined trustworthiness levels once the trusting agent has determined the criteria on which they are going to assign trustworthiness to the trusted agent.

If a given criterion was not expressed in clear terms to the trusted agent, then this fact should be taken into account when determining the correlation value of the interaction and, hence, the trustworthiness of the agent. If Clear$_{criterion}$ has a value of '0' for a particular criterion, this will result in the criterion not being taken into account when determining the trustworthiness of the trusted agent. On the other hand, if a given criterion was expressed in clear terms for the trusted agent, then it needs to be taken into account when determining the trustworthiness of the trusted agent. Assigning a value of '1' to Clear$_{criterion}$ will result in the criterion being taken into account. In order for Clear$_{criterion}$ to have an effect on the corresponding Commit$_{criterion}$ in an interaction, we multiply the Commit$_{criterion}$ value by Clear$_{criterion}$.

Similarly, we argue that if a given criterion was important or more important to the trusting agent, then its correlation value should be given more influence in determining the final trustworthiness value of the trusted agent. In order for the influence of a criterion (Inf$_{criterion}$) to have an effect on the Commit$_{criterion}$ of an interaction, we multiply the Commit$_{criterion}$ value by Inf$_{criterion}$.

Thus, if a value of '0' is assigned to Inf$_{criterion}$ (meaning that the criterion is unimportant in determining the outcome of the interaction), then the value of Commit$_{criterion}$ for that interaction will not be taken into account when determining trustworthiness of the trusted agent (as the resulting value is 0). On the other hand, if a value of '5' is assigned to Inf$_{criterion}$ (meaning that the criterion

Table 6.14   The maximum value and relative value of the correlation

The maximum value of $Corr_{qualities}$	$MaxCorr_{qualities}$	sum of ($MaxCommit_{criterion c} \times$ $Clear_{criterion c} \times Inf_{criterion c}$) for all criteria
The relative value of $Corr_{qualities}$	$RelCorr_{qualities}$	$Corr_{services}/MaxCorr_{qualities}$

was very important to the trusted agent), then the value of $Commit_{criterion}$ for that criterion will be given the highest Influence when determining trustworthiness. If a value of '4' is assigned to $Inf_{criterion}$ (meaning that the criterion was important to the trusted agent), then the value of $Commit_{criterion}$ for that criterion will be given more weight than a criterion whose *Influence* is '0' when determining trustworthiness. Such an $Inf_{criterion}$ would have less weight than a criterion assigned an *Influence* of '5' when determining trustworthiness.

6.7.5 Maximum Value and Relative Value

The maximum value of each component in the CCCI metrics ($Commit_{criterion c}$, $Clear_{criterion c}$ and $Inf_{criterion c}$) is 5. Therefore, for *N* criteria, the maximum value of $Corr_{qualities}$ is as shown in Table 6.14.

The *relative value* is the calculated value, that is, the actual value of the $Corr_{qualities}$ as a proportion of the $MaxCorr_{qualities}$.

6.7.6 Example

Let us consider the example of the interaction between the two logistic companies (East Logistics and West Warehouse) again. Let us assume that

- West Warehouse has allocated a warehouse space of $5000\,m^2$.
- This space was allocated for the duration of, say, 40 days.

West Warehouse has not fulfilled any of the criteria that East Logistics was looking for. Let us assume that West Warehouse considers the second criterion to be important and the first criterion to be very important in determining the outcome of the interaction. Furthermore, let us also assume that the influence of these criteria have been explicitly communicated to East Logistics. We now derive the maximum possible correlation value for the interaction. The maximum possible value of $Commit_{space}$ for the interaction is '5' (since the maximum possible correlation value for any criterion is '5'). In other words, if West Warehouse delivers the best possible service to East Logistics, we will get a value '5' for $Commit_{space}$. Since East Logistics considers this criterion to be very 'important', it has assigned the maximum possible value of '5' to Inf_{space}. The maximum possible value of $Clear_{space}$ for this interaction can be '5' (since East Logistics and West Warehouse both agree that the criterion was communicated to West Warehouse in 'very clear' terms).

The maximum possible value of $Commit_{days}$ for this interaction is '5' (since the maximum possible value of the correlation value for the criterion is '5'). If West Warehouse provides the best possible service to East Logistics, we will get a value of '5' for $Commit_{days}$. East Logistics considers this criterion to be 'important'. Therefore, the value of Inf_{days} is '4'. The value of $Clear_{days}$ for this interaction is '5' (since East Logistics and West Warehouse both agree that the criterion was communicated to West Warehouse in 'very clear' terms).

Therefore, using Equation 6.1, we can calculate the value of $MaxCorr_{qualities}$ as shown below:

$$MaxCorr_{qualities} = (MaxCommit_{space} * Inf_{space} * Clear_{space}) + (MaxCommit_{days} * Inf_{days} * Clear_{days})$$

$$= (5 * 5 * 5) + (5 * 4 * 5)$$

$$= 225$$

6.8 Trustworthiness Values and Corr$_{\text{qualities}}$

The value of Corr$_{\text{qualities}}$ depends on the service of the trusted agent in the interaction. If the trusted agent behaves in a way that was initially expected by the trusting agent, then the value of Corr$_{\text{qualities}}$ will be high. Conversely, if the trusted agent does not behave in a way that was initially agreed to by both the interacting agents, then the value of Corr$_{\text{qualities}}$ will be low. Corr$_{\text{qualities}}$ quantifies and communicates the service of the trusted agent in the interaction.

The best possible service or the expected service of the trusted agent is possible only when the trusted agent abides by what it had initially agreed for in respect of all the criteria in the interaction. This would mean a scenario where the trusted agent provides the best possible service. In the case where the trusted agent abides by and fulfils all the criteria in the interaction as it had initially agreed, we would get the maximum possible correlation value for that interaction. Thus, we can define the *maximum possible correlation value* (*Max*Corr$_{\text{qualities}}$) for an interaction as *a numeric value that quantifies the maximum possible concurrence between what the trusting agent expected and what it received from the trusted agent in its interaction.* Alternatively, it is the correlation value for an interaction when the service of the trusted agent is equivalent to the expected service.

Therefore, the correlation value for the interaction between East Logistics and West Warehouse when West Warehouse acts in a way that it had initially agreed to is '225'. This is the maximum possible correlation value for the interaction. Therefore, if the correlation value for the interaction with West Warehouse is '225', then West Warehouse can be considered to be fully *trustworthy*.

The ratio between the actual service of West Warehouse and the best possible service of West Warehouse would be quantified to the extent to which West Warehouse can be regarded as trustworthy.

6.8.1 Derivation of Trustworthiness Values

In order to determine the trustworthiness of the trusted agent, we need to express the correlation value for the interaction (Corr$_{\text{qualities}}$) relative to the maximum possible correlation value for the interaction (*Max*Corr$_{\text{qualities}}$). Additionally, a greater concurrence between (Corr$_{\text{qualities}}$) and (*Max*Corr$_{\text{qualities}}$) implies a better performance by the trusted agent and results in a higher trustworthiness value.

We define the *relative correlation value* (*Rel*Corr$_{\text{qualities}}$) as *a numeric value that quantifies the degree of* concurrence *between the correlation value for an interaction as a proportion of the maximum possible correlation value for the interaction.*

$$Rel\text{Corr}_{\text{qualities}} = (\text{Corr}_{\text{qualities}})/(Max\text{Corr}_{\text{qualities}})$$

The domain of $Rel\text{Corr}_{\text{qualities}}$ is $[0-1]$.

Equation 6.2: Expression of the relative correlation value

RelCorr$_{\text{qualities}}$ is a measure that quantifies the honesty, competency, calibre or service of the trusted agent in an interaction with the trusting agent. It shows the extent to which the trusted agent abides by what it had initially agreed to do and hence denotes the extent to which the trusted agent can be relied upon to perform a given action. Since trustworthiness denotes the amount of trust that can be reposed in the trusted agent, RelCorr$_{\text{qualities}}$ denotes the trustworthiness of the trusted agent.

In this method, trustworthiness has to be expressed on the scale of $[0-5]$, as previously discussed in Chapter 3. In order to express trustworthiness in the range $[0-5]$, we need to multiply RelCorr$_{\text{qualities}}$ by a factor of '5'.

$$\text{Trustworthiness} = 5 * (Rel\text{Corr}_{\text{qualities}})$$

Equation 6.3: Expression of trustworthiness in terms of the relative correlation value

By substituting the expression for *relative correlation value* from Equation 6.2 into Equation 6.3 we arrive at

$$\text{Trustworthiness} = 5 * ((\text{Corr}_{\text{qualities}})/(Max\text{Corr}_{\text{qualities}}))$$

Equation 6.4: Trustworthiness expressed in terms of $\text{Corr}_{\text{qualities}}$ and $Max\text{Corr}_{\text{qualities}}$

By substituting the expressions for $\text{Corr}_{\text{qualities}}$ and $Max\text{Corr}_{\text{qualities}}$ in the above equation, we derive the following:

$$\text{Trustworthiness} = 5 * \left(\frac{\sum_{C=1}^{N} \text{Commit}_{\text{criterion } c} * \text{Clear}_{\text{criterion } c} * \text{Inf}_{\text{criterion } c}}{\sum_{C=1}^{N} Max\text{Commit}_{\text{criterion } c} * \text{Clear}_{\text{criterion } c} * \text{Inf}_{\text{criterion } c}} \right)$$

Equation 6.5: Trustworthiness expressed in terms of $\text{Commit}_{\text{criterion } c}$, $\text{Clear}_{\text{criterion } c}$, $\text{Inf}_{\text{criterion } c}$ and their corresponding *maximum values*

Since the maximum possible correlation value for a given commit criterion is 5, Equation 6.5 can also be written as

$$\text{Trustworthiness} = 5 * \left(\frac{\sum_{C=1}^{N} \text{Commit}_{\text{criterion } c} * \text{Clear}_{\text{criterion } c} * \text{Inf}_{\text{criterion } c}}{\sum_{C=1}^{N} 5 * \text{Clear}_{\text{criterion } c} * \text{Inf}_{\text{criterion } c}} \right)$$

Equation 6.6: Equation 6.5 modified showing the *maximum possible correlation value* for a criterion of 5

The trustworthiness value that we obtain using Equation 6.5 or 6.6 will be a real number in the range [0–5].

6.8.2 Example

We can apply Equation 6.6 (the modified Equation 6.5) to the example in Section 6.4.7 of the interaction between West Warehouse and East Logistics where

$$\text{Corr}_{\text{qualities}} = (\text{Commit}_{\text{space}} * \text{Inf}_{\text{space}} * \text{Clear}_{\text{space}}) + (\text{Commit}_{\text{days}} * \text{Inf}_{\text{days}} * \text{Clear}_{\text{days}})$$

$$Max\text{Corr}_{\text{qualities}} = (Max\text{Commit}_{\text{space}} * \text{Inf}_{\text{space}} * \text{Clear}_{\text{space}}) + (Max\text{Commit}_{\text{days}} * \text{Inf}_{\text{days}} * \text{Clear}_{\text{days}})$$

Table 6.15 shows the actual values for $\text{Commit}_{\text{criterion } c}$, $\text{Inf}_{\text{criterion } c}$ and $\text{Clear}_{\text{criterion } c}$ for each criterion. The table also shows the maximum possible scores for $\text{Commit}_{\text{criterion } c}$ and $\text{Corr}_{\text{qualities}}$.

Table 6.15 Overview of the CCCI measure results for West Warehouse

Criterion c	$\text{Commit}_{\text{criterion } c}$		$\text{Inf}_{\text{criterion } c}$	$\text{Clear}_{\text{criterion } c}$	$\text{Corr}_{\text{qualities}}$	
	Actual	Max	Actual	Actual	Actual	Max
Space	4	5	5	5	100	125
Days	3	5	4	5	60	100
				Total	160	225

From Table 6.15, we conclude that in this case

$$Rel\text{Corr}_{\text{qualities}} = \text{Corr}_{\text{qualities}}/Max\text{Corr}_{\text{qualities}} = 160/225$$

Therefore,

$$\text{Trustworthiness} = 5 * Rel\text{Corr}_{\text{qualities}} = 3.5(\text{approximately})$$

Therefore, the visual representation of this trustworthiness value using the ordinal scale of trustworthiness introduced in Chapter 3 would be 3.5 stars (★★★⯪) and this corresponds to trustworthiness level 4.

According to the semantics for this level, the trusted agent (in this case, West Warehouse) is largely trustworthy.

6.9 Summary

The CCCI methodology for a trustworthiness measure provides four metrics, and defines the *maximum possible correlation* value for a business service interaction, $\text{Corr}_{\text{qualities}}$, and the maximum possible values of $\text{Commit}_{\text{criteria}\,c}$, $\text{Clear}_{\text{criteria}\,c}$ and $\text{Inf}_{\text{criteria}\,c}$. The *relative correlation value* is determined by the ratio of the correlation value and the maximum possible correlation value with respect to the trustworthiness scale. The user-defined trustworthiness value is in the range [0–5], so the trustworthiness value is obtained by multiplying the relative correlation value by a factor of 5.

The following important concepts were presented in this chapter:

- A conceptual framework for measurement of quality and trust.
- The four major steps of the conceptual framework for measurement of quality and trust, namely, the following:
  1. Obtain context-associated domain knowledge
  2. Identify the quality aspects from the knowledge domain
  3. Develop Quality Assessment Criteria for each quality aspect
  4. Measure the quality and trust against Quality Assessment Criteria through CCCI metrics.
- The use of CCCI metrics.
- The term *expectation* refers to what the trusting agent expects from its interaction with the trusted agent.
- The *correlation* ($\text{Corr}_{\text{qualities}}$) between the trusted agent's expected service and its actual service in an interaction is used as the basis for determining the trustworthiness of the trusted agent.
- The *correlation between the expected and actual service for a given criterion in an interaction* ($\text{Commit}_{\text{criterion}\,c}$) is a numeric value that quantifies the extent of concurrence between what the trusting agent expected of the trusted agent for that criterion in that interaction and what it actually got for that criterion.
- The *clarity of a criterion* ($\text{Clear}_{\text{criterion}\,c}$) is a metric used to express whether or not a criterion, based on which the trusting agent is going to evaluate the trusted agent, has been made explicitly clear to the trusted agent prior to the interaction.
- The *influence or the importance of a criterion* ($\text{Inf}_{\text{criterion}\,c}$) in an interaction is determined on the basis of the perception of the trusting agent. Therefore, the influence of each criterion in an interaction may vary depending on the following:
  - From the perspective of the trusting agent, some criteria may be more important than others in determining the outcome of the interaction.
  - One trusting agent may have a completely different perception from another trusting agent of the importance of each criterion in the same type of interaction with the trusted agent.
- The *maximum possible correlation value* ($Max\text{Corr}_{\text{qualities}}$) for an interaction as a numeric value that quantifies the maximum possible concurrence between what the trusting agent expected and what it received from the trusted agent in its interaction.

- We define *relative correlation value* (RelCorr$_{qualities}$) as a numeric value that quantifies the degree of concurrence between the correlation value for an interaction and the maximum possible correlation value for the interaction.
- RelCorr$_{qualities}$ denotes the trustworthiness of the trusted agent. In order to express a trustworthiness value in the range [0–5], we multiply RelCorr$_{qualities}$ by a factor of '5'.

7

Trustworthiness Systems

7.1 Introduction

In this chapter, we examine some well-known, successful e-business portals to demonstrate the advantages of using trust systems. The establishment of trustworthiness-based services helps to create business value through building trust relationships and consumer confidence.

We demonstrate how well-known e-business providers have begun to adopt trustworthiness technology and systems to build their customer relationships, raise their company's reputation and profits and attract millions of users from around the world. These sites include Amazon, Yahoo, Epinions, eBay, BizRate and CNet, to name a few. We shall see how they carry out the measurement of trustworthiness of agents, quality of services and quality of products. At the end, we provide an overall summary of the technology they have used. Note that in this chapter the trustworthiness measure is used interchangeably with the trustworthiness rating.

7.2 Amazon's Trustworthiness Systems

Amazon.com is a very well-known online bookstore, which now also sells many other consumer products such as games and computer software. Amazon.com provides two different kinds of trustworthiness rating systems: trustworthiness of transaction partners and trustworthiness of products. These systems help buyers make purchasing decisions.

7.2.1 Trustworthiness of Transaction Partners

Buyers and sellers carry out *buy and sell* transactions: Amazon calls them *transaction partners*.

The trustworthiness of sellers is measured by the quality of service provided by the seller (Figure 7.1). The sellers include 'new-product sellers' and 'used-product sellers' (or second-hand products). The trustworthiness system provides customers with a view of the products being sold by different sellers and the seller's trustworthiness level or rating [1–4].

The overall trustworthiness value of sellers does not consider seller's feedback on ratings; it only considers the buyer's trustworthiness rating or the measure of, and by, the seller (http://www.amazon.com/gp/help/seller/feedback-popup.html, Table 7.1).

Apart from assigning trust levels using a 5-star system to sellers or buyers, the transaction partners can leave comments or feedback to substantiate their trustworthiness ratings. All comments or feedback can be viewed on the Amazon site (Figure 7.2).

Apart from giving a trust value (5-star rating) and comments, the Amazon 'Safe Buying Guarantee' (next to the sellers) indicates that the condition of the item bought, and its timely delivery is guaranteed under the Amazon.com A-to-Z Guarantee. This guarantee provided by Amazon helps

Trust and Reputation for Service-Oriented Environments Elizabeth Chang, Tharam Dillon and Farookh Hussain
© 2006 John Wiley & Sons, Ltd

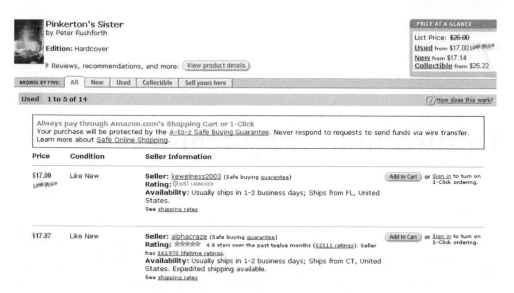

Figure 7.1 Amazon's seller trustworthiness rating. Reproduced by permission of Amazon.com. (c) 2004 Amazon.com, Inc. All Rights Reserved

Table 7.1 Amazon's three-level rating scale for reviews

Semantics for feedback	Matching stars
Positive	5 or 4 stars
Neutral	3 stars
Negative	2 stars or 1 star

protect the buyer, in case the seller tries to cheat by not delivering the products or delivering materially different products. This information and reviews about the sellers can influence a customer's decision to buy with confidence and can increase online purchases.

7.2.2 Trustworthiness of the Products

Amazon has two kinds of product reviews:

- Customer review
- Spotlight review.

Customer reviews allow every Amazon Registered User to review and rate a product on the basis of his/her experience with the quality of the product that is purchased from Amazon.com (Table 7.2). Amazon distinguishes two kinds of products:

- Electronic goods (games, software or electronics products, and so on)
- Other goods (books, apparel, and so on).

For electronic goods, only ratings are provided while other goods contain both reviews and ratings.

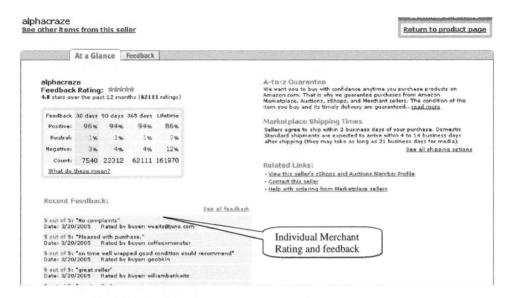

Figure 7.2 Amazon's feedback rating and statistics of total feedback. Reproduced by permission of Amazon.com. (c) 2004 Amazon.com, Inc. All Rights Reserved

Table 7.2 Amazon's five-level rating scale for products

Product quality semantics	Scales
Hated it	1 star
Do not like it	2 stars
It is ok	3 stars
Like it	4 stars
Loved it	5 stars

Customers can vote on whether they found a review to be useful, and reviewers whose reviews get the most positive votes are classified as 'Top Reviewers', (http://www.amazon.com/exec/obidos/subst/community/reviewers-faq.html).

This implies that people first get to read the most helpful reviews and then the reviews by all the customers, providing for a table of responses. The most helpful reviews are then easy to understand. As a result, *Top Reviewers* are sought after for opinions.

7.2.3 Rating of Customer Reviews

Any of Amazon's transaction partners and online users can view the comments or feedback and can vote if they found a particular review to be helpful. All reviews given by the users are stored in Amazon's centralized system. A user can view all the comments or feedback by all reviewers as well as the reviewers' information. The system has not yet implemented a mechanism for the justification of the review's rating.

These reviews, along with the information about the reviewers, provide customers with a guide about the product being sold and whether such a product is worth buying. This provides the customer with confidence by allowing them to incorporate the feedback into their purchasing decision. For an example of this, see Figure 7.3.

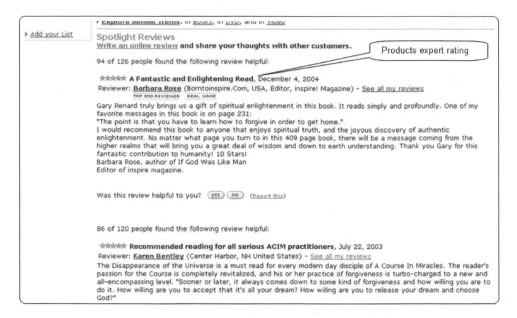

Figure 7.3   An example of Amazon's product rating and review. Reproduced by permission of Amazon.com. (c) 2004 Amazon.com, Inc. All Rights Reserved

7.3 Yahoo's Trustworthiness Systems

Yahoo's product search allows customers to compare and contrast prices and features for products and merchants [5-8].

7.3.1 Trustworthiness of Merchants

Yahoo's *merchant rating system* is an integral part of Yahoo's business strategy, which provides a boost to consumer confidence and loyalty to Yahoo.

Yahoo uses a six-level trustworthiness scale for merchant rating as shown in Figures 7.4 and 7.5. Each merchant rating has two parts:

- An overall rating
- A five-criterion rating.

Figure 7.4   Yahoo's six-level trustworthiness scale for rating merchants. Reproduced with permission of Yahoo! Inc. © 2005 by Yahoo! Inc. YAHOO! and the YAHOO! logo are trademarks of Yahoo! Inc

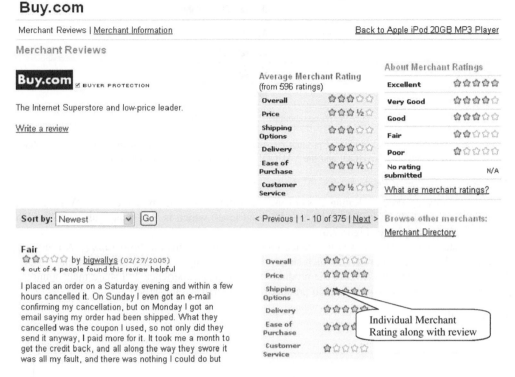

Figure 7.5 Yahoo's merchant trustworthiness. Reproduced with permission of Yahoo! Inc. © 2005 by Yahoo! Inc. YAHOO! and the YAHOO! logo are trademarks of Yahoo! Inc

The *overall rating* collects feedback from all shoppers providing feedback about the merchant and aggregates the total five-criterion ratings to determine an overall rating for that merchant. It permits ratings by customers (weighted * 2) and others and calculates the *average merchant rating* using a weighted average of total users' responses.

A '*Not Yet Rated*' icon is used when a merchant has less than five posted ratings.

Yahoo also offers a 'buyer protection programme' much like the 'safe buying guarantee' provided by Amazon. Merchants must earn and maintain a minimum 3-star average merchant rating. Yahoo protects its buyers in the event of 'payment for an item that is not received' or 'receiving an item that is materially different from that what is described'.

7.3.2 Rating User Reviews

In Yahoo, users can vote if they have found a particular review to be helpful using a two-category scale:

- Helpful
- Not helpful.

All reviews given by the users are stored in Yahoo's centralized system and can be accessed easily by users. Yahoo has not incorporated sophisticated *quality of review* rating procedures; however, it counts the total number of people who found the service helpful against the total number of people giving feedback.

Figure 7.6   Yahoo's six-level quality of product measurement scale. Reproduced with permission of Yahoo! Inc. © 2005 by Yahoo! Inc. YAHOO! and the YAHOO! logo are trademarks of Yahoo! Inc

7.3.3 Rating Quality of Products

Yahoo's Quality of Product Rating System uses a 5-star system from 'Excellent', 'Very good', 'Good', 'Fair', and 'Poor' and includes an additional level, which is 'No rating submitted' which indicates unknown quality (Figure7.6).

All registered users are allowed to rate the products, which forms the *'user rating'*. A user may choose to 'Quick Rate' a product, which simply involves giving it a rating between 1 and 5 stars. They may also 'write a review' for the product to detail their experience and share their opinions.

In Figure 7.7 for a *user review*, each product rating contains four parts:

- An overall rating
- Average product rating

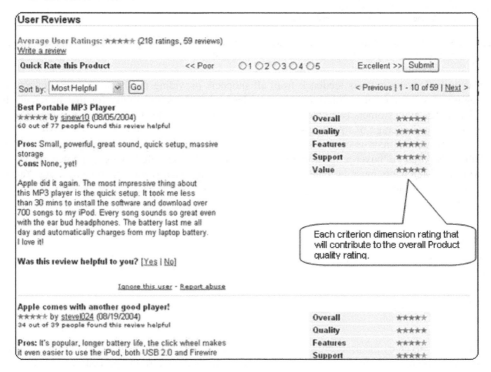

Figure 7.7   Yahoo's user review. Reproduced with permission of Yahoo! Inc. © 2005 by Yahoo! Inc. YAHOO! and the YAHOO! logo are trademarks of Yahoo! Inc

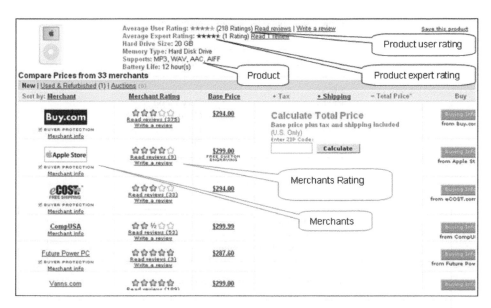

Figure 7.8 Yahoo's overall product recommendation system. Reproduced with permission of Yahoo! Inc. ©
2005 by Yahoo! Inc. YAHOO! and the YAHOO! logo are trademarks of Yahoo! Inc

- Criterion rating
- Expert rating.

An *'overall product rating'* is displayed in search results.

'Criterion rating' addresses the specific quality measure of the product, and all registered users
are allowed to rate the products, which form the *'user rating'*.

'Average product rating' is simply an average of the total number of posted ratings for that
product.

'Expert rating' is provided by Yahoo's known experts or trusted agents. These expert ratings
inspire higher confidence among users than user ratings.

In Figure 7.8, an overview of Yahoo's recommendation system is shown. A product along with
its ratings is compared along with the rating of merchants selling this product.

7.4 Epinions.com's Trustworthiness System

Epinions.com is a consumer reviews portal on the Internet. It offers product ratings, merchant
ratings, review ratings, as well as reviewer ratings. The objective is to provide reliable sources that
provide valuable consumer insight, unbiased advice, in-depth product evaluations and personalized
recommendations to the public, aiming to help customers make informed buying decisions [9, 10].

Epinion's main objective is to target advertising opportunities to reach shoppers who are research-
ing products and making buying decisions in every major commercial category. They help sellers
and service providers to capture the customers' needs by learning what people like and dislike about
shopping, how they use the Internet in their everyday lives and what would make them prefer to
buy online rather than off-line.

7.4.1 Trustworthiness of Stores

Epinion presents customers with trusted stores to help prevent possible fraud. Epinion's store
trustworthiness rating contains five levels (Table 7.3).

Table 7.3   Epinion's five-level rating scale
for stores

Level of Trust for store	Semantics
5 ticks	Excellent
4 ticks	Very Good
3 ticks	Good
2 ticks	Fair
1 tick	Poor

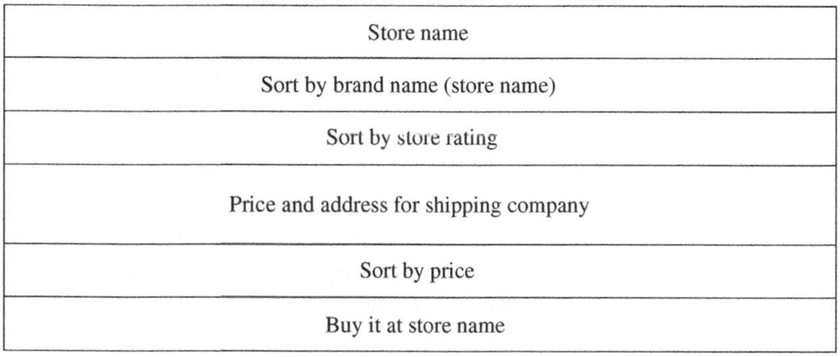

Store name
Sort by brand name (store name)
Sort by store rating
Price and address for shipping company
Sort by price
Buy it at store name

Figure 7.9   Epinions store comparison

Epinion's store trustworthiness rating contains three parts.

- An overall store rating is calculated as an average of all the reviews.
- A four-criterion rating is given for each store.
- Individual reviews can all be viewed for any merchant.

Epinion's store trustworthiness system also provides a list of different stores that contain the same or a similar product, a list of prices from different stores about a product and ratings of each store (Figure 7.9).

Epinion uses a five-level trustworthiness scale for rating stores with 1–5 ticks (5 being the highest). It is also possible to see the individual store rating and reviews by various customers. Here, people can talk about their experience with a particular store and what they thought of its service (Figure 7.10).

7.4.2 Trustworthiness of Reviewers

Epinion creates a network of reviewers. Reviewers can review and rate each other's trustworthiness. Each reviewer (as a trusting agent) can build a trusted virtual community with trusted members (trusted agents). It allows each new member to create its own trusted community (Figures 7.11 and 7.12).

Networked reviewers are assets of Epinion and are known as a *web of trust*. The web of trust system ensures that the opinions of trusted members (trusted agents) are always at the top of a member's list.

In Epinion, the web of trust system builds a hierarchy of agents who trust each other.

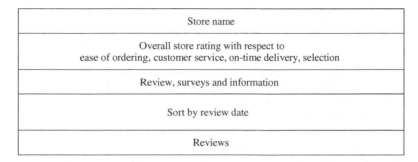

Figure 7.10   Epinions trusted store rating

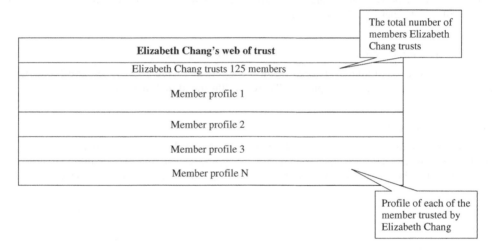

Figure 7.11   Elizabeth Chang as a trusting agent in Epinion

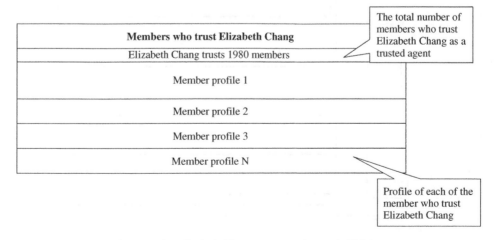

Figure 7.12   Elizabeth Chang as a trusted agent in Epinion

The trusting agent establishes trust relationships with trusted agents using four procedures:

(a) Looking at the trusted agent's previous records. If the agent has consistently given good advice to you, you are more likely to believe the agent's suggestions in the future.
(b) Looking at the trusted agent's reviews. If many people consistently found the agent's review helpful and trustworthy, then maybe the agent is giving good advice.
(c) Checking the trusted agent's profile pages.
(d) Looking at the trusted agent's webs of trust.

Once the above procedure is completed and the trusting agent is satisfied with the four-procedure output, the trusting agent may add the trusted agent to the trusting agent's web of trust. Hence, the trusting agent builds a virtual community with trusted agents. Note that the trusting agent will also encounter untrustworthy agents, which the trusting agent will also record in their community. This record will affect other agent's webs of trust.

7.4.3 Trustworthiness of Reviews

Epinions.com's *review rating* rates each member's opinion (Figure 7.13). There are four criteria with a four-level rating scale for opinions. These are illustrated in Table 7.4.

7.4.4 Rating Products

Epinion has a *product ratings system* that ranks all of the products in a category or subcategory and the ranking is based on their overall star rating. Note that in Epinion, different products have different rating methods. For example, a book has only a rating and no reviews, whereas

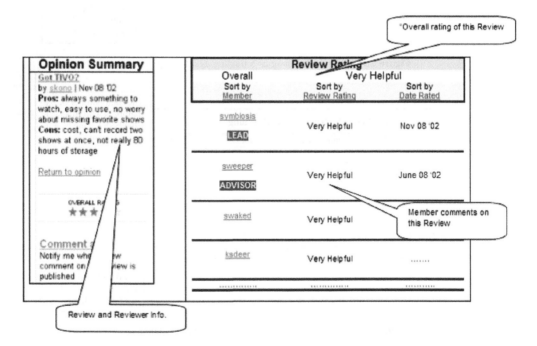

Figure 7.13   Schematic diagram of Epinion's review rating

Table 7.4   Epinion's four-level trustworthiness scale for review rating

Review rating level	Semantics	Descriptions
1.	Not helpful	The review is off-topic, inaccurate, offensive or copied from another source.
2.	Somewhat helpful	The review is poorly presented or somewhat inaccurate.
3.	Most helpful	The review is accurate and well presented.
4.	Very helpful	The review is exceptionally accurate, contains a significant amount of useful information about the subject and is very well presented.

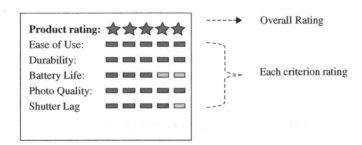

Figure 7.14   Each product rating and its criterion dimensions

electronic goods have reviews and ratings. Some products have a criterion rating, a review as well as an overall rating. The reviews help agents talk about their experience with a particular product and what they liked and disliked about it, whereas ratings give only a brief quality rating of a product.

An overall star rating is calculated on the basis of the following criteria:

- Overall product rating, with extra weight given to high-quality reviews
- Number of reviews about the product
- The date of review of the product.

Figure 7.14 shows an example of a product rating and its criteria and Figure 7.15 illustrates an overall review of the same product.

7.5 eBay.com's Trustworthiness Systems

eBay is a virtual online marketplace. It provides an international platform for anyone to trade anything. It sells goods and services for diverse individuals, business players and service providers. Currently, it handles a volume of 25 million items of sale with new items added everyday [11-14].

7.5.1 Trustworthiness of an eBay Member

eBay offers *reputation rating systems* to help its prospective users to make informed decisions to assist in their purchasing preferences. eBay *only* considers the feedback from eBay transaction partners (eBay sellers or eBay buyers) and calculates trust for eBay's trusted members.

Feedback is made up of comments and ratings left by other eBay members who have traded online. These comments and ratings are valuable indicators of the reputation of buyers or sellers on eBay. They are included along with an overall feedback score, in a *member profile*.

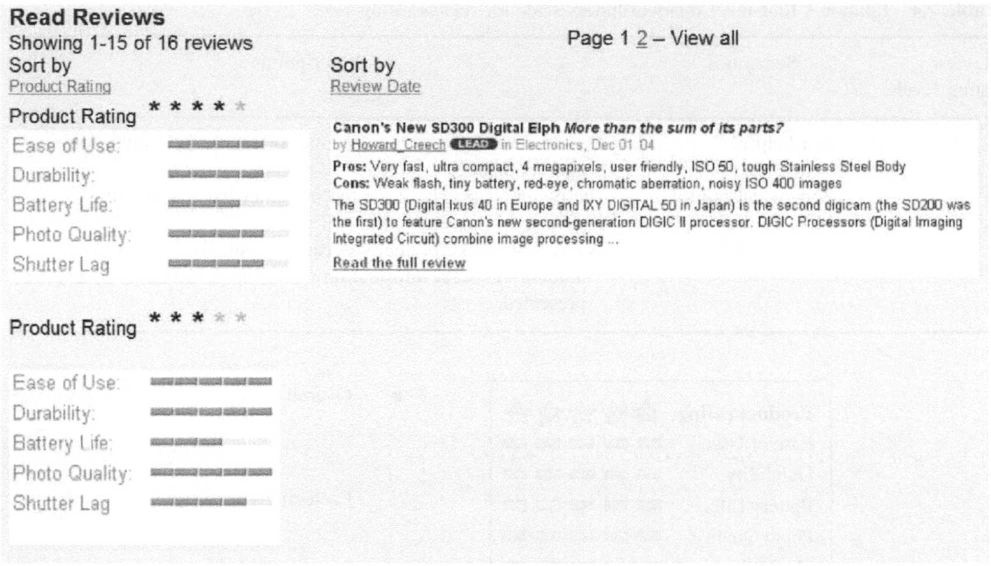

Read Reviews

Showing 1-15 of 16 reviews

Page 1 2 – View all

Sort by
Product Rating

Sort by
Review Date

Product Rating ★ ★ ★ ★ ☆

Ease of Use:

Durability:

Battery Life:

Photo Quality:

Shutter Lag:

Canon's New SD300 Digital Elph *More than the sum of its parts?*
by Howard_Creech (LEAD) in Electronics, Dec 01 '04

Pros: Very fast, ultra compact, 4 megapixels, user friendly, ISO 50, tough Stainless Steel Body
Cons: Weak flash, tiny battery, red-eye, chromatic aberration, noisy ISO 400 images

The SD300 (Digital Ixus 40 in Europe and IXY DIGITAL 50 in Japan) is the second digicam (the SD200 was the first) to feature Canon's new second-generation DIGIC II processor. DIGIC Processors (Digital Imaging Integrated Circuit) combine image processing ...

Read the full review

Product Rating ★ ★ ★ ☆ ☆

Ease of Use:

Durability:

Battery Life:

Photo Quality:

Shutter Lag:

Figure 7.15   Schematic diagram of Epinion's product rating system

In eBay, reputation is determined using members feedback after their interaction with other members. Such feedback is categorized as positive, negative, or neutral ratings (http://pages.eBay.com.au/help/feedback/evaluating-feedback.html).

An example schematic diagram of an eBay member, *'eBayRose'*, is shown in Figure 7.16.

A total eBay score is calculated for the number of eBay members who are satisfied doing business with a particular member. It is calculated on the basis of the difference between the total number of members who left a positive rating and the total number of members who left a negative rating. In the example shown in Figure 7.16, the feedback score is $3531 - 3 = 3528$.

Positive feedback is given as the calculated percentage of the positive over the total, which is 3531/3534, that is, 99.9 %. If a member has had several transactions with *eBayRose* and leaves more than one positive rating, they will be counted only once in the total.

Feedback score:	3528
Positive feedback:	99.9%
Members who left positive feedback:	3531
Members who left negative feedback:	3
All positive feedback received:	6587

Recent Ratings:

		Past 1 month	Past 6 months	Past 12 months
⊕	Positive	392	1833	2805
⊡	Neutral	1	1	1
⊖	Negative	1	1	1

Figure 7.16   Schematic diagram of eBay's trustworthiness rating system

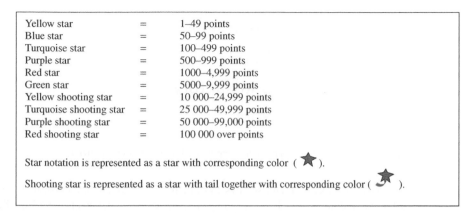

Yellow star	=	1–49 points
Blue star	=	50–99 points
Turquoise star	=	100–499 points
Purple star	=	500–999 points
Red star	=	1000–4,999 points
Green star	=	5000–9,999 points
Yellow shooting star	=	10 000–24,999 points
Turquoise shooting star	=	25 000–49,999 points
Purple shooting star	=	50 000–99,000 points
Red shooting star	=	100 000 over points

Star notation is represented as a star with corresponding color ( ).

Shooting star is represented as a star with tail together with corresponding color ( ).

Figure 7.17 Schematic diagram of eBay's positive feedback stars

On the basis of the total points scored (between 1 to over 100 000 points), the eBay member is assigned stars as shown in the schematic diagram given in Fig 7.17. As the number of positive feedback scores increase, a higher star level is assigned.

On the basis of the profile and reviews of an eBay member, a buyer can decide if a seller is genuine and reliable and confidently place an order. This can further influence the decision of a buyer in favour of someone with a good record. In eBay, the online user can also visit each comment or opinion or feedback made by each individual member. For further details, please see eBay member profile and rating on http://pages.eBay.com.au/services/buyandsell/powersellers.html.

7.6 BizRate.com's Trustworthiness Systems

BizRate.com is a shopping search engine that has a business objective to list every store and every product around the world. By listing the price, availability and ratings information, shoppers can confidently buy anything that is available for sale anywhere. BizRate.com data indicates that they have an index of over 30 million product offerings from more than 40 000 stores.

BizRate.com uses a proprietary shopping search and rating algorithm termed *ShopRank*, which is determined by weighting price, popularity and availability of products against the reputations of merchants that sell them. BizRate collects feedback about products and merchants from more than one million online buyers and sellers each month [15, 16].

7.6.1 Trustworthiness of Merchants

Merchants can be found in the *Compare Prices and Stores* section for any product. This section lists the various merchants who carry particular products along with their price, availability and merchant rating. This rating is performed on a four-level scale as illustrated in Figure 7.18.

The 'smiley scale' can quickly help find stores with the level of quality required at the right price. For example, a consumer may question whether it is worth saving $10 on a bouquet of flowers when the merchant has a red sad-face for 'on-time delivery'?

BizRate needs a minimum of 20 surveys in the previous 90 days to regard the rating as valid. Once determined, the ratings are not static as they can be changed on the basis of some valid reason. These fluctuations might be caused by some evidence of suspicious review activity or the store may have changed its business model, resulting in changes to services or products.

A *customer certified logo* provided by BizRate increases trust in a particular merchant. To gain this certificate, the store has to participate in a two-dimensional process. Firstly, the store should

Figure 7.18   BizRate merchant rating. Reproduced by permission of Bizrate

support BizRate by collecting feedback and surveys from their customers as they finish their deal and secondly, customers should give at least 'a rating of satisfactory or better on the 12 quality rating dimensions of service'. If the store satisfies these criteria, it gets this certificate, which is displayed along with its rating by icon. This is illustrated in Figure 7.19.

The BizRate.com rating system hence allows them to claim that they have the most reliable and recent ratings possible. The latest user feedback is taken from their point-of-sale survey network and combined with the latest feedback from BizRate members to determine store ratings.

7.6.2 Rating Products

All registered users can rate and review a product. Their ratings collectively form the *overall product rating* shown as 1–5 stars (5 stars being the highest). These stars are classified on the rating scale shown in Figure 7.20.

Many products have a further breakdown of the *overall product rating* into the pros and cons of the product, the reviews and comments of users are also accessible. On the basis of this information,

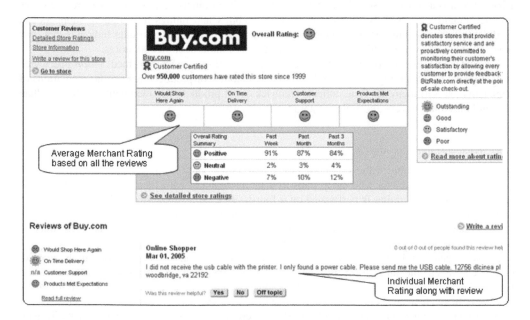

Figure 7.19   Merchant rating by individual customers. Reproduced by permission of Bizrate

Figure 7.20 Product rating scale. Reproduced by permission of Bizrate

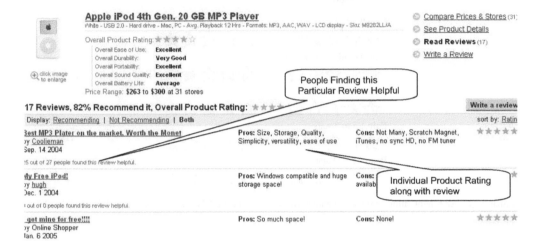

Figure 7.21 Individual product quality rating. Reproduced by permission of Bizrate

users can decide if the product is suitable for their needs, which then allows them to actually search for a suitable merchant. This is illustrated in Figure 7.21.

7.6.3 Rating Reviews

Users can vote if they find a particular review helpful or not. All reviews given by the users are stored in the system and can be accessed easily to display reviews and reviewer information along with the other reviews provided. However, this process is incomplete in the sense that it does not provide a rating for the review or the reviewer, but it just has information about how many people found the review helpful or otherwise. The system has not yet implemented a justification of the review rating.

7.7 CNet.com's Trustworthiness Systems

CNet attempts to provide expert and unbiased advice on technology products and services to inform users and expedite purchasing. By integrating an extensive directory of more than 200 000 computer, technology and consumer electronic products with editorial content, downloads, trends, reviews and price comparisons, CNet attempts to provide users with the most up-to-date and efficient shopping resources on the web [17, 18].

7.7.1 Trustworthiness of Merchants

The CNet merchant rating system is termed *certified store rating* and a store can get this rating if it participates in the CNet certified store programme. All the stores in this network have to adhere to the CNet code of conduct and a strict set of quality service guidelines and CNet certified rating criteria. These ratings are indicated by the stars next to the stores sites. Stores can receive a rating from half a star to five stars depending on compliance, and these ratings are updated weekly. Registered users can rate a store on the basis of their experience and write a review for the store. New stores are tagged as *not yet rated* (Figures 7.22 and 7.23).

Rating Scale

Excellent 4.5 to 5 stars

Very Good 3.5 to 4 stars

Good 2.5 to 3 stars

Fair 1.5 to 2 stars

Poor 0.5 to 1 star

Figure 7.22 Merchant rating scale. Used with permission from CNET Networks, Inc. Copyright © 1995–2005 CNET

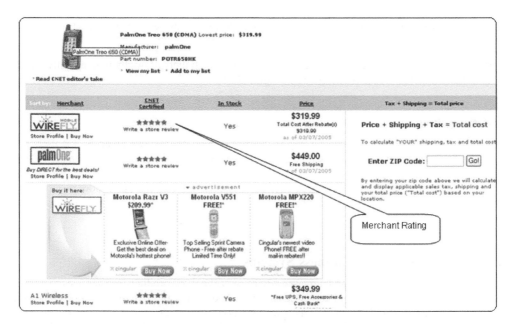

Figure 7.23 Certified merchant rating. Used with permission from CNET Networks, Inc. Copyright © 1995–2005 CNET

Figure 7.24 Merchant rating criteria. Used with permission from CNET Networks, Inc. Copyright © 1995–2005 CNET

7.7.2 Store Rating Criteria

There are four categories that are used as criteria in deciding the rating for a store.

Site functionality – Maximum 1 star
Store standards – Maximum 1 star
Order fulfilment – Maximum 1 star
Customer feedback – Maximum 2 stars

Each category above is further decomposed into several criteria that the store has to satisfy. The overall number of stars for a store is equal to the sum of the stars for each category and these are discussed in more detail in Chapter 11.

An illustration of the merchant rating criteria can be found in Figure 7.24.

7.7.3 Rating Products

CNet product rating is done by an expert panel of editors. CNet editors are special employees of CNet with expertise in giving reviews of products. CNet indicates that their editors offer hands-on experience. They seek out the products, note where they see them, touch them and test them. Most of the products they review are provided to them by manufacturers for trial. Once the review is complete, the product is returned and with the consent of the seller the review is published.

There are five categories used as generic rating criteria for all products:

1. Set up
2. Design
3. Features
4. Performance
5. Service and support.

These categories change for different products, for example, laptops may have battery life taken into consideration while cameras may have image quality considered. In each of the above categories, there are further criteria for measurement. These criteria could be the same for all categories or vary depending on the type of product. For example, for a camera, the criteria for each of the above categories are as follows:

- Point-and-shoot
- Midrange
- Semipro.

In order to calculate the quality of the product, CNet has set coefficients that weight each criterion differently as a percentage of the total. The overall scale for quality is between 0 and 10. On the basis of the input from the above criteria, the rating is calculated using the coefficient set for the particular product. For example, a Mainstream Notebook rating is calculated as follows:

$$\text{Rating} = [(\text{design} * 0.2) + (\text{features} * 0.3) + (\text{performance} * 0.1) + (\text{battery life} * 0.2)$$
$$+ (\text{service and support} * 0.2)]/10$$

The outcome of this equation is a number in the range 1 to 10 as shown in Table 7.5.

Along with editor ratings, CNet also lets its registered members express their personal views on any article or product. User ratings allow the users to rate a product and express their views in the form of pros and cons. These ratings collectively form the *average user rating*. This is illustrated in Figure 7.25.

7.7.4 Rating Reviews

Users can also vote if they find a particular review to be helpful or not. All reviews given by the users are stored in the system and can be accessed easily to display the reviews and the reviewer information along with the other reviews offered by that particular reviewer. However, it is incomplete in the sense that it does not provide a rating for the review or the reviewer, but it provides just information about how many people found it helpful or otherwise. This is illustrated in Figure 7.26.

Table 7.5 Product quality scale

Rating	Explanation
1–1.9	Abysmal
2–2.9	Terrible
3–3.9	Very poor
4–4.9	Poor
5–5.9	Mediocre
6–6.9	Fair
7–7.9	Good
8–8.9	Very Good
9–9.9	Excellent
10	Perfect

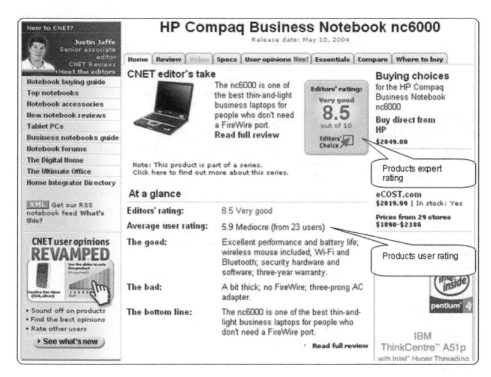

Figure 7.25 CNet's product ratings. Used with permission from CNET Networks, Inc. Copyright © 1995–2005 CNET

7.8 Review of Trustworthiness Systems

7.8.1 Commonality of Functions of Trustworthiness Systems

The analysis of the existing trust rating systems has shown that they generally have a number of common characteristics including the following:

- Trustworthiness and reputation of a seller, buyer or provider.
- Review and ratings of the product.
- The rating system processes the request of buyers and generates a list of recommendations of the preferred products or the providers list.
- The more familiar users becomes with the system, the more comfortable they are purchasing online.
- All these sites give transparent validation and publish their validation logic.

The above survey is summarized in Table 7.6. The solid dots indicate complete ratings, while empty dots indicate incomplete implementation of the rating.

7.8.2 Review of Existing Trustworthiness Measurement

In comparing and contrasting the above trustworthiness systems, we find that these large e-business operators have deployed different levels of sophistication for trustworthiness systems. Both Amazon and Yahoo, at the time of writing this book, had very simple seller, product and review trustworthiness rating systems. They have similar seller/merchant trustworthiness rating methods, which

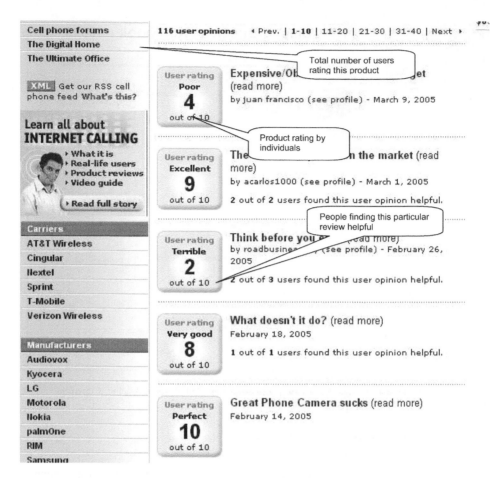

Figure 7.26 Individual product ratings and reviews. Used with permission from CNET Networks, Inc. Copyright © 1995–2005 CNET

Table 7.6 Summaries of ratings

	Amazon	Yahoo	Epinion	eBay	BizRate	CNet
Rating merchants	●	●	●	●	●	●
Rating products	●	●	●		●	●
Rating customers				●		
Rating reviews	○	○	●		○	○
Rating reviewers			●			

Note that ● represents a sufficient rating system and a ○ represents a preliminary system only.

consider an overall rating, average rating and aggregation of each including a criterion rating. Both, along with BizRate and CNet, rate reviews in a simple way by having two levels: 'helpful' and 'not helpful'. However, the average user may find that the simplicity of this provides an ease of use. Epinion's review rating is carried out on a four-level scale: 'very helpful', 'helpful', somewhat helpful' and 'not helpful'. Epinion offers trusting agents and trusted agents virtual community trustworthiness systems based on the context of how helpful a member finds an agent's reviews. In eBay, only transaction partners can rate each other. This trust value remains with them for their lifetime. BizRate uses a slightly more complicated formula than simple averages to rate the merchants from the individual trust values but is mostly similar to Amazon and Yahoo. CNet, on the other hand, includes the additional feature of having a panel of expert editors who give their opinions after testing the product themselves. This additional feature inspires more confidence in reviews than customer reviews that are more or less like those of unknown agents.

7.8.3 Weakness of Existing Trustworthiness Measures

As we can see, although the above systems are simple, they have all attracted millions of users to utilize them.

However they have limitations, including the following:

(a) There is no standard rating scale, for example, 2 levels, 5 levels or 6 levels are used.
(b) There is no standard visual representation, for example, stars, ticks or smiling faces are used.
(c) There is no standard quality of service measure.
(d) The context of trust is limited.
(e) The criterion measure is limited and not sufficient to represent quality of service.
(f) Time, during which trust can change, has not yet been considered.
(g) Trustworthiness prediction has not yet been considered.
(h) There is a lack of a proper aggregation method that helps the weighting of different levels of opinions and reviewers.
(i) The current trustworthiness systems provide limited functionality, for example, they have not yet considered quality of service measures based on the service agreement.
(j) There is a lack of a systematic approach to trustworthiness measures.

7.9 CCCI for Trustworthiness of E-service

The following case demonstrates how a CCCI methodology can be applied for trustworthiness measurement in a logistics service network. This methodology is being utilized for logistics network services and is developed and hosted by the Centre for Extended Enterprise and Business Intelligence at www.ceebi.curtin.edu.au and www.logistics.curtin.edu.au.

7.9.1 Example of Logistics Network Service

An example of trustworthiness for a logistics network service can be found in Figure 7.27.

Figures 7.28(a and b) best illustrate trustworthiness in logistics services. For example, on 12 February, Mr Wong has a consignment that needs to be transferred from Melbourne to Western Australia. This consignment consists of 10 boxes of fragile items including computer, vases and glasses that needed to be delivered to Mr Ping in Perth. Another 10 boxes, mainly clothing, should be delivered to Mr Xin in Albany.

Mr Wong is looking for a company that can provide a packing service for the fragile items and door-to-door pick-up/delivery service for those 20 boxes. He expects this consignment to arrive in Perth and Albany on 18 February. In addition, Mr Wong prefers to deal with the potential service provider that offers an online tracking and tracing facility and signature-based proof of delivery service, which enables him to know the status of his consignment at any time.

After searching the market, Mr Wong chooses Fast Delivery (FD) Pty Ltd as his trusted agent to handle his consignment.

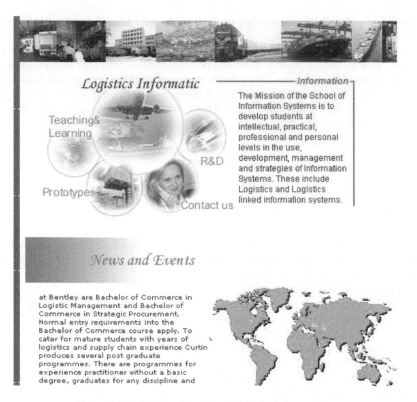

Figure 7.27   Trustworthiness centre for logistics services

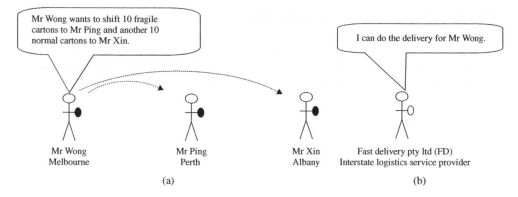

(a)   (b)

Figure 7.28   (a) Example of a logistics service and request, (b) Example of a logistics service provider

The following summary highlights the key points in the contract between Mr Wong and FD.

—   The total cost is AU$1500.00 (AU$900 for the delivery service from Melbourne to Perth and
    additional AU$600 for the delivery service from Perth to Albany). All sales taxes are included
    as are packing charges, door-to-door delivery service charges from Melbourne to Perth and to
    Albany, along with transit insurance.

— Packing and picking time is on the same day, which is 10 February. FD will pack the fragile items and collect the consignment at Mr Wong's office in Melbourne.
— Ten boxes of fragile items will be delivered to Mr Ping in Perth in the morning between 7 and 9 am on 18 February in an intact status.
— Ten boxes of clothing will be delivered to Mr Xin in the afternoon between 4–6 pm on 18 February.
— Mr Wong must be able to track and trace his consignment at any time by using the Easytrack system offered by FD.
— FD will offer signature-based proof of delivery service once Mr Ping and Mr Xin receive their respective shipments.
— Money back guarantee: if the consignment is not delivered on time or is damaged, FD will, at Mr Wong's request, give a replacement service or a second service free of charge. Mr Wong's refund will not be applicable if on-time delivery is not made owing to circumstances such as inclement weather, industrial disputes or things beyond the logistic provider's control.

On the basis of the above agreements, Mr Wong informed and promised his clients, Mr Ping and Mr Xin, on the delivery time on the same day when the contract between Mr Wong and FD was signed. However, FD did not fully fulfil its commitment. Mr Wong was told that Mr Xin did not receive 10 boxes of clothing until the 18th of March because of the fact that FD's partner in Perth was unable to dispatch its vehicle to deliver the consignment from Perth to Albany. Moreover, FD failed to offer an online track and trace service owing to a system breakdown, when Mr Wong needed to know where his consignment was. Therefore, Mr Wong got a replacement of the service free of charge and worth about AU$600 according to the 'Money back guarantee' agreement.

7.9.2 Application of CCCI

As described in Chapter 6, the primary function of the CCCI metrics is to seek the gap or difference between an original commitment from the service provider and the actual service delivered from the service provider. In CCCI methodology, the *criterion* would be the *original commitment* of the trusted agent set out in the 'terms and conditions'. We can obtain these through their service agreement. The trustworthiness measure is determined through the correlation of what was committed and what was actually delivered.

We, therefore, use the concepts and definitions found in Table 7.7.

The correlation of the service is calculated using the formulae found in Table 7.8. (For details, please refer to the discussion of this in Chapter 6.)

Table 7.7 CCCI metrics for correlation of quality of service (Section 6.3.2)

1	$Commit_{criterion\,c}$	The *commitment to the criterion* represents the fulfilment of each commitment (each criterion) of the trusted agent.
2	$Clear_{criterion\,c}$	The *clarity of the criterion* represents the clarity of each commitment (each criterion) in the service agreement (contract), and whether it is understood in the same way by both parties.
3	$Inf_{criterion\,c}$	The *influence of the criterion* represents the importance of each commitment (each criterion) that affects the trustworthiness determination.
4	$Corr_{service}$	To measure the fulfilment of the mutually agreed service, we calculate the *correlation* of the *committed services* from the trusted agent against the *delivered services* of the trusted agent.

Table 7.8 CCCI formula (Trustworthiness expressed in terms of $\text{Commit}_{\text{criterion }c}$, $\text{Clear}_{\text{criterion }c}$, $\text{Inf}_{\text{criterion }c}$ and their corresponding maximum values; repeated from Equation 6.5 in Chapter 6)

$$\text{Trustworthiness} = 5 * \left(\frac{\sum_{C=1}^{N} \text{Commit}_{\text{criterion }c} * \text{Clear}_{\text{criterion }c} * \text{Inf}_{\text{criterion }c}}{\sum_{C=1}^{N} \text{Max Commit}_{\text{criterion }c} * \text{Clear}_{\text{criterion }c} * \text{Inf}_{\text{criterion }c}} \right)$$

7.9.3 Define the Criteria

From Mr Wong's (trusting peer) perspective, the criteria to be utilized to measure the service quality of FD were as follows:

Agreed price (AU$600 + AU$900)
On-time pick-up service
On-time delivery service
Intact delivery
Tracking and tracing capability
Signature based proof of delivery
Money back guarantee (if a service is badly done, a free service in the future is provided free of charge).

7.9.4 Importance of Criterion (Inf$_{\text{criterion}}$)

The importance of the criteria for measurement in this case includes on-time service, intact delivery, online tracking and trace capability, agreed costs, signature-based proof of delivery and a money back guarantee. These are elaborated in this section.

7.9.4.1 On-time Service

On-time service in this case includes on-time pick up and delivery. Nevertheless, it is important to distinguish between on-time delivery and on-time pick up as their impact on Mr Wong is quite different. In addition, this distinction enables Mr Wong to measure service more accurately.

In terms of on-time pick up, Mr Wong expected that the potential supplier (a logistics company) should pick up this consignment on the particular time agreed to by both parties. Obviously, any late pick-up service would disrupt Mr Wong's daily schedule and cause inconvenience for Mr Wong. Regarding on-time delivery, any late delivery to Perth or Albany would negatively affect the business relationship between Mr Wong and his clients (Mr Ping and Mr Xin) since Mr Wong has also promised the delivery time to them and any delay additionally would affect them. In other words, Mr Wong focused more on on-time delivery than the on-time pick-up service. Accordingly, Mr Wong assigned the value of importance for on-time delivery and on-time pick-up as 5 and 3 respectively.

7.9.4.2 Intact Delivery

Owing to the fact that this consignment consists of 10 boxes of fragile items, intact delivery becomes one of the key criteria to measure the performance of the service provider. Receiving fragile items such as a computer, vases and glasses in a broken condition does not make business sense. Hence, Mr Wong gave the value of importance for this criterion as 5.

7.9.4.3 Online Tracking and Tracing Capability

Online track and trace ability advantages Mr Wong by allowing him to gain instant consignment status information. This may help him notify business partners in the event of any delay. Thus, the value of importance in terms of tracking and tracing capability was 4.

7.9.4.4 Agreed Costs

An increment in the fixed costs with the same service requirement and standard agreed to by both sides of the service would affect Mr Wong's perception of the service. Therefore, Mr Wong gave a value of 3 for this criterion.

7.9.4.5 Signature-based Proof of Delivery

Signature-based proof of delivery was one of the key factors for Mr Wong to select the service provider. Mr Wong considered this criterion to be important as he wants to know when the consignment is delivered and to whom. In the case that Mr Ping or Mr Xin was unable to receive the shipment by themselves (refer to Figure 7.28) and appointed someone else on their behalf to receive the consignment, signature-based proof of delivery allows the consignment carrier (logistics provider) and Mr Wong, Ping and Xin to determine and confirm who received the cargo. Consequently, the value of importance for this criterion was 4.

7.9.4.6 Money Back Guarantee

Before the business interaction occurred, the trustworthiness between the trusting peer and the trusted peer was 0 since both parties to the service were new to each other. Thus, the money back guarantee item would make financial sense for Mr Wong in the case where FD failed to fulfil its commitment. Accordingly, the money back guarantee was given a value of 4 in terms of its importance.

7.9.5 Clarity of Criterion (Clear$_{criterion}$)

In this case, both criteria were written very clearly in the contract, and hence, Clear$_{criterion}$ for each criterion was 5. The trusting peer (Mr Wong) and the trusted peer (FD) understood their responsibilities for this business transaction without any ambiguity.

Table 7.9 shows the importance and clarity for each criterion. On the basis of the actual performance (Commit$_{criterion}$) of FD, together with the importance and clarity of criteria mentioned previously, the Corr$_{criterion}$ can be worked out in a straightforward manner. The trustworthiness value can be obtained. Table 7.10 summarizes Commit$_{criterion}$, Inf$_{criterion}$, Clear$_{criterion}$ and Corr$_{criterion}$ for each criterion selected by the trusting peer.

Table 7.9 Importance and clarity of each criterion

Criterion	Influence	Clear
Intact delivery	5	5
On-time delivery	5	5
Agreed costs	3	5
Tracking and tracing capability	4	5
Signature-based proof of delivery	4	5
On-time pick-up service	3	5
Money back guarantee	3	5

Table 7.10   Summary of Commit$_{criterion}$, Inf$_{criterion}$, Clear$_{criterion}$ and Corr$_{criterion}$

Criterions	Commit$_{criterion}$		Inf$_{criterion}$	Clear$_{criterion}$	Corr$_{criterion}$	
	Actual	Max	Actual	Actual	Actual	Max
Intact delivery	5	5	5	5	125	125
On-time delivery	2.5	5	5	5	62.5	125
Agreed costs	5	5	3	5	75	75
Tracking and tracing capability	0	5	4	5	0	100
Signature-based proof of delivery	5	5	4	5	100	100
On-time pick-up service	5	5	3	5	75	75
Money back guarantee	5	5	3	5	75	75
				Total	512.5	675

7.9.6 Correlation of the Criterion

$$RelCorr_{service} = Corr_{service}/Max_{service} = 512.5/675$$

$$Trustworthiness = 5^* RelCorr_{service} = 3.796296296$$

This trustworthiness value indicates that the trusted agent is largely trustworthy, which predicts a higher possibility of occurrence for the next potential business transaction between the trusted agent and the trusting agent under the same circumstances. More precisely, the trusting agent (Mr Wong) would choose the trusted agent (FD) to handle another consignment if the criteria, Inf$_{criterion}$ and Clear$_{criterion}$ are the same as in the previous case.

7.9.7 Exceptions

In this case, the trusting agent (Mr Wong) selected only seven criteria to measure the quality of service and examined the trustworthiness towards the trusted agent (FD) by applying CCCI methodology. However, other than the criteria illustrated, there are a number of criteria that affect the trustworthiness between the trusting agent and the trusted agent either positively or negatively. Therefore, it is essential to group these unselected ones but vital criteria as exceptions in terms of the criteria for this case.

7.9.7.1 Technology

In this case, Mr Wong emphasized signature-based proof of delivery and online track and trace technologies. Yet, there are numerous technologies available to facilitate the information flow as well as goods flow. For instance, transportation management system (TMS) can be used to improve transportation performance and warehouse management system (WMS) can be used to optimize warehouse operation. A study by Lewis and Talalayevsky [19] examined how the evolution of information technology has allowed the largest users of logistics services to focus on their core competencies and contract out logistics. Sophisticated technologies applied by service providers enable them to improve the efficiency of information flow, and ultimately influence their performance, which determines the trustworthiness between a trusting agent and a trusted agent.

7.9.7.2 Hands-on Experience and Knowledge

It is vital that operational departments have hands-on experience and knowledge to meet their customers' needs. For example, some service providers can consolidate consignments from different suppliers with the ability to send them forward as one shipment. In Mr Wong's case, the hands-on experience and knowledge over packing and handling fragile items strengthened the trustworthiness to some extent.

7.9.7.3 Consistency in Services

Consistency in service is vital in both B2B and B2C circumstances; and customers prefer to deal with companies that are able to provide consistent services. Evidence suggests that inconsistency in service tends to be detrimental to trust [20]. For instance, in peak season, a freight forwarder ABC provides cargo space for 300 kg in each flight operated by ABC from Perth to Melbourne every day, contrasting with the cargo space provided by another freight forwarder XYZ, which varies from 100 to 500 kg. ABC's customer, who has 300 kg of air cargo that needs to transferred from Perth to Melbourne, would trust freight forwarder ABC instead of freight forwarder XYZ to handle their goods, as freight forwarder ABC provides more certainty in terms of cargo space in the aircraft.

7.9.7.4 Flexibility

Customers are more willing to trust service providers who are flexible in customer service. Flexibility reflects the service providers' capability to amend and adjust their services in order to meet customers' special requirements. A provider that can bring forward schedules (deliver earlier) or postpone shipping at the customer's request displays flexibility. An example of this is where a manufacturer is unable to pass their consignment to carriers at an agreed time, yet the consignment still needs to be delivered to the destination on time. The likelihood of enhancing trustworthiness is higher if the carrier can demonstrate their flexibility by adjusting transportation and distribution plans to achieve the manufacturer's expectation, within the same providers' limits.

7.9.7.5 Reputation

Reputation has a great deal to do with trust. An individual is more willing to commit to another if the other person holds a reputation for cooperative behaviour [21]. The same mechanism operates among firms and serves to check for misbehaviour, thereby building trust, especially in long-term relationships. Reputation for fairness should have a positive effect on a buyer's level of trust in a third-party provider. This reputation for fairness is built on the edifice of reliable and consistent behaviour over time [22]. A logistics provider's reputation for fairness should be transferable across buyers and enhance the level of trust toward the provider.

7.9.7.6 Communication

Communication has been described as 'the glue that holds together the channel of distribution' ([23], p.38). Other than in a formal manner, communication can be informal and behind the scenes, which greatly enhances trust and coordinative behaviour [21]. Anderson and Narus [24] and Morgan and Hunt [25] have empirically demonstrated that communication leads to trust. Additionally, communication results in trust in an indirect way by eliminating misunderstandings. Evidence suggests that unsatisfactory performance is not just affected by a service provider's ability but depends also on the extent that the trusting agent and trusted agent understand each other.

7.10 Summary

Reputable sellers and trusted service providers have major and unsurpassed advantage in product, services or information sales when compared to others. We have studied seven well-known e-business providers to illustrate how the trustworthiness technology and system is reshaping the world of e-commerce, e-business and e-services. Even when e-commerce is considered to be in a slump, these companies are still increasing profits and serving millions of users around the world. These sites include Amazon, Yahoo, Epinion, eBay, BizRate, and CNet. We also glanced at a further detailed example of trustworthiness of e-service via logistics examples to demonstrate the

CCCI trustworthiness measure for e-services. The above e-business portals are chosen because they represent the new generation of intelligent business models and operations.

Each of the above e-business portals uses its own language, terms and measurement scales to describe trust and trust relationships. This field is still very new. However, standards will emerge in the near future, as this intelligent business tool develops further.

These trustworthiness technologies and systems are not just limited to e-businesses. They should also apply to brick and mortar companies. Building a trust relationship with customers is an intelligent tool for all types of businesses, and not just for dot.com businesses. The challenge in e-business is the open and distributed environment, where communication is often anonymous and 'short-cut'. Therefore, remote users or providers often receive incomplete information about an order, a user, a service description, and so on. The problem of gaining inefficient information is solved through trust technologies as well as trustworthiness systems. *Trustworthiness technologies and systems* give an online user the same experience as in the real world in respect of trust and reputation in the business situation.

The following important concepts were presented in this chapter:

- The trustworthiness systems
- Trustworthiness measure for rating agents, sellers, reviews, reviewers, and products
- The weakness of existing measures.

References

[1] http://www.amazon.com, accessed (2005)
[2] http://www.amazon.com/gp/help/seller/feedback-popup.html, accessed 2005.
[3] http://www.amazon.com/exec/obidos/subst/community/reviewers-faq.html, accessed (2005).
[4] http://www.amazon.com/exec/obidos/tg/browse/-/537868/pop-up/ref=olp_wa_1/, accessed (2005).
[5] http://www.yahoo.com, accessed (2005).
[6] http://help.yahoo.com/help/us/shop/shop-77.html, accessed (2005).
[7] http://help.yahoo.com/help/us/shop/shop-06.html, accessed (2005).
[8] http://help.yahoo.com/shop/shop-68.html, accessed (2005).
[9] http://www.epinions.com, accessed (2005).
[10] http://www.epinions.com/help, accessed (2005).
[11] http://www.eBay.com, accessed (2005).
[12] http://pages.eBay.com.au/help/feedback/evaluating-feedback.html, accessed (2005).
[13] http://pages.eBay.com.au/services/buyandsell/powersellers.html, accessed (2005).
[14] http://pages.eBay.com/help/feedback/reputation-stars.htmlBottom of Form, accessed (2005).
[15] http://www.BizRate.com/content/ratings_guide.html, accessed (2005).
[16] http://www.BizRate.com, accessed (2005).
[17] http://reviews.CNet.com, accessed (2005).
[18] http://reviews.CNet.com/4002-5_7-5100969.html, accessed (2005).
[19] Lewis I. & Talalayevsky A. (2000) 'Third-party logistics: Leveraging information technology', *Journal of Business Logistics*, vol. 21, no. 2, pp. 173–185, [Retrieved:February 25, 2005, from ProQuest 5000].
[20] Stock J.R. & Lambert D.M. (2001) *'Strategic logistics management'*, 4th ed, McGraw-Hill, Boston, USA.
[21] Pruitt D.G. (1981) *'Negotiation behaviour'*, Academic Press Inc., New York, USA.
[22] Ganesan S. (1994) 'Determinants of long-term orientation in buyer-seller relationships', *Journal of Marketing*, vol. 58, no. 2, pp. 1–19, [Retrieved: February 25, 2005, from ProQuest 5000].
[23] Mohr J. & Nevin J.R. (1990) 'Communication strategies in marketing channels: A theoretical perspective', *Journal of Marketing*, vol. **57**, no. 4, pp. 36–51, [Retrieved: February 25, 2005, from ProQuest 5000].
[24] Anderson J.C. & Narus J.A. (1990) 'A model of distributor firm and manufacturer firm working partnerships', *Journal of Marketing*, vol. 54, pp. 42–58.
[25] Morgan R.M. & Hunt S.D. (1994) 'The commitment-trust theory of organizational commitment', *Journal of Marketing*, vol. 58, no. 3, pp. 20–38, [Retrieved:February 23, 2005, from ProQuest 5000].

8

Reputation Concepts
and the Reputation Model

8.1 Introduction

Reputation has a profound impact on the trusting agent and the trusted agent in business interactions. Moral, ethical and legal guidelines are implemented as a result of the promotion of fair trading practices, honesty from all parties, consumer protection legislation, service quality assessment, and assurance for customers, e-businesses and service-oriented environments.

In order to understand this leading edge technology, we first need to understand the concept of reputation along with reputation query and the three inner relationships of reputation, which are the recommendation trust relationship, the reputation query relationship and the third-party trust relationship. We also need to distinguish between third-party agents, recommendation agents, known agents, referred agents, unknown agents, malicious agents and untrusted agents. We shall examine first-, second- and third-hand opinions, the trustworthiness of recommendation agents to provide opinions (also known as *credibility* of recommendation agent) and the trustworthiness of the opinion. We learn how to modify the trustworthiness value of the recommendation agent to provide an opinion, and finally we present the reputation model for a better and more adequate agent, product and service assessment. The above ethical fundamentals are the foundations of a customer's perceptions of business and service integrity.

Some technologies (Chapter 11) treat human networks (both users who are registered on their website and customers of the site) as trustworthy, and others (outsiders or unknown agents) as untrustworthy. The questions of trust in networked environments include issues concerning interactive relationships within these environments, such as someone who is known can be untrustworthy, and someone who is known may be trustworthy in one *context* (e.g., work on projects) and untrustworthy in another (e.g., promotion as they become competitors in this *context*).

Another question that could arise is, 'If it is someone we do not know, is he/she trustworthy or untrustworthy'? It cannot simply be stated that *unknown people* are all untrustworthy or malicious, especially with billions of existing Internet users and a large client or user population that is unknown. This is unlike the physical world, where our immediate social sphere is much smaller (e.g., in an organization of 100 people, all of which are *known* agents to each other, it might be assumed that 2 % of them are malicious or untrustworthy). However, in the virtual world, our reach to known and unknown agents is beyond comprehension in conventional trust relationships.

Another problem remains in relation to trust and reputation where agents can sometimes be inconsistent over time. How do we manage the need to distinguish malicious agents from untrustworthy agents? Our understanding of *trustworthy agents* is that they may actually be untrustworthy

Trust and Reputation for Service-Oriented Environments   Elizabeth Chang, Tharam Dillon and Farookh Hussain
© 2006 John Wiley & Sons, Ltd

in some time slots and contexts. Our final considerations may be the trust approach toward unknown agents and the level of trust we place in third-party opinions regardless of whether they are known to us or not.

In this chapter, we take these concerns into consideration. We explain the complex concepts and terminologies one by one. We define *reputation* in a form suitable for service-oriented environments and e-business. Existing definitions will also be reviewed. A new set of terms that comes with the notion of reputation will be examined. We also focus on the differences between trust and reputation, including distinguishing between trust *relationships* and *reputation relationships*.

The key complication in reputation and reputation relationships is the involvement of *third-party agents*. We take the reader step by step through the complexities of the *reputation relationship*, including the *reputation query relationship, recommendation trust relationships* and *third -party trust relationships*. We also separate third-party trust agents into known agents, referred agents, unknown agents and malicious agents. Furthermore, we explain the need to consider first-, second- and third-hand opinions and how a *reputation value* should be aggregated. We also develop *reputation models* using these advanced reputation concepts, all of which are for the purpose of a more adequate and fair assessment of agents, products, services and quality assurance for end users and consumers.

8.2 Reputation in Literature

Reputation has been widely used in different disciplines like sociology and economics [1, 2] and psychology [3, 4]. In the area of computing, the concept of reputation has been applied to multi-agent systems [5–14]. In this section, we review some of the definitions of reputation as given in the relevant literature and discuss their shortcomings when applied to service-oriented network environments. Sabater and Sierra [7] defined reputation as an *'Opinion or view of one about something'*. Abdul-Rahman and Hailes [9] defined reputation as *'an expectation about an Agent's behaviour based on information about or its past behaviour'*. Mui *et al.* [14] defined reputation as the *'perception that an agent creates through past actions about its intentions and norms'*. Note that the above definitions have not fully considered the time factor or context in detail in their reputation definition and method, such as that the reputation of a given agent has a time frame (time slot) and the reputation context may or may not be the same in the next instance of time. Context is important when defining reputation. It is also crucial to identify those who are best placed or eligible to vouch for an agent's reputation. For example, a university may have a good reputation in engineering but not in medicine. Miztal [15] notes that 'Reputation helps us to manage the complexity of social life by singling out trustworthy people in whose interest it is to meet promises'. This definition of reputation focuses on the purpose of reputation as a means of finding trustworthy people. Although this is correct, it does not mention whose reputation is under consideration, at what given point in time, in what context and, more importantly, who is eligible to vouch for the reputation.

8.3 Advanced Reputation Concepts

8.3.1 Reputation

In the following definitions, we provide a base level definition as well as an advanced definition of reputation (Figure 8.1).

The base level definition (Definition 1) is at a simple level for the understanding of trust and reputation. It is easier for computation and implementation in trust and reputation systems. The advanced definition (Definition 2) is for the next generation of trust and reputation systems, which is a current research topic, and currently has no fully implemented system available.

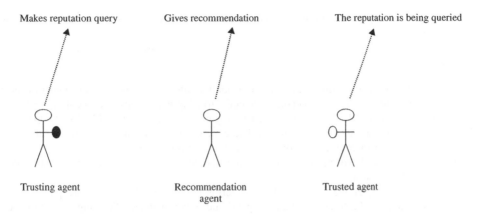

Makes reputation query   Gives recommendation   The reputation is being queried

Trusting agent   Recommendation   Trusted agent
                                agent

Figure 8.1   Three key entities in reputation definition

Definition 1 – Basic Reputation Concept

Definition 1a: In service-oriented environments, we define *agent reputation* as an aggregation of the recommendations from all of the third-party recommendation agents, in response to the trusting agent's reputation query about the *quality* of the trusted agent.

The definition also applies to the reputation of the *quality of product* (QoP) *and quality of service* (QoS).

Definition 1b: In service-oriented environments, we define *product reputation* as an aggregation of the recommendations from all of the third-party recommendation agents, in response to the trusting agent's reputation query about the Quality of product (QoP).

Definition 1c: In service-oriented environments, we define *service reputation* as an aggregation of the recommendations from all of the third-party recommendation agents, in response to the trusting agent's reputation query about the Quality of the Service (QoS).

Definition 2 – Advanced Reputation Concept

Definition 2a: In service-oriented environments, we define *agent reputation* as an aggregation of the recommendations from all of the third-party recommendation agents and their first-, second- and third-hand opinions as well as the trustworthiness of the recommendation agent in giving correct recommendations to the trusting agent about the quality of the trusted agent.

Definition 2b: In service-oriented environments, we define *service reputation* as an aggregation of the recommendations from all of the third-party recommendation agents and their first-, second- and third-hand opinions as well as the trustworthiness of the recommendation agent in giving correct recommendations to the trusting agent about the Quality of the Service (QoS).

Definition 2c: In service-oriented environments, we define *product reputation* as an aggregation of the recommendations from all of the third-party recommendation agents and their first-, second- and third-hand opinions as well as the trustworthiness of the recommendation agent in giving correct recommendations to the trusting agent about the Quality of the Product (QoP).

Fundamentally, reputation is an aggregated value of the recommendations about the trustworthiness of a *trusted agent* (such as a trusted agent or QoP and services). The reputation value is not assigned but only aggregated by the trusting agent.

There are four primary concepts that should be clearly understood in the reputation definition:

(1) *Reputation*: aggregation of all the recommendations from the third-party recommendation agents about the quality of the trusted agent.
(2) *Recommendation (including opinion and recommendation value)*: submitted by the third-party recommendation agents (recommenders).

(3) *Recommendation agent*: submit the recommendation or opinion to the trusting agent or respond to a reputation query.
(4) *Reputation query*: a query made by the trusting agent about the trusted agent in a given context and time slot.

The terms 'reputation query', 'third-party recommendation agents', 'first-, second- and third-hand opinions', 'trustworthiness of recommendation agent', 'trusting agent', 'trusted agent' are essential when defining reputation. These new terms can be regarded as the building blocks of reputation, particularly in service-oriented environments. These new terms introduced in the definition of reputation make a fundamental distinction between trust and reputation.

In this chapter, we will pay more attention to the new terms, new concepts and the difference between trust and reputation.

Note that in this chapter we use the following terms interchangeably: 'the third-party recommender' is a synonym of the 'recommendation agent' and 'recommender'. 'Making a recommendation' is synonymous with 'giving an opinion', and 'recommendation' is synonymous with 'opinion'. Trustworthiness of recommendation agent to give an opinion and credibility of recommendation agent are synonymous and used interchangeably.

8.3.2 Third-party Recommendation Agents

In many large websites, the concepts of third-party agents and third-party recommendation agents are interchangeable. However, as we shall see in this section, they are not quite the same.

Third-party agent: In service-oriented network environments, all other agents with the exception of the trusting agent and the trusted agent in a given relationship are referred to as third-party agents. They could be non-anonymous agents, pseudo-anonymous agents and anonymous agents.

Third-party recommendation agent: In service-oriented network environments, third-party recommendation agents are third-party agents who provide a recommendation, feedback or opinion. Not all third-party agents give recommendations; therefore, third-party recommendation agents are a subset of the third-party agents. Third-party recommendation agents can be thought of as a subset of third-party agents.

Sometimes we use the term *recommendation agents* or *recommender* to refer to *third-party recommendation agents*.

8.3.3 Reputation Query

A *reputation query* is the enquiry made for a specific context and time slot regarding a trusted agent, product or service. It may include context ID, context description, context time and so on. The issues to consider are the following:

- In regard to a trusted agent, the enquiry is about the trustworthiness of the trusted agent.
- In regard to a product, the enquiry is about the QoP.
- In regard to a service, the enquiry is a request for feedback about the QoS.

The term reputation query is also interchangeable with 'asking opinions', 'getting recommendations', 'calling for referees', or 'an invitation for feedback'.

The reputation query has the following characteristics:

- Each *reputation query* is associated with one context and a given time slot. For example, when someone asks for a person's reference, such as 'Does Jo's work experience in the last 5 years have relevance to our company's primary operations?', the *context* is Jo's experience or history, as well as the company's primary operation, that is, Internet hosting. The time slot is the last five years. The reputation query is considered reasonable if it states clearly what the trusting agent is seeking.

- The *context* in the reputation query is specific to an agent, a product or a service. For example, it can be about sellers, shops, trusted agents, service providers, products or particular services.
- The *context* is a key component in reputation query
- Each *context* in reputation query could be associated with a context name, quality aspects and quality assessment criteria.
- The context has many aspects. Each aspect may have many criteria. If a reputation query or reference is stated as 'Does Jo fit into our company?', we could easily ask about what aspects would be important in this context, 'cooking', 'marking', or 'managing business development' and so on.
- Dynamic factors are involved in a reputation query. The trustworthiness assigned by an agent to another agent gathers increasing uncertainty with the passage of time. For example, Alice knew Bob was trustworthy in 1990 in the context of 'storing a file'. Faro enquires about Bob's reputation in 2005. Since Alice does not know anything about Bob's reputation in the last 15 years, Alice should not reply to the reputation query, and if she did, how can she now ensure that Bob is trustworthy, untrustworthy or very trustworthy? Therefore, the impact of her reply will be reduced. We also need to consider that, in the reputation query, time specification together with context is a necessary requirement.
- A reputation query may involve positive activities or negative activities. For example, one may be seeking good qualities of people or good qualities of service or product, or one may be seeking negative comment or comments on weaknesses.
- A reputation query is associated with exactly one trusted agent, also known as the *reputation-queried agent*. For example, a reputation query or request for reference such as 'I would like to know the reputation of Charlie, Toby and Lily', actually contains a few queries. However, *no one* can give one opinion or one recommendation about all three people in one instance.

As mentioned in Chapter 1, an agent could be a seller, a service provider, or a buyer. The agent, who made the reputation query, is also called the *reputation-querying agent*.

- The reputation query is responded to by zero-to-many third-party agents. It may be associated with zero-to-many (0:M) *third-party agents* who respond to the reputation query. A reputation query may result in no response at all. No agents may come forward to offer their opinions or recommendations. However, the more the opinions collected from third parties, the better the assessment made by the trusting peer about the trusted peer.
- Similar reputation queries may be posted by many trusting agents. For example, it may be possible that several agents want to know details about a new combined business and IT solution called 'trust and reputation in service-oriented environments'. These trusting agents may not know each other and not everyone receives exactly the same number of third-party agent replies. Therefore, these trusting agents will each have different third-party agent replies. Some third parties may even offer the same opinion twice to different trusting agents about the same query.

8.3.4 Trusting Agent in a Reputation

A *trusting agent* in a reputation query is an agent who poses a query, requests recommendation or asks for feedback or seeks references about the trusted agent or entity from third-party agents on the network.

If a trusting agent wants to carry out a transaction with a trusted agent, but he/she has not interacted with the trusted agent previously, the trusting agent is to ask for recommendations from other agents (third parties) about their past experiences with the trusted agent, in order to decide whether or not to carry out a transaction with the trusted agent or choose to use the trusted agent.

8.3.5 Trusted Agent in a Reputation

A trusted agent could be an agent, a product, a service, a software agent, a website, and so on, that the public are interested in and is available for comments over a service-oriented network environment.

Trusted agent is the agent whose reputation or quality is being queried by the trusting agent. We use the term *trusted agent* as referring to *an agent, a service or a product*.

The trusted agent in a reputation query is also known as the *reputation-queried agent* whose reputation is under consideration or evaluation. The trusting agent may request a reputation query before entering into an interaction or transaction with the trusted agent or choosing the trusted agent. The trusting agent can do this by sending or broadcasting their request to the third-party agents on the network.

8.3.6 First-, Second- and Third-hand Opinions

The terms 'Opinion' and 'recommendation' are used interchangeably. It *must* come from a third party. It is the belief of the third-party agent about the trusted agent or product or service.

Reputation is an aggregated value of the entire *third-party recommendation agents'* opinions or recommendations. There are a number of distinct characteristics in third-party opinions:

(1) Not every *third-party agent*'s opinion is of equal importance. For a more precise assessment of reputation, we should divide opinions or recommendations into three categories: first-, second-and third-hand opinions.
(2) Not every *third-party agent* gives the correct opinion. There may be intentional misleading opinions provided or agents may be incapable of giving the right opinion. For example, a first- hand opinion may not be a true opinion and a known agent may not be trustworthy. We may trust them in one context, but not in every context. For a more precise measure of the reputation, we should consider the trustworthiness of the opinions, especially for the first- and second-hand opinions, and trustworthiness of agents in giving the right opinion.

Figure 8.2 provides examples of first-, second- and third-hand opinions:

- First-hand opinion (direct opinion e.g., Budi says to Alice that Liz is trustworthy)
- Second-hand opinion (indirect opinion e.g., Bob says to Alice that Sarah said Liz is trustworthy)
- Third-hand opinion (public opinion e.g., Joe says to Alice that he believes that Liz is partially trustworthy)

The following points need further explanation in the Example in Figure 8.2:

(a) Alice has a trust relationship with Bob and Budi only; she has no direct trust relationship with Sarah, Jo or Liz. Now Alice wants to form a trust relationship or interact with Liz, so she asks for a reputation query about Liz from all of the agents in the network.
(b) Budi and Sarah have or have had interactions or a trust relationship with Liz.
(c) Sarah had had trust relationships with Bob and Liz, but not with Alice and Jo.
(d) Jo may know Liz, but does not know anybody else.
(e) Bob, Budi, Sarah and Jo are all third-party agents, from the trusting agent's (Alice's) perspective.
(f) All third-party agents can reply to Alice's reputation query when Alice asks about the reputation of Liz.
(g) The trusted agent Liz may also receive a reputation query, but Liz should not reply for herself.

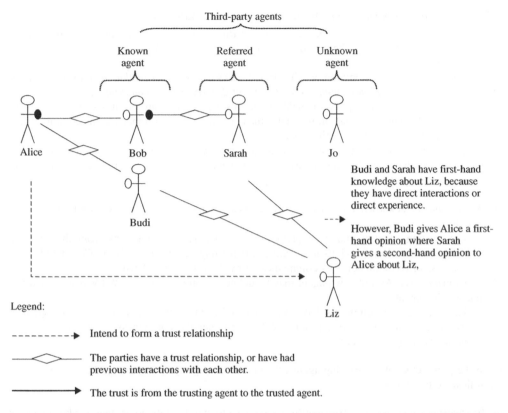

Figure 8.2 Sarah and Budi had first-hand knowledge about Liz; however, from Alice's point of view, Budi gives first-hand opinion and Bob gives second-hand opinion, because Bob says Sarah knows Liz

Weighting the Opinions

- By default, first-hand opinion is weighted higher than second- and third-hand opinions.
- A second-hand opinion is an indirect opinion. If Bob had replied to the reputation query about Liz, his opinion is an indirect opinion, that is, from his friend Sarah.

  The weight for second-hand opinion is customer defined. Normally, for the quality of the assessment, the distinction between first- and second-hand opinion is important; however, the challenge is to be able to identify the first- and second-hand opinions in the open anonymous network because the agent may not be honest when giving a recommendation.
- A third-hand opinion is a public opinion. Those who have not had direct or indirect experience with Liz can give opinions here.

  Most trust and reputation systems do not implement first-, second- or third-hand opinions; this is because of the cost involved in their implementations. This is also true in the physical world. In the virtual world, the majority of reputation systems are targeted first-hand opinions, with no separation concerning first-, second- and third-hand opinions, In this case, the assessment of the reputation may not be adequate. To overcome these problems, assembling large volumes of data is necessary, or it could be by restricting queries to first-hand opinions only, as well as monitoring the recommendation agents feedback or behaviour and validating the trustworthiness of the recommendation agent in giving correct recommendations.

8.3.7 Trustworthiness of Opinion or Witness Trustworthiness Value (WTV)

The trustworthiness of opinion is referred to as the trustworthiness of the opinion of the recommendation of agent. It can also be called as *Witness Trustworthiness Value (WTV)* or the credibility of the opinion.

The trustworthiness of an opinion or WTV is defined as *a numeric value that denotes or quantifies the correctness of an opinion communicated by the third-party recommendation agent.*

The trustworthiness of an opinion is held by the trusting agent about the third-party recommendation agent or the witness agent and it denotes the correctness of recommendations communicated by the third-party agent to the trusting agent.

From the perspective of the trusting agents, the third-party agents are classified into the following two types based on the correctness of recommendations communicated by the third-party recommendation agent.

(1) *Known agents*: A known agent may be defined as an agent whose WTV is known to the trusting agent.

   The trusting agent would have previously solicited recommendations from the agent in question and based on previous interaction/s the trusting agent knows the WTV (or rather the correctness of recommendations communicated by the third-party Agent).

(2) *Unknown agents*: An unknown agent may be defined as an agent whose WTV is not known to the trusting agent.

   The trusting agent might not have previously solicited recommendations from the agent in question and hence does not know the WTV (or rather the correctness of recommendations communicated by that agent).

From the perspective of the trusting agents, the known third-party agents are further classified into the following two types,

a. *Trustworthy known agents*: Trustworthy known agents may be defined as agents whose witness trustworthiness is within a specified range or value. The specified value or range of WTVs for a third-party agent to be considered as trustworthy is usually specified by the trusting agent.

   In business terms, a trustworthy known agent is an agent from whom the trusting agent had previously taken recommendations and the WTV of this agent based on these previous recommendations falls within a given range of WTVs. The trusting agent considers each agent falling within this range as a trustworthy known agent or a trustworthy witness agent or a trustworthy recommender or a credible agent.

b. *Untrustworthy known agents*: Untrustworthy known agents may be defined as agents whose witness trustworthiness is outside the specified range of WTVs.

   In business terms, an untrustworthy known agent is an agent from whom the trusting agent had previously taken recommendations and the WTV of this agent based on these previous recommendations does not fall within a given range of WTVs. The trusting agent considers each agent falling within this range as an untrustworthy known agent or an untrustworthy witness agent or an untrustworthy recommender or a non credible agent.

In this book, the concept 'giving an opinion or a recommendation' is interchangeable with 'writing a review' or 'offering feedback' about the trusted agent or product or service. Therefore, we sometimes treat the above words as the same and use them interchangeably.

8.3.8 Quality of Agent, Product or Services

The quality of an agent is assessed by a set of simple and well-defined criteria.

   The quality of product is also assessed by a set of simple and well-defined criteria.

What is a well-defined criterion? A well-defined criterion may be defined as a criterion that can be extracted from a service-level agreement, sales contract or mutual agreement. It can also be derived from customer service requirements or best practice in the service industry as well as national or international standards.

QoS is commonly assessed by the fulfilment of the service agreement, which clearly states the terms and conditions of the service. These terms and conditions imply the responsibility or commitment of the service provider.

In the section above, we have introduced all the concepts that are relevant to reputation. In the next section, we shall discuss the reputation relationship.

8.4 Reputation Relationship

The nature of the third-party agents and their involvement in the reputation process complicates the reputation relationship. This is also true in the physical world. Dealing with *binary* relationships is much easier than dealing with *ternary* relationships where a third party is involved, but normally in the physical world, the third-party agent or agents are well known to both agents. However, in the virtual world, there are millions of third-party agents that neither party may know. In this section, we address some of these concerns.

Importantly, we must consider the following issues:

- What are the interconnections between the three types of agents?
- How do we reveal the intertwined inner relationships that surround third-party recommendation agents?

As we can see, these challenges of reputation relationships do not exist in a trust relationship. This is because the trust relationship is a direct binary relationship between trusting agent and trusted agent, whereas, the reputation relationship involves an extra agent, the third-party recommendation agent. Therefore, in this next section, we reveal the more complicated relationships and their effect on the business and commercial world.

8.4.1 Definition of the Reputation Relationship

Reputation relationship definition: We define a *reputation relationship* in a service-oriented environment as an association between the trusting agent, third-party recommenders and trusted agents that revolves around the reputation query.

The term reputation query is essential when defining the reputation relationship because if there is no reputation query the trusting agent will not be associated with the third-party agent (in the sense of soliciting recommendations from the third-party agent) and the trusted agent (in the sense of soliciting recommendations about the trusted agent). Similarly, the only reason why the third-party agent is communicating recommendations about the trusted agent to the trusting agent is because of the reputation query issued by the trusting agent. It can be clearly seen that the reputation relationship between the third-party agent, trusting agent and the trusted agent is centred around the reputation query.

The reputation relationship is much more complex than the trust relationship in view of the following [16]:

- There is an extra agent known as the *third-party agent* involved in the relationship.
- The third-party agent engages in three major activities in the reputation process: having a trust relationship with the trusted agent, giving opinions about the trusted agent to the trusting agent, and being available for trustworthiness of opinion assessment being carried out by the trusting agent.

- The third-party agent's recommendation is the key to the quality of the reputation assessment for the trusted agent.
- The nature of the third-party agents, which we may or may not know, creates the uncertainty about the trustworthiness of their opinion in the reputation relationship.
- The method of dealing with a third-party agent's uncertainty, dishonesty or trustworthiness of opinion is a big challenge in the reputation process.

The complexity lies in the involvement of the *third-party recommendation agent*, which creates a greater complexity in the *reputation relationship* over the *trust relationship*.

In the following text, we shall reveal the inner complexity of the reputation relationship step by step.

8.4.2 Agents in the Reputation Relationship

In order to explain the reputation relationship, it is necessary to have an overview of the agents and their activities in the reputation process.

In the conceptual view of the reputation relationship in Figure 8.3, we see that there are three entities involved, namely, the trusting agent, the trusted agent and the third-party recommender. Each agent has its own duties in the reputation process, which are explained in detail in the following text.

From the trusting agent's point of view:

- The trusting agent (e.g., Alex) needs to issue a clear *reputation query* about the *trusted agent* (e.g., Liz), in the specific *context* (e.g., warehousing) and specific *aspect of context* (e.g., cold

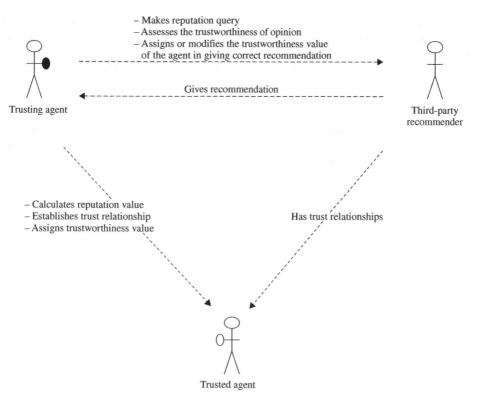

Figure 8.3   Agents and their activities in the reputation relationship

storage rental, price, discount, etc.), or even *criteria* (e.g., room condition, JIT service, auto tracking etc.) and time (e.g., Dec, 2004) to the third-party agents on the network.
- The trusting agent needs to collect as many recommendations as possible from the third-party agents who give a recommendation.
- Before the trusting agent aggregates the total recommendation value, he/she may carry out an assessment on the trustworthiness of the opinions or recommendations, and classify first-, second- and third-hand opinions for a more adequate reputation measure.
- The trusting agent needs to assign or modify and maintain a record of the trustworthiness of the recommender in giving correct opinions for a more adequate reputation measurement in the future.

From the third-party agent's point of view:

- Third-party agents give honest opinions about their experience or trust relationship with the trusted agent. They could give first-, second- or third-hand opinions.
  However, there is no guarantee that all third-party agents will give correct opinions. This is a situation that the trusting agent has to accept.

From the trusted agent's point of view:

- The trusted agents should not make recommendations about themselves. The trusted agents do not contribute to the reputation process. Therefore, they are in a vulnerable situation.

8.4.3 The Inner Relationships within Reputation Relationship

In Figure 8.3, we can see that there are clearly three distinct, independent relationships that exist: The recommendation trust relationship, giving of the recommendation or opinion and the third-party trust relationship. These are explained further (Figure 8.4).

Recommendation trust relationship establishment is a strategy from the trusting agent. The purpose is to help assess the trustworthiness of opinion that is offered by the third-party agents.

Gives recommendation or opinion is an event, and this event binds the third-party agent, their opinions and the trusting agent's reputation query into an association relationship. This shall be further explored in this chapter.

The third-party trust relationship is established between the third-party agent and the trusted agent. The trust value that the third-party agent assigns to the trusted agent becomes the recommendation value when the third-party agent makes a recommendation.

8.4.4 Three Agents and Three Inner Relationships

The three entities involved in the reputation relationships are the following:

- Trusting agent
- Third-party recommendation agents
- Trusted agent.

Three important sub-relationships are as follows:

- *Recommendation trust relationship*   (relationship between trusting agent and third-party recommendation agent to provide an opinion)
- *Third-party trust relationship*   (relationship between third-party recommendation agent and trusted agent)

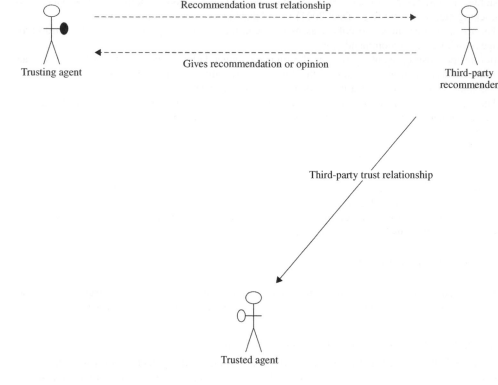

Figure 8.4   There are three inner relationships within reputation relationship

- *Reputation query relationship* (relationship between reputation query and third-party recommendation agent)

These three inner relationships form the entire reputation relationship. The three inner relationships are known as *reputation query relationship, recommendation relationship* and *third-party trust relationship* (8.5).

   In this section, we have seen how the third-party agents and their involvement complicate the reputation relationship, the interconnection between the three types of agents, and their three intertwined inner relationships. In the following sections, we shall further explore these three inner relationships in detail.

8.5 Recommendation Trust Relationship

In this section, we focus on the *recommendation trust relationship*, and its difference compared to the *trust relationship* introduced in Chapter 2.

8.5.1 Definition

The recommendation trust relationship is defined as an association between the trusting agent and the third-party recommendation agent in the context of communicating recommendations about a trusted agent in a particular context and at a given point in time.

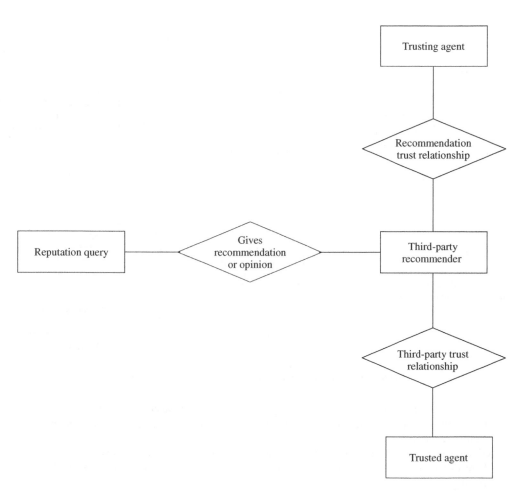

Figure 8.5 A high-level view of the reputation relationship with the three major entities and three inner relationships

The recommendation trust relationship can be regarded as an instance of a trust relationship in which the context of the relationship is 'communicating recommendations'. It is qualified by a trustworthiness value for the third-party recommendation agent – which denotes the quality of the recommendation by the third-party recommendation agent in giving a correct recommendation to the trusting agent.

8.5.2 Recommendation Trust Relationship Diagram

The layout of the recommendation relationship diagram (Figure 8.6) is similar to the trust diagram introduced in Chapter 2. However, the name of each entity in the *recommendation trust relationship* diagram is different from that in the trust diagram except for the 'trusting agent'.

In the next section, we introduce the difference between the trust diagram and the recommendation relationship diagram.

Here, we note the following:

(1) There is a M:M (many-to-many) relationship between the third-party recommendation agent and the trusting agent. A trusting agent can have many third-party recommendation agents, and

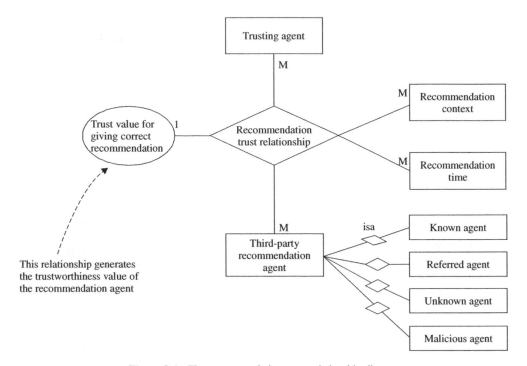

Figure 8.6 The recommendation trust relationship diagram

a third-party recommendation agent can reply to many trusting agents' calls. Each recommendation trust has a recommendation context where the trusting agent trusts the recommendation agent's recommendation.

(2) Given a particular third-party recommendation agent and a particular context and a time slot, there can be one trust value for the trustworthiness of the recommendation, assigned by the trusting agent.

(3) There is a M:M:M:M:1 (many-to-many-to-one) relationship between the trusting agent, third-party recommendation agents, recommendation contexts and time slot to the recommendation trust values.

(4) The recommendation relationship generates the trust value of the recommendation agent in giving correct opinions.

8.5.3 The Difference between Trust and Recommendation Relationship

Despite the similarity of the above diagram to the trust model described in Chapter 2, there are fundamental differences between the recommendation relationship diagram and the trust model. They are as follows:

(1) It is a recommendation relationship for which the trusting agent receives the recommendation from the third-party agents, and the context of the relationship only considers the recommendation, and nothing else. This is a major difference between the trust and the recommendation relationship.

(2) The trust value from the recommendation relationship *only* represents the trustworthiness of the third-party recommendation agent's willingness and capability to give a correct recommendation in a specified context and time slot.

(3) From the trusting agent's point of view, third-party recommendation agents are of different types namely, known agents (have a trust relationship with the trusting agent), referred agents (an agent referred by a known agent), unknown agents and malicious agents. Therefore, the reputation relationship is not the same as the trust relationship. In the trust model, the trusted agent is a known agent only, and trust refers to the trustworthiness of the trusted agent. In the reputation relationship, it refers to the trustworthiness of the recommendation or opinion.

(4) There must be a 'recommendation context' and 'time' in which the trusting agent requests a particular third-party recommendation agent's opinion on that 'context', for example, recommendations about where to buy a low-end database server, a high-end application server, JIT services, and so on.

There are two challenges in the reputation relationship :

- Determining trustworthiness of recommendation agents
- Determining trustworthiness of the opinions or recommendations.

8.5.4 Trustworthiness Value for Recommendation Agent

The recommendation relationship is important in addressing the trustworthiness of third-party recommendation agents in giving the correct recommendation. In other words, the recommendation relationship depicts whether the third-party has credibility in giving a correct recommendation.

8.6 Third-party Trust Relationship

In this section, we will formally define and elucidate the *third-party trust relationship*. Additionally, the difference between the *third-party trust relationship* and the *trust relationship* introduced in Chapter 2 is enumerated.

8.6.1 Definition

The *third-party trust relationship* is defined as the bond or association between the third-party recommendation agent and the trusted agent in the context and time specified in the reputation query.

From the perspective of the trusting agent, who initiates the whole process of reputation computation, this relationship is between the third-party recommendation agent and the trusted agent. From the perspective of the trusting agent in the reputation relationship, the trusted agent in the reputation relationship plays the role of the trusted agent in the third-party trust relationship and the third-party recommendation agent, in the reputation relationship plays the role of the trusting agent in the third-party trust relationship. Therefore, this relationship may be called as *third-party trust relationship*.

This relationship is not from the trusting agent to the trusted agent, but it is about a third-party agent involved in a trust relationship with the trusted agent. The trusting agent has no direct relationship (at the time of making the reputation query) with the trusted agent. Therefore, we call this relationship a third-party trust relationship.

This relationship is a similar to the trust relationship explained in Chapter 2, except that the trusting agent is a third-party recommendation agent.

8.6.2 Third-party Trust Relationship Diagram

Each third-party recommendation agent on the network could be a trusting agent. If they are trusting agents, they will have trusted agents. Figure 8.7 depicts this relationship.

There are two differences between this *third-party trust relationship* and the trust model described in Chapter 2, namely, the third-party recommendation agent is the trusting agent (Figure 8.7), and

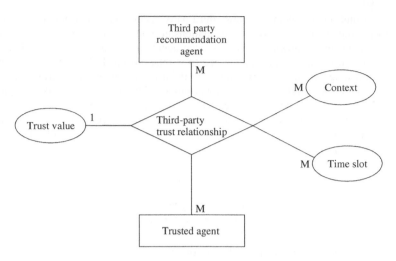

Figure 8.7   The third-party trust relationship diagram

the trusted agent is adjacent to the trusted agent (agent may imply product or service, and so on, whereas in the trust model of Chapter 2, only trusted agents were involved).

Figure 8.7 depicts the relationship between third-party recommendation agents and the trusted agent.

Here, we note the following:

(1) There is a M:M (many-to-many) relationship between the third-party recommendation agent and trusted agent. A third-party recommendation agent can have many trusted agents, and a trusted agent can be trusted by many third-party recommendation agents. Each trust relationship is context and time dependent. This is the same as in the trust model.
(2) Given a particular third-party agent and a particular trusted agent engaged in a trust relationship for a given context and in a given time slot, there can be one trust value. This is the same as in the trust model.
(3) There is a M:M:M:M:1 (many-to-many-to-one) relationship between third-party recommendation agents, trusted agents, contexts and time slots and the third-party *trust value*.
(4) The *trust value* is single valued attribute in each of the trust relationships. This is the same as in the trust model.

Not all the agents in the network have a trust relationship with other agents. Also, not all the third-party recommendation agents have a trust relationship with the trusted agent. If they do, they could voice their opinions about the trusted agent. However, if they do not, they *should not* voice their opinions, since these opinions could be misleading. We call these opinions *malicious opinions*. In Chapters 9 and 10, we shall learn how to detect *malicious opinions*.

As mentioned in the definition of reputation, a trusted agent may be an agent, a product or a service. This is shown in Figure 8.8.

8.6.3 The Difference between Trust and Recommendation Relationship
There is a close similarity between the above diagram and the trust model described in Chapter 2. However, there are some differences between this diagram and the trust model.

(1) The relationship is a third-party trust relationship, and is not with the trusting agent directly. The trusting agent in this relationship is the third-party agent.

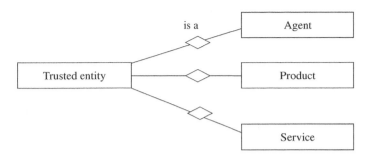

Figure 8.8 The trusted agent in Figure 8.7 could be an agent, a product or a service

(2) The trust value from the recommendation agent to the trusted agent is context and time dependent. When this value is passed on as a recommendation, it becomes a *recommendation value*.
(3) The trusted agent in reputation could be an agent, a product or a service.
(4) The *context* in the third-party trust relationship may be different from the *context* in the
recommendation relationship. There may be no relationship between these *contexts*.

8.6.4 Third-party Trust and Recommendation

Any third-party agent could be a trusting agent in a case where the third-party agent has a direct
interaction with the trusted agent. Therefore, there should be a trustworthiness value assigned by the
third-party agent to the trusted agent. However, when the third-party agent provides his/her opinion
about the trustworthiness of the trusted agent, this trustworthiness value becomes a recommendation
value.

In other words, the trustworthiness value assigned by the trusting agent to the trusted agent
becomes the recommendation or opinion of the trusted agent when the third-party agent vouches
or conveys this to other agents.

For example, Sarah and Liz have directly interacted with each other and the context of their
interaction was 'storing files'. Since Alice does not know Liz and wants to use some of the
processing power of Liz's computer, Alice asks Sarah about her perceived trustworthiness of Liz.
All Sarah can convey to Alice is the extent to which she regards that Liz can be trusted in the
context of storing of a file. She has no idea of the trustworthiness of Liz in the context of sharing
her processing power with others. Hence, she cannot advise Alice on the trustworthiness of Liz for
sharing her processing power.

8.7 Reputation Query Relationship

In this section, we introduce the *reputation query relationship* and the difference when compared
with the trust relationship introduced in Chapter 2. It is one of the key relationships in the reputation
model.

8.7.1 Definition

The *reputation query relationship* is defined as the association between the reputation query entity
and the third-party agent in the reputation relationship that signifies the request from the trusting
agent for the recommendation and the reply from the third-party agent.

In this section, we introduce the *reputation query relationship* and the difference when compared
with the trust relationship introduced in Chapter 2. It is one of the key relationships in the reputation
model.

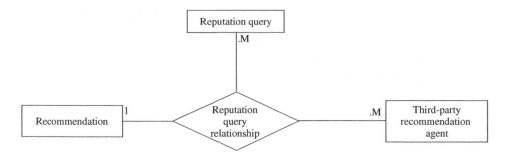

Figure 8.9   The reputation query relationship diagram

Table 8.1   Example of reputation query

Reputation query ID#	Context	Trusted agent	Trusting agent
100	Supervising PhD students	Joe	Farookh
101	Proofreading English	Sonya	Farookh
102	Thai cooking	Popi	Farookh

Unlike a trust relationship, the reputation query relationship involves and signifies the request and the exchange of the recommendation. The reputation query relationship will result in a collection of repute values that are recommended by the third-party recommendation agents, which can be aggregated by the trusting agent to form the reputation of the trusted agent.

8.7.2 Reputation Query Relationship

The reputation query relationship is represented in Figure 8.19.

A *reputation query* is represented as an entity because an 'event' can be represented as an entity. It also has multiple attributes including reputation query ID, reputation context and a time slot associated with each reputation query. Table 8.1 gives an example of the reputation query.

The recommendation is represented as an entity because it has multiple attributes, namely, the recommendation or opinion, first-, or second- or third-hand opinions, recommendation value, and so on.

Table 8.2 gives some examples of recommendations.

Recommendation is unlike a trust value, which has only a single value. The relationship represents the recommendation from the third-party recommendation agent in reply to a reputation query or the call for opinion or feedback about the trusted agent's willingness and capability to deliver QoS or QoP in a given context and time slot.

Table 8.2   Examples of recommendations

Reputation query ID #	Recommendation agent	Opinion	First-hand opinion	Second-hand opinion	Third-hand opinion	Recommendation value
100	Bob	Like it	Yes	No	No	4
100	Liz	Hate it	No	No	Yes	1
100	Leo	Not bad	No	Yes	No	3
102	Leo	Very good	Yes	No	No	5

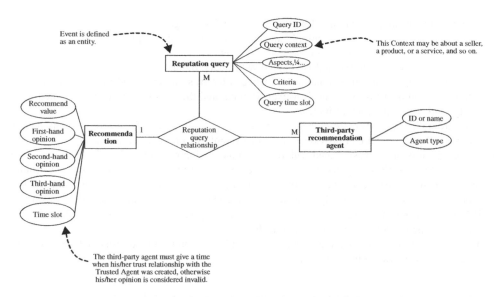

Figure 8.10 Complete view of reputation query relationship

The reputation query relationship has the following characteristics (Figure 8.10):

(1) A reputation query relationship involves third-party agents who might or might not be trust-worthy. They offer their opinions or recommendations in reply to the reputation query. This relationship is NOT a trust relationship.
(2) It collects recommendations or opinions and recommendations of trust values and these can be used to generate a reputation value for the trusted agent. Unlike the trust value that is assigned by the trusting agent itself, reputation is an aggregated value of third-party opinions or third-party recommendations.
(3) There is a M:M:1 (many-to-many-to-one) relationship between reputation query, the recommendation agents and the reputation value.
(4) In each reputation query, there may be zero-to-many (0:M) recommendation agents who are willing to give their opinions. The more the replies regarding the same reputation query context and within the time slot specified, the better the reputation value calculated.
(5) Each recommendation agent can reply to zero-to-many (0:M) reputation queries.
(6) If there is no reputation query, there will be no recommendation value or reputation generated for the trusted agent. If no recommendation agent replies to the reputation query, there is also no reputation value for the trusted agent. Therefore, recommendation as an agent may have 0..1 values, whereas reputation query and recommendation agents have 0:M cardinality in their relationships.
(7) There must be a query context and time slot specified in each reputation query.
(8) There will be an aggregation of all the recommendations and recommendation values. This aggregation is according to the time slot, from start to end time, and all the recommendation agents' replies.

8.7.3 The Difference between Trust and Reputation Query Relationship
There are fundamental differences between the reputation query relationship and the trust model described in Chapter 2:

(1) The reputation query relationship is not a trust relationship. It is a general association relation-ship. There is no trusting agent and no trusted agent.

(2) The recommendation is submitted by the recommendation agents.

(3) The recommendation contains multiple attributes. Therefore, it is represented as an entity.

(4) Reputation query is mainly about the context. However, context is a complicated term as it contains aspects and criteria, and is coupled with query ID and time slot.

(5) There is a simple association relationship between the trusting agent and the reputation query.

(6) The reputation query is an event. However, in relational theory, an event can be defined as an entity, and it has multiple attributes.

(7) The cardinality of (0:M) represents that a reputation query may have from zero-to-many replies, and a third-party agent can provide zero or many replies to reputation queries.

(8) The cardinality of (0:1) associated with the recommendation represents that, for each reputation query, with each third-party agent, there may be no reply or no opinion submitted or one opinion only. This means that, each third-party agent, if he/she gives a recommendation to a particular reputation query, he/she can only give one recommendation, or one value, not two. For example, if Liz gives a recommendation about Patricia, she can give only one opinion, such as 5 stars. However, she cannot give another value to Patricia.

(9) The *context* in the reputation query may be stored in a recommendation trust database for later evaluation of trustworthiness of opinion.

8.8 Trustworthiness of Third-party Recommendation Agents

In the introduction of this chapter, we noted that people may have doubts when using the Internet. The doubt is worsened when people have to interact or carry out a transaction with anonymous, pseudo-anonymous and non-anonymous users/agents on the Internet. How to balance the expanding use of the Internet in social and commercial environments with the perceived lack of trust is a critical question. The trigger of this difficulty is the management of the third-party agent.

In this section, we address the following questions:

- If someone is known, could he/she be untrustworthy?
- If someone is known and trustworthy, could he/she be untrustworthy in another *context*?
- If someone is unknown, could he/she be trustworthy?
- Could we assume all *unknown people* are all untrustworthy?
- How can we differentiate the unknown agent from the untrustworthy agent (known agent but not trustworthy) and malicious agents? What is the impact if we treat them as the same?
- How do we trust an unknown agent?

The above questions are the key issues for reputation technology. If they are addressed properly, they will be the key to success for reputation technology.

8.8.1 Four Types of Third-party Agents

There are four types of third-party agents from the trusting agent's point of view:

(a) *Known agents* – Agents who are known to the trusting agent and include trustworthy and untrustworthy agents. They are non-anonymous or pseudo-anonymous agents.

(b) *Referred agents* – Agents who are not known to the trusting agent directly and are only engaged as a result of references provided by known agents. They could be non-anonymous or pseudo-anonymous agents.

(c) *Unknown agents* – Agents with no referral or direct interaction with the trusting agent. They are anonymous agents.

(d) *Malicious agents* – Agents who intentionally disrupt trust relationships, businesses or services, and are discovered by the trusting agent through interaction. They could be anonymous, non-anonymous or pseudo-anonymous.

Third-party agents become important to the trusting agent if they come forward and respond to the reputation query and offer their opinions about the trusted agent, product or service. This, in turn, creates a special relationship between them, known as a *recommendation relationship*. A strong relationship implies more trustworthiness of the third-party recommendation agent in giving correct opinions. This is similar to the situation in the physical world.

Examples of third-party agents in a service-oriented network environment may include customers, service providers, buyers, sellers or online shoppers. These also include untrusted agents whom we know and who have very low trustworthiness values, or unknown agents whom we do not know, or malicious agents who intentionally make errors and give untrue feedback.

8.8.2 Eligibility to be a Recommendation Agent

Any agents in the open service network environment can give their opinions about another agent, service, service provider or product. However, when a reputation query is made, the following agents should not give their opinion or recommendation:

- Malicious agents (if they do, their opinion should be disregarded)
- Trusting agent (agent who made the reputation query)
- Trusted agent (agent whose reputation is being considered).

Figure 8.11 depicts the agents who are *not* eligible to vouch for trust values or give recommendation.

8.8.3 Known Agent as a Recommendation Agent

Known agent: In service-oriented network environments, a known agent means he/she has had a direct trust relationship, or direct interaction, or direct experience with the trusting agent, in a particular context and time and there is trust value assigned by the trusting agent. The agent could be trustworthy, partially trustworthy, or untrustworthy, and so on, and it is context dependent.

If *a known agent* gives a first-hand opinion, followed by the assessment of the trustworthiness of the recommendation, and if it is very trustworthy, the opinion is then considered as more important than any other type of opinion (Figure 8.12).

If *a known agent* gives a first-hand opinion, but the assessment shows that the trustworthiness of the recommendation is very low or the recommendation is untrustworthy, even if this is a first-hand

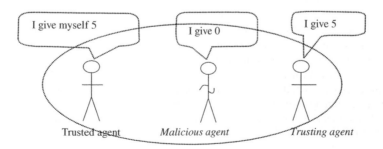

Figure 8.11 Malicious agents, the trusting agent, and the trusted agent himself/herself are not eligible to vouch for their opinions about the trusted agent

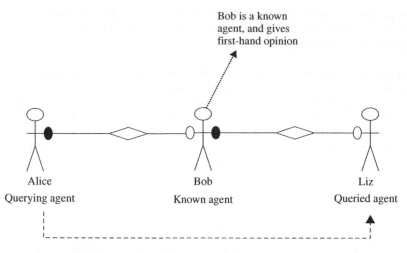

Figure 8.12 A first-hand opinion

opinion, the weight of consideration of this opinion will be reduced and the weight ratio is reflected by the trustworthiness level of the third-party recommendation agent (to provide an opinion) who provides it.

If a *known agent* gives a second- or third-hand opinion, and it is a trustworthy recommendation agent, even if this is a known and trustworthy agent, the weight of consideration of this opinion will be reduced and the weight ratio is reflected by the opinion ratio that is defined by the reputation system.

The trustworthiness of a known recommendation agent with first-hand opinion is considered as very important when aggregating all the recommendations from third-party agents. The weight assigned to this category of recommendation is user defined or can be customized and it is an organizational decision. Some reputation systems, as shown in Chapter 11, may only partially consider these.

8.8.4 Dual Relationships Associated With a Known Agent

Trustworthiness of the recommendation agent is different from the *trustworthiness of a trusted agent*. The former relates to the context giving a 'recommendation' only, but the latter may relate to many other contexts, such as different services. However, this could be easily confused. In this section, we provide a detailed analysis of these two kinds of trustworthiness.

Between a trusting agent and a known agent, a dual relationship can coexist. But they are not the same and should have minor or no influence on each other.

For example, Alice and Bob have had dual relationships for the last 10 years and have both benefited from these dual relationships (Figure 8.13).

The trusting agent could have trust relationships with many of the *recommendation agents*, who are known agents. When a trusting agent carries out a reputation query of the known agents as well as of other third-party agents, there is a *recommendation trust relationship*, not a *trust relationship*, and it has little or nothing to do with their trust relationships in other contexts. See Figure 8.14.

Figure 8.14 represents two entities with two different relationships: on the left, it is a recommendation trust relationship, and on the right, it is a normal trust relationship, which is the same as introduced in Chapter 2

Note also that the *recommendation context* and time slot are different from the *trust context* and time slot.

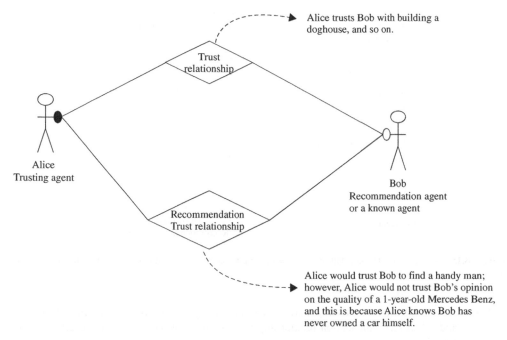

Figure 8.13 Dual relationships between the trusting agent and the recommendation agent

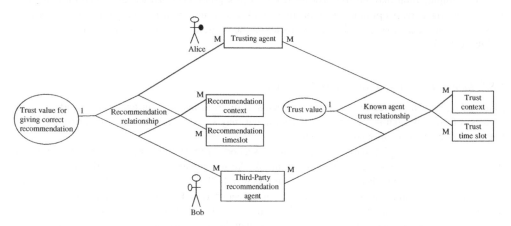

Figure 8.14 An alternative view of dual relationships with relationship details

- In the recommendation trust relationship, the *recommendation context* contains the *context of the opinion* that the recommendation agent has given, and the trust value represents the trustworthiness of the recommendation agent in giving a correct opinion or represents the credibility of the recommendation or trustworthiness of the opinion.
- In the trust relationship, the *trust context* contains the *context in which* the trusting agent holds a trust value for the trusted agent.

If a recommendation agent constantly gives correct recommendations, opinions or feedback, the *trustworthiness of the recommendation agent* is high and the *trustworthiness opinion* is rated high.

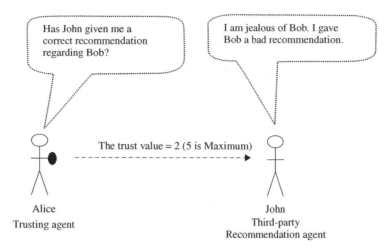

Figure 8.15   A recommendation relationship may determine the trustworthiness of the third-party recommendation agent

Note that a good *trust relationship* may or may not strengthen the *recommendation trust relationship*, and a bad recommendation relationship may or may not have a damaging effect on the trust relationship.

For example, John may be trustworthy from Alice's point view. However, Alice may not trust John in giving correct opinions, as John may think everyone could be competing with him (e.g., IT service), so he always gives bad opinions about the reputation-queried agents. This is illustrated in Figure 8.15.

8.8.5 Referred Agent as a Recommendation Agent

Referred agent: In service-oriented environments, a referred agent is an agent whose identity and trust is referred by a known agent. A referred agent may be indirectly known to the trusting agent but there is no trust relationship between them.

A referred agent's opinion may provide a first-, second- or third-hand opinion or recommendation. It is an indirect opinion or recommendation and the known agent passes it on to the trusting agent.

Referred agents and their opinions are given less weight than known trustworthy recommendation agents with first-hand opinions. The weight of consideration for the referred agent and his/her opinion also reflects the trustworthiness level of the known agent who referred the agent that provided the first- or second-hand opinion. The weight given to the referred agent and his/her opinion is user defined and it is an organizational decision.

If a referred agent provides a second-hand opinion (an indirect opinion), the weight of the opinion would be even less than that of a referred agent with first-hand opinion Figure 8.16.

If an agent is a referred *referred agent* (multiple referred agent), he/she is counted as a referred agent. For example, a trusting agent, Alice, who does not know about the trustworthiness of the reputation-queried agent Liz puts out a reputation query. Bob does not know Liz either, so he passes the reputation query to Sarah. Sarah does not know Liz either and passes the query to Joe. Joe has a trust relationship with Liz, so Joe is a referred–referred agent, but with first-hand opinion.

It is important to note that all agents on the network can vouch for their opinion. For each context, each agent should only vouch once; however, the opinion can be extended, but there cannot be multiple responses by one agent.

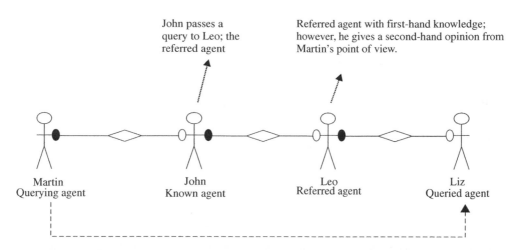

Figure 8.16 A referred agent gives second-hand opinion and their trustworthiness is transited by the known agent

8.8.6 Unknown Recommendation Agents

Unknown agent: In service-oriented environments, the unknown agent is an agent whose identity is unknown or undefined. An unknown agent may be new to the network or may have had no direct interaction with other agents.

The characteristics of the unknown agent include the following:

(a) The unknown agent may be an honest agent, or may be a dishonest agent (malicious agent).
(b) When an unknown agent joins the network or is new to an environment, by default, the trustworthiness is regarded as unknown and his/her reputation is also regarded as unknown.
(c) The trust value for the unknown agent is unknown. The unknown agent may be untrustworthy or trustworthy.
(d) They may or may not be anonymous.

By default, we should not assume that the unknown agent is an untrustworthy agent. This is especially important in the service-oriented network environment. Otherwise, there will be no business on the Internet.

In Figure 8.17, unknown agent A does not know the trusted agent Liz, untrusted agent B is a known agent but untrustworthy, and malicious and unknown agent C may wish to disrupt the assessment by giving a bad opinion about the trusted agent. Unknown Agent D knows Liz but is unknown to the trusting agent Alice. Malicious agent E knows Liz and has negative feelings toward her. Therefore, he offers a bad opinion of Liz. This example shows that unknown agents could give correct *or* wrong opinions; therefore the larger the sample for data collection, the more efficacy is achieved in the reputation assessment, as it might filter spurious opinions.

For simplicity, the unknown agent's opinion can be considered as a third-hand opinion, as there is no trustworthiness value that helps the judgement of the opinion. In some situations, a third-hand opinion or unknown agent is totally discarded in the aggregation of reputation values; however, in many other situations, such as involving political decisions, the third-hand opinions and unknown agents are considered important, however with a different weighting calculation than the first- and second-hand opinions. The weight ratio is user defined and it is an organizational decision.

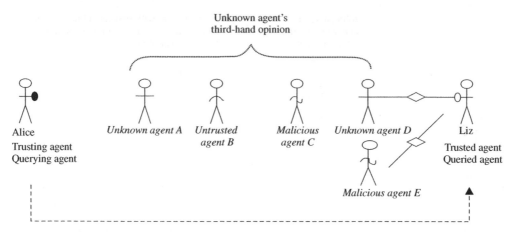

Figure 8.17 An unknown agent may give first-, second- or third-hand opinions (unknown agent may be a malicious agent or may be an honest agent)

8.8.7 Malicious Agents

Malicious agent: In service-oriented environments, the malicious agent is the agent whose identity is unknown or undefined, and such an agent is one who is discovered to be untruthful and makes intentional mistakes and errors in order to disrupt the operation or business or provides misleading opinions to misguide the community.

Malicious agents have the following characteristics:

(a) They may be unknown agents.
(b) They may be known agents who have changed their behaviour.
(c) They might be intruders and may have stolen the identities of other agents.
(d) They intentionally try to disrupt the service or business.
(e) They could be anonymous, non-anonymous or pseudo-anonymous agents.

To detect a malicious agent, one could use a well-kept recording system, through data mining, pattern matching or behaviour monitoring and JIT reporting systems.

8.8.8 Untrusted Agents

Untrusted agent: In service-oriented environments, the untrusted agent is the agent whose identity is known to the trusting agent and whose trustworthiness value is very low.

An untrusted agents is an agent with a very low trustworthiness value, and with whom the trusting agent has had experience and regards as untrustworthy.

The characteristics of an untrusted agent are as follows:

(a) An untrusted agent is a known agent with a very low trustworthiness value assigned to him/her by the trusting agent.
(b) There is a trust relationship existing; however, the relationship is not good because the trusting agent does not believe this agent has the willingness and capability to deliver the quality of recommendation in a given context and given time slot.
(c) He/she could be a non-anonymous or pseudo-anonymous agent.

8.8.9 Coalition and Collusion

Known and referred agents can form a *chain of a trust relationship*, which provides *transitive* trust between each other that could form a closed community.

In order to counter the possibility of soliciting recommendations from unknown third-party agents, those agents whom the trusting agents do not know but are referred by known agents are called *referred agents* and their opinions are taken into a different level of consideration.

This chain of a trust relationship forms a *coalition* and all agents within this coalition know each other either directly or indirectly, and they could help each other by sharing recommendations and any feedback about their experiences.

Forming a coalition could help the closed community in a positive way; however, it is important to note that some coalitions may act negatively to outsiders, exclude anyone whom they do not like or are unknown, squeeze the market, attack competitors, may work together forcing in the one direction they want to go and give unfair trading or assessments on the quality of other businesses, agents or organizations. This becomes a *collusion* that has destructive impact in the service-oriented environment and networked economy.

To address this issue, we shall carefully distinguish the type of recommendation agent, keep records of their recommendation performance, carefully examine the trustworthiness of their opinions, calibrate, validate and update their trustworthiness values, and not simply assume that known agents are all trustworthy, nor that all unknown agents are untrustworthy.

8.8.10 Trustworthiness of the Recommendation Agent

In the virtual world, we live in an open community rather than in a closed community, and the population of unknown agents is too large to ignore or treat all of them as untrustworthy or malicious. If we do that, it will result in inadequate assessment and evaluation of the quality of agents, products or service. What we need is an automated system that could tell us what to trust, how much we should trust, and what is the likely trust level in future interactions. In considering the quality assessment of reputation, *the trustworthiness of the recommendation agent in giving correct opinions* is important and very necessary.

The trustworthiness of the recommendation agent is an *overall rating* of the recommendation agent; it is an aggregated rating from the trustworthiness of the opinions. Opinion is context based, and each context is associated with a trustworthiness of opinion on that context. Therefore, trustworthiness of the recommendation agent is assessed by the trustworthiness of each of the opinions given and it is context based.

As can be seen in Table 8.3 the top field represents the trustworthiness of the recommendation agent. It is an average of all the values for trustworthiness of opinions. On the left, the recommendation context defined by the trusting agent is listed; in the middle, Patricia's opinion is listed; and on the right, we see the evaluation of the trustworthiness of each opinion that the recommendation agent provided for the trusting agent. This example is used to demonstrate the difference between the *trustworthiness of recommendation agent* and *trustworthiness of opinion*.

The trustworthiness of opinion is synonymous with the trustworthiness of recommendation. Key concepts that must be considered together are 'recommendation agent' and 'recommendation or opinion'. We must be clear that we consider nothing else but the '*recommendation* or *opinion*' that the '*recommendation agent*' made. As explained in the previous section, there is little or no influence from other trust relationships that the parties hold.

8.9 Trustworthiness of the Opinion

From previous sections, we come to to the following conclusions:

- A trustworthy agent may not give a correct recommendation/opinion.
- A trustworthy agent has no credibility to give a correct recommendation.

Table 8.3 An example of trustworthiness of the recommendation agent

Recommendation agent: Patricia
Trustworthiness in giving correct recommendation: 4.5 (5 is maximum) ★ ★ ★ ★ ⯪

Recommendation context (from a reputation query)	Opinion or recommendation (from Patricia, the recommendation agent)	Trustworthiness of opinion Evaluation by trusting agent
Fishing in Perth	Fishing in April and May is good	★ ★ ★ ★ ⯪
Digital cameras	The battery life of Cannon S230 is very short. Recommend 2 out of 5	★ ★

- A first-hand opinion may not be trustworthy.
- An unknown agent may not always be untrustworthy or malicious.

In this section, we shall explain how the *trustworthiness of opinion* may be examined.

8.9.1 Credibility in Giving a Correct Opinion

The credibility of the agent in giving a correct opinion represents the trustworthiness of opinion.

A trustworthy agent for some other *context* may not give a correct recommendation, nor has the credibility to give the right recommendation and a first-hand opinion may not be trustworthy. For example, Alice and Bob have had a very good trust relationship for a long time, and Alice would lend her car or house to Bob, as Bob does not have a car or a house of his own. Bob's opinion on Alice's query would be considered as a first-hand opinion. However, when Alice wants to buy a Mercedes Benz, the first-hand opinion from Bob is not worth very much. Since Bob never owned a car, how could he possibly have a valid opinion about purchasing a Mercedes? In this case, Bob is a trustworthy agent; however, he has an inability to give a valid recommendation about purchasing a Mercedes (Figure 8.18).

In order to handle the trustworthiness of the opinion, we not only consider whether the opinion is first-, second- or third-hand, but we also consider the trustworthiness of each opinion in a particular context. We consider this issue by noting that we trust the opinion for a specific context and a specific time slot.

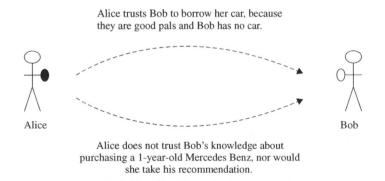

Alice trusts Bob to borrow her car, because they are good pals and Bob has no car.

Alice   Bob

Alice does not trust Bob's knowledge about purchasing a 1-year-old Mercedes Benz, nor would she take his recommendation.

Figure 8.18 By correlating the dual relationship, Alice can conclude Bob's credibility in giving correct opinion

8.9.2 Determining the Trustworthiness of Opinion

There are at least three ways to help determine the trustworthiness of the opinion:

- By correlating the dual relationships that coexist between the trusting agent and the recommendation agent, one could determine the credibility of the recommendation agent in giving a quality or trustworthy opinion.
- By keeping historical records of past recommendations for the recommendation context, and using the history to determine the trustworthiness of the opinion.
- Through the use of a scientific calculation to update the trustworthiness of the opinion, as it is compared with the actual trustworthiness obtained for the trusted agent for the interaction.

The trusting agent will consider the aggregated reputation about the trusted agent and use it as an initial trust value to carry out an interaction or a business deal. After this interaction, the trusting agent would derive his/her own trustworthiness value about the trusted agent. He/she then uses this value to evaluate how much difference there is between what the recommendation agents have recommended. He/she then reassigns the trustworthiness of the recommendation agents in giving a correct recommendation.

8.9.3 Dual Relationships

From the dual relationship diagram presented earlier, we note the following (Figure 8.19):

- Each reputation query has a context and a time slot.
- Each opinion or recommendation has a context and a time slot.
- Each trust relationship has a context and a time slot.

One way to judge the credibility of an agent in giving a correct opinion or recommendation is to correlate the *context* for which we trust the recommendation or the *context* for which the party had a trust relationship.

Let us consider another example. ABC Logistics Company is famous for its efficiency in local delivery but has very poor warehousing services. Suppose that DEF Warehouse wants a recommendation regarding another warehousing service in the city. Even though DEF Warehouse uses a lot of ABC's local delivery service and ABC's opinion counts as a first-hand opinion, DEF Warehouse would not trust the recommendation regarding another good warehousing service, from ABC Logistics.

Credibility in giving a good opinion is related to the domain expertise.

8.9.4 Recommendation Time Consideration

Let us consider an example. Charlie tells Toby that a Perth wedding reception hall is very good, as the offer includes many fantastic services relating to entertainment and food, and he was there when his uncle was getting married. But Charlie's uncle was married 10 years ago and the management has changed. Therefore, the time slot plays a key role in assessing the trustworthiness of the recommendation.

The above examples show that the trustworthiness of the opinion is different from the trustworthiness of an Agent. It is heavily contextual and time slot dependent.

The above examples also show that the trustworthiness of an opinion is judged by correlating the dual relationships, namely, the trust relationship and the recommendation trust relationship. The judgment is based on the *context and the time slot* in the *trust relationship* as well as the *context and time slot* in the *recommendation trust relationship*.

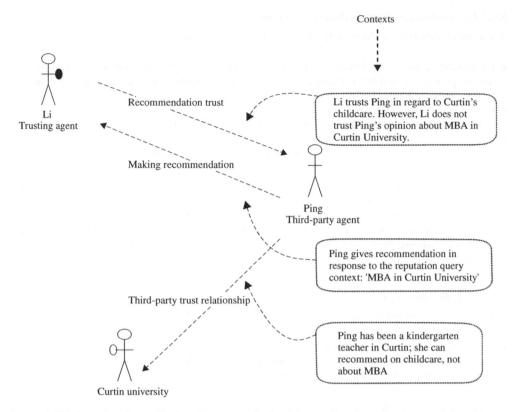

Figure 8.19 Assessing the credibility of recommender in giving quality of opinion by correlating the past recommendation record or database information

8.9.5 Utilize Recommendation History

To assess the credibility of the agent, we rely on past experiences or a record either from the recommendation relationship or trust relationship and context, to evaluate the *credibility* of the Agent in giving a correct opinion.

For example, a reputation query context is about an MBA Degree in Curtin University (trusted agent). The recommendation relationship that a trusting agent (e.g., Ms Li) has with the recommendation agents (e.g., Ms Ping) in the past was about childcare, and the third-party trust relationship that the recommendation agent (Ms Ping) has with the trusted agent (Curtin University) was childcare (Figure 8.19).

8.9.6 Scientific Methods

There are several well-known aggregation methods that could be applied to help determine the trustworthiness of opinion, such as employing a deterministic approach, utilizing the Bayesian approach, using a fuzzy systems approach (to be introduced in Chapter 10) or the Markov model, and so on (Chapter 12). Note that these methods were primarily used to aggregate the recommendation value for the reputation-queried agent (trusted agent). These methods also have a technique for progressively updating and modifying the trustworthiness of opinion using the error between the agent's recommendation and the actual trustworthiness after the interaction as the basis. These are explained in Chapters 10 and 12.

8.9.7 Aggregation of Recommendation to Generate Reputation Value

After understanding the different types of third-party recommendation agents and what they can do, we come to know that there are four key factors influencing the quality of the assessment of the reputation of the trusted agent:

(1) Trustworthiness of the opinion (take into consideration known and unknown Agents etc.)
(2) First-, second- and third-hand opinion
(3) Dynamic factors (reputation can increase or decrease as time passes)
(4) The aggregation method used

The reputation is an aggregation of the recommendations from all of the third-party recommendation agents. The reputation takes the above four key factors into consideration. The generated reputation value is an aggregated value that represents the total cohort of third-party agents' beliefs of the trustworthiness of the trusted agent or reputation-queried agent.

8.9.8 The Aggregation Within the Time slot

Note 1: Reputation is an aggregation of recommendations (Figure 8.20). Suppose Alice issues a reputation query for Bob's skills as a salesman over the last 5 years. A third-party recommendation agent, Sonya, may give the reply 'I knew Bob in 2001–2002, he sold $20 million dollars of goods and was an excellent salesman'. Jo may reply 'I know Bob's selling skills have been good in the last 6 months, July 2005–Dec 2005'. The time slots within which Sonya and Jo knew about Bob's skills as a salesman are within the last 5 years, and reflect *third-party trust relationship* time slots. These time slots may be different from the *reputation query* time slots but they are within the time slot of the reputation query.

Note 2: If an agent, Sonya, had an interaction with Bob between 2001 and 2002, she will only reply by giving one value. However, if she had two interactions with Bob, one between 2001 and 2002 and the other in 2004, she should only reply by giving one value (i.e., her average trust value of Bob), and should not give two values.

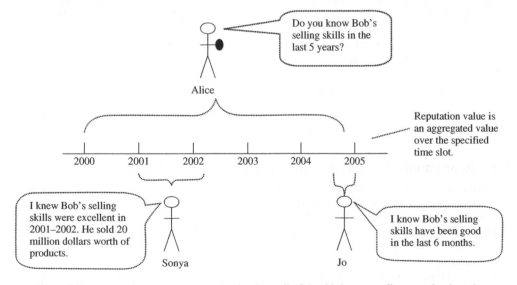

Figure 8.20 Reputation is an aggregated value from all of the third-party replies over the time slot

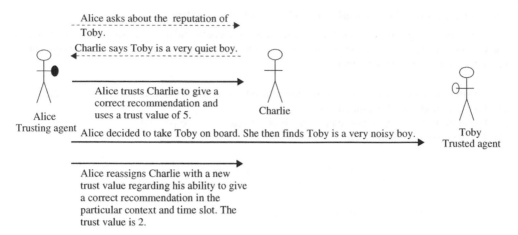

Figure 8.21   Calibrating the trust value for the recommendation agent to give a correct recommendation

The agent's *reputation query* contains the specification of what trusting agents are querying about. If it is about the quality of service of a trusted agent, it should contain a clear specification of the context (a context may contain several aspects) and a specified time slot. Figure 8.10 shows the essential attributes.

8.9.9 Updating the Trustworthiness of the Recommendation Agent

The opinion or recommendation from a third-party recommendation agent might be correct or incorrect. The trusting agent will know this after a direct interaction with the trusted agent. The trusting agent will then validate the trustworthiness value of the recommendation agent.

This process can be called the *calibration* of the recommendation or opinion. Only then will the agent know whether the recommendation was correct or not. This is then used to assign a trust value to the third-party recommendation agent. The trust value in the recommendation relationship is reflected by the *trustworthiness of the third-party recommendation agent in giving the correct recommendation*.

The trustworthiness of recommendation agents represents the trustworthiness of the opinions or recommendation and vice versa.

8.9.10 Process of Reassigning the Trustworthiness of Opinion to the Recommendation Agent

There is a four-step process when considering the trustworthiness of the recommendation agent in giving a correct opinion. This is demonstrated in Figure 8.21.

8.10 Reputation Model and Reputation Relationship Diagrams

The reputation relationship is built around the third-party recommendation agents who give recommendations or opinions to the trusting agent about the opinion of the trusted agent who are involved in the three relationships: reputation query relationship, recommendation relationship and third-party trust relationship. Their trustworthiness is assessed by the trusting agent on the basis of the willingness and capability to give the right information or correct recommendation in a given context and time slot.

In this section, we give a comprehensive overview of the reputation relationship and reputation model.

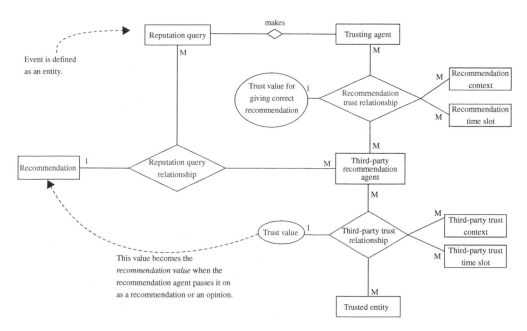

Figure 8.22 A high-level view of the reputation model

8.10.1 High-level View of Reputation Relationship Model

It is important to note the fundamentals of the reputation relationship, which are as follows (Figure 8.23):

a. It is built around third-party recommendation agents.
b. All three relationships involve context and time slots. Note that these *contexts* may or may not be the same in each of the relationships. Each context serves its own purpose in its own relationships.
c. Recommendation agents provide recommendations or opinions about the trustworthiness of the trusted agent.
d. Recommendation agents are involved in three relationships: the reputation query relationship, the recommendation relationship and the third-party trust relationship.
e. The trustworthiness of a recommendation agent's recommendation or opinion is assessed by the trusting agent on the basis of the willingness and capability to give the right information or correct opinion in a given recommendation context and time slot. This is represented by the recommendation relationship.

The top right-hand side of the diagram in Figure 8.22 contains the *recommendation relationship*. The bottom half of the diagram contains the *third-party trust relationship* and the left-hand side contains the *reputation query relationship*.

Another important feature of the reputation relationship is the type of the trustworthiness of recommendation agents. Note that the trustworthiness of the recommendation agent has nothing to do with whether the relationship is a known relationship, an unknown relationship, and so on, nor does it depend on whether the opinion is a first-, second- or third-hand opinion. However, they all have influence in the aggregation of the *final* reputation value.

In many existing recommendation systems, there is no distinction between first-, second- and third-hand opinions. There is only a minor consideration of the trustworthiness of the recommendation agent and how the trustworthiness of the recommendation agent is to be to updated

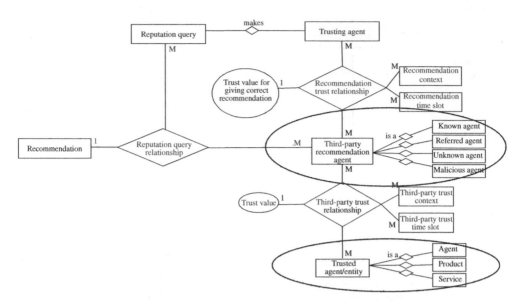

Figure 8.23   The entity level view of the reputation relationship (see circled areas)

or *calibrated*. However, as the trend of e-business develops, this advanced consideration will be crucial to the quality assessment and evaluation in the service-oriented environment.

8.10.2 Full Entity View in Reputation Relationship

In Figure 8.23, the complete entities that are involved in the reputation relationship are displayed in the circled parts. It highlights that the trustworthiness of the recommendation agent is inherited by all the sub-class entities. However, for the simplicity of calculations, we normally consider only the known agent's trustworthiness value, and not those of others.

In view of Figure 8.23, we could convert the *reputation relationship* to tabulated forms by using the entity–relationship concepts (high-level only), as shown in Tables 8.4–8.6

8.10.3 Attribute-level View of Reputation Relationship

In Figure 8.24, we give an overview of all the necessary attributes that are part of each major entity (see the circled parts)

8.10.4 Dual relationship View in Reputation

In Figure 8.25, we display the dual relationships that could exist in the reputation relationship. The circled part distinguishes that the recommendation trust relationship is a different trust relationship compared to a normal trust relationship.

Table 8.4   Sample of reputation query

Reputation query ID#	Context	Context time	Trusting agent
100	About Jo's mathematics ability	2004–2005	Mark
101	About Sonya's English skills	2004–2005	Mark
102	About HP Compaq desk-top PC	2004–2005	Toby

Table 8.5 Sample of recommendation

Reputation query ID #	Recommendation agent or ID #	Recommendation from recommendation agent	Time slot	First-, second-, third-opinion	Trustworthiness of recommendation agent
100	Faro	5	Oct 2004	First	4
100	Liz	5	Jan 2005	First	5
100	Leo	3	Feb 2005	Second	2
102	Liz	4	Mar 2005	First	5

Table 8.6 Sample of recommendation agents

Recommendation agent or ID #	Trusted agents ID #	Recommendation context	Recommendation time	Trustworthiness of opinion
Faro	A1000	Computing	2004	5
Liz	A1000	IT	2003	4
Liz	A1000	Project Management	2004	5
Leo	B1020	Logistics	2005	2

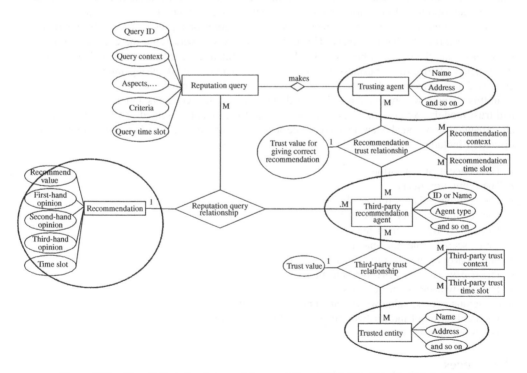

Figure 8.24 The attribute-level view of the reputation relationship (see the circled areas)

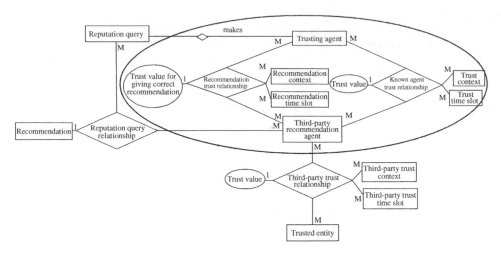

Figure 8.25   The view of dual relationship in the reputation relationship (see circled areas)

8.11 Conclusion

Reputation is a powerful tool that needs to be understood in distributed service-oriented environments and networks. This is because most people may have some doubts when using the Internet. This doubt is worsened when people have to interact or carry out a transaction with anonymous, pseudo-anonymous and non-anonymous users/agents over the Internet. If people do not trust the Internet or network, trying to make them trust an interaction or transaction over the Internet is a much more difficult thing to achieve. However, we are now experiencing the enormous impact of the Internet and the benefit it brings to us. How can we balance the dilemma of needing to embrace the benefits of electronic business environments, yet maintain trust during interactions within a context that is sometimes fraught with security and privacy concerns?

In this chapter, we presented and described the following key concepts and terms related to trust and trust models. We also elaborated on the original trust model presented in Chapter 2 to include an expanded conceptual view to incorporate a relationship model. The key issues that we have discussed include the following:

- The concept of reputation
- The reputation query
- The recommendation relationship
- The reputation query relationship
- Third-party trust relationships
- First-, second- and third-hand opinions
- Third-party agents, recommendation agents
- The trustworthiness of recommendation agents
- The trustworthiness of opinions
- Malicious agents, unknown Agents, untrusted agents
- The calibration of the trustworthiness of the recommendation agents
- The reputation model

References

[1] Celentani M., Fudenberg D., Levine D.K. & Pesendorfer W. (1966) '*Maintaining a Reputation Against a Long-lived Opponent*', Available: [http://citeseer.nj.nec.com/cache/papers/cs/25188/http:zSzzSzlevine. sscnet.ucla.eduzSzpaperszSzedit9.pdf/celentani66maintaining.pdf] (10/09/2003).

[2] Marimon R., Nicolini J.P. & Teles P. (2000) *'Competition and Reputation'*, Available: [http://www.utdt. edu/departamentos/economia/pdf-wp/WP002.pdf] (2004).

[3] Bromley D.B. (1993) *'Reputation, image and impression management'*, John Wiley & Sons, Chichester.

[4] Abelson I. (1970) *'Persuasion, How Opinion and Attitudes are Changed'*, Crosby Lockwood & Son, London.

[5] Esfandiari B. & Chandrasekaran S. (2003) *'On How Agents Make Friends: Mechanisms for Trust Acquisition'*, Available: [http://citeseer.nj.nec.com/cache/papers/cs/26840/http:zSzzSzwww.sce.carleton.cazSznet-managezSzpaperszSztrustworkshop.pdf/on-how-Agents-make.pdf] (10/10/2003).

[6] Yu B. & Singh M.P. (2003) *'An Evidential Model of Distributed Reputation Management'*, Available: [http://www-2.cs.cmu.edu/~byu/papers/p406-yu.pdf] (99/09/2003).

[7] Sabater J. & Sierra C. (2003) *'REGRET: A Reputation Model for Gregarious Societies'*, Available: [http:// citeseer.nj.nec.com/cache/papers/cs/22333/http:zSzzSzwww.iiia.csic.eszSzReportszSz2000zSz2000-06. pdf/sabater00regret.pdf] (15/09/2003).

[8] Pujol J.M., Sanguesa R. & Delgado J. (2003) *'Extracting Reputation in Multi Agent Systems by Means of Social Network Topology'*, Available: [http://citeseer.nj.nec.com/pujol02extracting.html] (09/10/2003).

[9] Abdul-Rahman A. & Hailes S. (2000) *'Supporting Trust in Virtual Communities'*, Available:[http://citeseer. nj.nec.com/cache/papers/cs/10496/http:zSzzSzwww-dept.cs.ucl.ac.ukzSzcgi-binzSzstaffzSzF.AbdulRah-manzSzpapers.plzQzhicss33.pdf/abdul-rahman00supporting.pdf] (30/11/2003).

[10] Cornelli F., Damiani E., di Vemarcati S., Paraboschi S. & Samarati P. (2003) *'Choosing Reputable Servants in a P2P Network'*, Available: [http://citeseer.nj.nec.com/cache/papers/cs/26951/http:zSzzSzseclab. crema.unimi.itzSzPaperszSzwww02.pdf/choosing-reputable-servents-in.pdf] (10/10/2003).

[11] Aberer K. & Despotovic Z. (2003) *'Managing trust in an Agent-2-Agent Information System'*, Available: [http://citeseer.nj.nec.com/cache/papers/cs/26315/http:zSzzSzwww.p-grid.orgzSzPaperszSzCIKM2001. pdf/aberer01managing.pdf] (10/10/2003).

[12] Xiong L. & Liu L. (2002) *'Building Trust in Decentralized Agent-to-Agent Electronic Communities'*, Available: [http://citeseer.nj.nec.com/cache/papers/cs/26940/http:zSzzSzdisl.cc.gatech.eduzSzAgentTrustz-SzpubzSzxiong02building.pdf/xiong02building.pdf] (10/04/2003).

[13] Lee S., Sherwood R. & Bhatacharjee B. (2003) *'Cooperative Agent groups in NICE'*, Available: [http:// citeseer.nj.nec.com/cache/papers/cs/27074/http:zSzzSzwww.ieee infocom.orgzSz2003zSzpaperszSz31_03. PDF/lee03cooperative.pdf] (30/10/2003).

[14] Mui L., Mohtashemi M. & Halberstadt A. (2002) *'A computational Model of Trust and Reputation'*, Available: [http://www.cnn.com/2002/WORLD/europe/10/04/world.cities/] (10/10/2003).

[15] Miztal B. (1996) *'Trust in Modern Societies'*, Polity Press, Cambridge, MA.

[16] Hussain F.K., Chang E. & Dillon T.S. (2004) 'State of art in trust in Agent-to-Agent communication', *Proceedings of the International Conference on Computer Applications*, Mynamar, pp. 99–104.

9

Reputation Ontology

9.1 Introduction

The growing development in trust and reputation technologies and systems in the 21st century has powerful social and economic implications for every organization and business entity. In the distributed networked economy, they will potentially bring an open, secure and transparent quality assessment and assurance for customers. A reputation system can offer recommendations to buyers or end-users for many kinds of trustworthiness evaluation from online or offline businesses (such as companies, retailers, public or private organizations, etc.), service providers (hotels, restaurants, bars, entertainments, city tours, logistics, etc.), brokers (dot.com sites, travel agents, service brokers, Yellow Pages, posters, etc.), to many kinds of commercial products and industrial suppliers as well as individual agents (such as buyers, customers, end-users, browsers, game players, etc.).

In the previous chapters, we introduced the concepts of reputation and the reputation relationship. We have given a great deal of attention to the differences between trust and reputation. One of the key issues in the discussion of reputation is that of the third-party recommendation agents and the complicated relationships generated among all the agents involved. We next explained what they could do and how we can either trust or distrust them and their opinions.

In this chapter, we introduce reputation ontologies. The reputation ontology is much larger and more complex than the trust ontology. We suggest that before studying this chapter, the reader should have already understood Chapters 4 and 8.

As explained in Chapter 4, an ontology represents a set of concepts that are commonly shared and agreed to by all parties in a particular domain. In this chapter, we shall introduce the reputation ontology and the ontological manifestation of inner relationships known as the *ontology for trust recommendation*, third-party trust and the reputation query. We apply this generic set of ontologies to the specific ontology known as the *agent reputation ontology* and service and product reputation ontologies.

We provide a brief review of reputation ontology databases, and later in the chapter, we present seven level scales used for the reputation measure. At the end of the chapter, we introduce and present the fuzzy and dynamic nature of reputation.

9.2 Reputation Ontology

The most current reputation technologies are still in their infancy. Many organizational trust and reputation systems use simple measures and mechanisms. However, despite using simple versions of reputation measures, the literature study has shown that these basic trust and reputation technologies have already been able to generate sufficient trust among the buyers and suppliers, generate substantial interest among online users as well as facilitate success in business volumes [1].

Trust and Reputation for Service-Oriented Environments   Elizabeth Chang, Tharam Dillon and Farookh Hussain
© 2006 John Wiley & Sons, Ltd

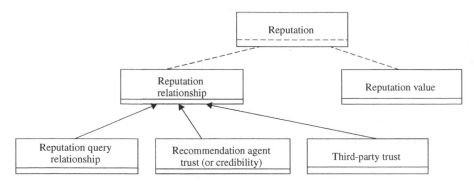

Figure 9.1   Conceptualization of the concept 'reputation' and its three inner relationships

For example, the well-known successful online web business eBay is currently in the process of updating its trust and reputation systems, as well as other business intelligence tools, by teaming up with epinions.com.

In this section, we represent the ontology on two levels; namely, a simple reputation ontology, for which a simple reputation measure is currently widely used and an advanced reputation ontology. An advanced reputation measure is the next generation of trust and reputation systems that will eventually be adopted in the service-oriented environment.

9.2.1 High-level Reputation Ontology

From Chapter 8, we learnt that reputation is '*an aggregation of recommendations from the third-party recommendation agents, in response to the trusting agent's reputation query for the quality of the trusted agent*'.

An advanced definition of reputation would be '*an aggregation of recommendations from first-, second- and third-hand opinions as well as the trustworthiness of the third-party recommendation agent to provide a recommendation opinion in response to the trusting agent's reputation query for the quality of the trusted agent or service or product*'.

It is important to appreciate and understand the overall manifestation of the reputation ontology from concept to application. To appreciate this manifestation fully, in a service-oriented environment we define that a *high-level reputation ontology* is the conceptualization of a high-level view of reputation and its three inner relationships. A diagrammatic representation of this high-level ontology is given in Figure 9.1.

In Figure 9.1, we see that the reputation relationship is composed of three inner relationships, namely, reputation query, recommendation trust for opinion and third-party trust.

- On the upper level, a defined concept of reputation and reputation of a trusted agent is generated though reputation relationship and aggregation of reputation values.
- On the top left, a reputation relationship involves three inner relationships that together form the reputation relationship.
- The reputation relationship is very complex. It contains a trust relationship (known as *third-party trust*) and a recommendation agent trust relationship to provide an opinion and reputation query; see Figure 9.2.

Regarding the reputation relationship, the following may be noted:

- For the trusted agent, there are two trusting agents. One is the recommendation agent and the other one is the forthcoming trusting agent who is currently seeking a reference or reputation inquiry about the trusted agent.

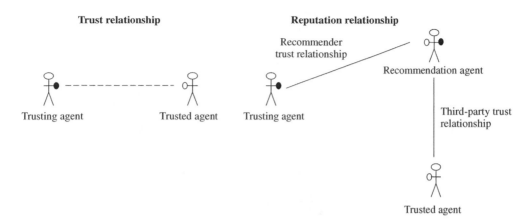

Figure 9.2 A comparison between trust relationships and reputation relationships

- For the recommendation agent, he or she is the trusting agent in regard to the trusted agent, and is the trusted agent from trusting agent's (reputation query agent) point of view for providing an opinion.
- The solid line represents the trust relationships and dotted line represents an 'intention to form a trust relationship'.

The three inner relationships of the reputation relationship are described in further detail in Figure 9.3 for the reputation query relationship, in Figure 9.4 for the recommendation trust relationship and in Figure 9.5 for the third-party trust relationship.

9.2.2 The Ontological View of Reputation Query

Before we start ontology presentation, let us review the ontology presentation notations introduced in Chapter 4. There are five key ontology notations and they are listed in Table 9.1.

Reputation Query Ontology: In a service-oriented environment, the reputation query ontology is defined as *the conceptualization of the reputation query relationship,* that is, the association of the

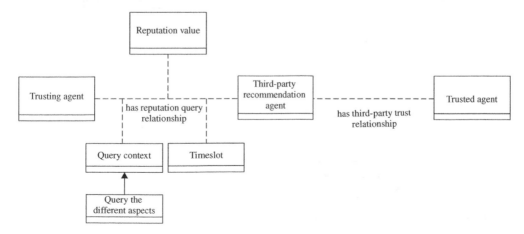

Figure 9.3 Reputation query ontology concept hierarchy (one of the three inner relationships of reputation)

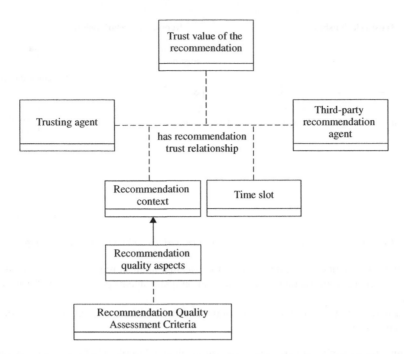

Figure 9.4   Recommendation agent trust ontology concept (one of the three inner relationships of reputation)

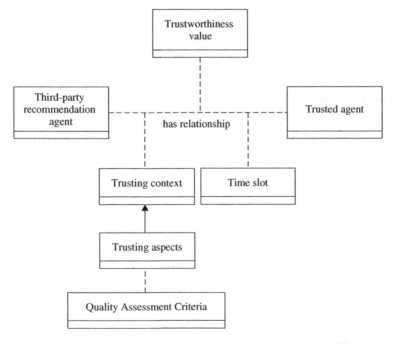

Figure 9.5   Third-party trust ontology concept (one of the three inner relationships of reputation)

Table 9.1 Ontology presentation notations

Ontology notation	Semantics of the notation
<<Concept>>	Double-field box represents the *ontological concept*.
(circle-line symbol)	Circle-line represents the instance of ontological concept.
- - - - - - - - - -	A dotted line represents ontology concept relation as M:M relationship, that is, a concept is closely related to another concept.
——————→	A line with a solid arrow represents *the upper–lower layer concept or 'generic and specific' concept.*
——————⇒	A line with an open arrow represents *the 'upper–lower part of ' concept.*

trusting agent and the third-party agent that signifies *a request* or *enquiry* about the quality of the trusted agent in a particular context and time slot from the trusting agent and the recommendations from the third-party agents.

The ontological representation of the concept 'reputation query' and its relationship with other concepts is given in Figure 9.3.

In the reputation query ontology concept hierarchy diagram (Figure 9.3), we see that there is a similarity between this and the trust ontology. However, this is only at the hierarchy level, and not at the semantic level.

As in the top left of the diagram, a reputation contains a trusting agent, a reputation requesting agent (not compulsory), and other compulsory components, such as a trusted agent, whose reputation is provided by a third-party recommendation agent, who responds to the reputation query, and the query context and time slot of the reputation inquiry.

The context in a reputation query for an example such as 'the management skill' refers to the quality aspects such as the control of budget, handling pressure, meeting deadlines, team building, and dealing with politics. These aspects are the basis on which the criteria for the quality measures are formed. Similar to the trust ontology, each aspect can have one or more criteria that require measurement. Furthermore, you could specify for the time slot that you require, for example, 'only the last 5 years experience is required'. Any time space or period contains any number of time slots and each time slot contains a number of time spots, and so on.

The concept of a trusted agent in this case is the same as that in the trust concept introduced in Chapter 2 and in the trust ontology introduced in Chapter 4.

Note that there is no 'quality assessment' concept involved in the concept of 'reputation query', because the concept of 'reputation query' generates a 'reputation value', not 'trust' or 'trustworthiness' value.

The most important concept in reputation is that of the third-party recommendation agents. These are also known by the term recommenders. They could be known agents, unknown agents or malicious agents. Trustworthiness is high if an agent is known. However, a known agent may also be untrustworthy. A trustworthy agent may also give an incorrect recommendation, which is called *an unintentional error*. A known agent may also give intentional errors, such as purposely damaging the trusted agent or reputation queried agent, and so on. A first-hand opinion is always more important than second- and third-hand opinions. However, in anonymous networks, this is hard to capture, so this feature has not been fully implemented in current technology. If it is, it will surely increase the quality of the assessment and trust and reputation measurement. The trustworthiness of the recommendation agent is derived through the recommendation relationship.

9.2.3 The Ontological View of Recommendation Agent Trust

Recommendation agent trust ontology: In a service-oriented environment, the *recommendation agent trust* ontology is defined as *the conceptualization* of the *trust* involved in the *recommendation trust relationship* which is the association between the trusting agent and the third-party recommendation agent in the context of the quality of recommendation on a particular context and time slot. This ontological representation of the concept 'recommendation trust' and its relationship with other concepts are given in Figure 9.4.

From the recommendation agent trust ontology concept hierarchy diagram (Figure 9.4), we note the following:

The recommendation agent trust is represented by the recommendation relationship and a trust value for the recommendation, which is assigned by the trusting agent. This ontological presentation is quite similar in structure to the trust diagram. The two major differences are as follows:

- The third-party recommender in this diagram is the trusted agent. The underlying meaning is similar to that of the trusted agent. However, here, the only concern is regarding the credibility to give a correct recommendation. The trusting agent is interested only in this context and not in any other context or aspects.
- The other difference is that this relationship also generates the trust value for the opinion. The opinion is based on the context. The context specifies the opinion content. In most cases, a trustworthy agent would give a trustworthy opinion. However, for a detailed quality assessment, we make the clear distinction that a trustworthy agent may not give a correct recommendation, whether it is intentional or unintentional. However, this may create challenges in the implementation of technology.

In the middle of the ontological hierarchy, we also see that a context can be represented by aspects and an aspect can be represented by criteria. The criteria are used to measure the quality of the agent (the trustworthiness of the agent) and his/her opinion (the trustworthiness of the opinion). The assessment methods can be correlations or regressions. This is introduced in Chapters 10 and 11.

Note that, after the assessment, the trustworthiness value and the trustworthiness of opinion are reassigned to the recommendation by the trusting agent. In many currently used trust and reputation technologies, the implementation of these features has been minimal because of specific implementation limitations. These limitations are generally evident in applied business situations and will not be elaborated upon here.

9.2.4 The Ontological View of Third-party Trust

From the diagram in Figure 9.5 of the third-party trust ontology concept hierarchy, we see there is a similarity in structure to the trust ontology concept hierarchy. This is correct. The only difference is that the name of the trusting agent is changed to the third-party recommendation agent or recommender.

Third-party Trust Ontology: In a service-oriented environment, the third-party trust ontology is defined as *the conceptualization* of the *trust* in the *third-party trust relationships*. This represents the trust between the third-party recommendation agent and the trusted agent and it is context and time dependence. An ontological representation of the concept 'third-party trust' and its relationship with other concepts is given in Figure 9.5.

Note that in the above definition, the trust context at a time slot may or may not be the same as specified in the reputation query.

9.3 Basic and Advanced Reputation Ontology

It is important to understand the fundamentals of the basic reputation ontology and the advanced reputation ontology. This includes an understanding of the reputation of a trusted agent and how it is aggregated in both ontologies. It also requires us to understand the significant differences using statistical methods in reputation ontology.

9.3.1 Basic Reputation Ontology

Reputation is about developing the measure of trustworthiness from third-party agents' recommendations and not by the trusting agents' own interactions and opinions. This is because the trusted agent is unknown to the trusting agent.

Basic Reputation Ontology: In a service-oriented environment, the basic reputation ontology of an agent or a service or a product is defined as the conceptualization of the reputation of a trusted agent, which is determined by simple aggregation methods of recommendations or opinions that are recommended or provided by the third-party recommendation agents in response to a particular reputation query.

The reputation value is known as the *reputation of the trusted agent*. It is an aggregated trust value obtained from all the recommendation agents who responded to a reputation query.

There are several methods used to aggregate the feedback that will be introduced in the next chapter. However, we will assume a key premise in the basic reputation of the trusted agent as outlined below:

Basic Reputation of an Agent $= \bigcup$ (Recommendation Value)

where we define $\bigcup$ as an operator for combining the recommendation value.

A graphical view of the basic reputation ontology is shown in Figure 9.6 through the use of ontology presentation notation.

Table 9.2 is a formula table for the basic reputation ontology.

9.3.2 Advanced Reputation Ontology

Advanced reputation measurement methodologies utilize more sophisticated statistical methods that impact on the accuracy of any reputation measure, thus influencing the quality and moral hazards of service-oriented environments.

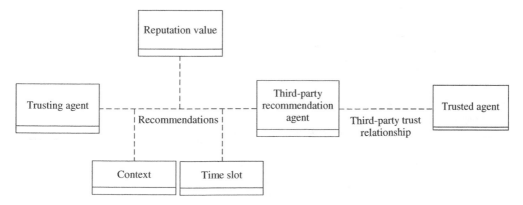

Figure 9.6   Ontology representation of basic reputation concept

Table 9.2   Formal axiom table of the basic reputation ontology

Formula name	Formula of basic reputation value
Concept	Reputation value
Inferred attribute	Basic value
Formula	Basic value = ∪(recommendation value)
Description	Basic reputation value of the trusted agent
Variable	Recommendation value
Ad hoc binary relation	Query about trusted agent

Advanced Reputation Ontology: In a service-oriented environment, the advanced reputation ontology is defined *as the conceptualization* of the *reputation* of the *trusted agent (or a service or a product)*, which is determined by advanced aggregation methods (the responses of the *third-party recommendation agents* weighted by the trustworthiness value of the third-party recommendation agent and the trustworthiness value of the opinion as well as the ranking of the first-, second- and third-hand opinions and time elapsed factors).

Advanced Reputation of the Trusted Agent = ∪ (Recommendation Value * Trustworthiness of opinion * Perceived first-, second- and third-hand opinion * Time elapsed factor)

where we define ∪ as an operator for combining and taking into account the trustworthiness of the recommendation agent's opinion, ratio of first-hand, second-hand and third-hand opinions and time factors. This advanced aggregation formula will enable the system to eliminate recommendations that are not trustworthy, self-recommendations and recommendations that are malicious.

A graphical view of the advanced reputation ontology is shown in Figure 9.7 through the use of the ontology presentation notation.

Table 9.3 is a formula table for the advanced reputation ontology.

9.3.3 Issues with Basic Reputation

With the simple (or basic) reputation measure, three problems could be created:

(a) It may end up without a normal distribution in statistical analysis, such as 99 % of third-party recommendation agents giving 'positive' or 'trustworthy' ratings to 99 % of agents (see eBay example in Figure 9.8). Chapter 11 gives more details on this. Only a small fraction of third-party recommendation agents tend to give bad recommendations in current systems. For more details on this, see Chapter 11.

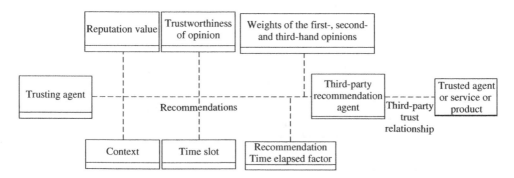

Figure 9.7   Ontology for advanced reputation of the trusted agent

Table 9.3 Formal axiom table of the advanced reputation ontology

Formula name	Formula of advanced reputation value
Concepts	Reputation value, trustworthiness of opinion, time slot, first-hand opinion, second-hand opinion, third-hand opinion
Inferred attribute	Advanced value
Formula	Advanced value = ∪ (recommendation value * trustworthiness value * first-hand opinion value, second-hand opinion value, third-hand opinion value * time elapsed factor)
Description	Advanced reputation value of the trusted agent
Variables	Recommendation value, trustworthiness value, first-hand opinion value, second-hand opinion value, third-hand opinion value, time elapsed factor
Ad hoc binary relation	Query about trusted agent

```
Member profile: Farookh Khadeer Hussain (2100)

Feedback score: 2100
Positive feedback: 99.95%

Members who gave positive feedback: 2100
Members who gave negative feedback: 10

Total positive feedback received: 2700


Recent ratings:              Positive    Neutral    Negative

Past month                     1000          0           0

Past 6 months                  1100          0           0

Past 12 months                  600          0          10
```

Figure 9.8 Description of e-Bay's rating for merchants

(b) It may create doubt on the accuracy and adequacy of the reputation measure itself, such as the truthfulness of the reputation rating and the depth of the criteria addressed in the reputation.

(c) Trust and reputation change over time. But this reputation measure may not address the issue of the dynamic nature of trust and reputation. A simple 'one value for the lifetime' is not convincing, as many assumptions may not be explored and explained clearly to the end customer and end-user.

Therefore, there is a need to use a more sophisticated measurement method for reputation. This is introduced in the next section.

9.4 Trustworthiness of Opinion Ontology

9.4.1 Opinions in Reputation

The most crucial factor for reputation measurement is the validation of trustworthiness of opinion (ToO) or the recommendation provided by the third-party recommendation agents.

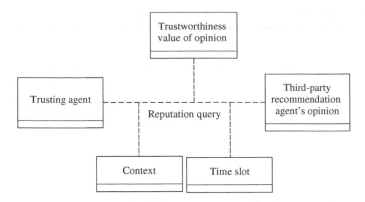

Figure 9.9 Ontology representation of the concept 'trustworthiness of opinion'

Table 9.4 Formal axiom table of the trustworthiness of opinion ontology

Formula name	Formula of trustworthiness of opinion value
Concepts	Trustworthiness of opinion in giving correct recommendation
Inferred attribute	Trustworthiness value of opinion
Formula	Trustworthiness value of opinion $= f$(actual trust value found on interaction – recommendation value)
Description	Trust value of recommendation agent in giving correct opinion
Variables	Actual trust value found on interaction, recommendation value
Ad hoc binary relation	Query about trusted agent

9.4.2 Ontology for Trustworthiness of Opinion

In Chapter 4, we introduced the opinion trust ontology as the following trust tuple:

Review trust (receiver, reviewer, review or feedback, assessment criteria, time slot and trustworthiness of each assessment criterion)

The graphical view of the ToO ontology is shown in Figure 9.9 through the use of ontology presentation notation.

Table 9.4 illustrates the trustworthiness of opinion ontology.

9.4.3 Validation of the Opinion during the Calculation of Reputation

There are two processes in evaluating the ToO. Firstly, while aggregating the recommendation value, we need to consider the ToO (credibility of the opinion). Then, after the aggregation process, the trusting agent will conduct a monitoring of performance of the trusted agent or entity, and find out the real trustworthiness value of the trusted agent and use this value to compare what value the recommendation agent has been given in the past, and if there are any discrepancies, the trusting agent shall readjust the ToO of the recommendation agent, and this in turn will affect the overall rating of the recommendation agent's trustworthiness in a right opinion.

The process of considering the credibility of the agent in giving the correct opinion during the calculation of reputation is carried out through the reputation databases. There is no calculation of the ToO of the recommendation agent during the reputation measure of the trusted agent. This is because the reputation measure is about the trusted agent, and not about the recommendation agent. However, the evaluation and adjustment of the ToO is done after the measurement of reputation and trustworthiness calculation of the trusted agent by the trusting agent after an interaction between the two.

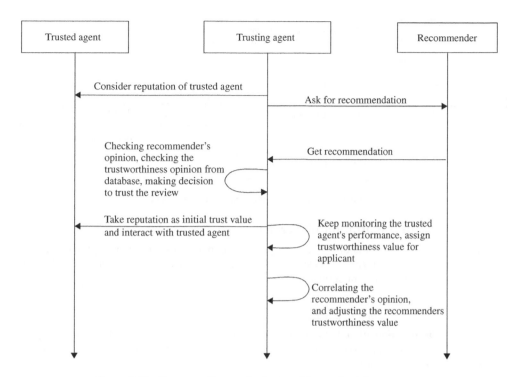

Figure 9.10   Sequence diagram for trustworthiness of opinion measures

9.4.4 Validation and Adjustment of the Trustworthiness of Opinion after Reputation

The process of validation and adjustment of the trustworthiness of opinion is done when the reputation determination is finished, performance data of the trusted agent is collected and real trustworthiness is derived for the trusted agent. Then this newly derived trustworthiness value (from the trusting agent) is compared with the recommendation or opinion submitted from the recommendation agent. If there are discrepancies, the trusting agent can decide whether to use the recommendation again or not, or simply modify the reputation database he or she had and use it for the future reputation process. Thus in Table 9.4 we note that new trustworthiness value of opinion for a partial recommender is a function f (actual trust value found on interaction − recommendation value).

Figure 9.10 details the processes generated by the trusting agent in the validation and adjustment of the trustworthiness of the trusting agents based on recommendations from the recommending agents. Note the continual feedback loops to progressively aggregate each recommendation as they are returned from the agents on the network.

9.5 Ontology for Reputation of an Agent

9.5.1 Ontology for Reputation of an Agent

Ontology for Reputation of an Agent: In the service-oriented environment, the ontology for reputation of an agent is defined as the conceptualization of the reputation of a trusted agent.

9.5.2 Conceptual View of the Ontology for Reputation of Agent

The graphical view of the ontology for reputation of an agent is shown in Figure 9.11 through the use of ontology presentation notation.

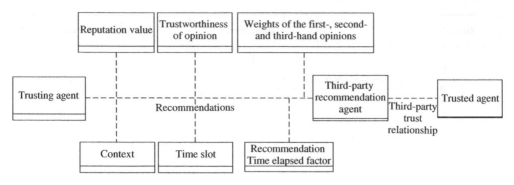

Figure 9.11   Ontology for reputation of an agent

9.6 Ontology for Reputation of Service

The ontology for the reputation of services has potential implications for the large growing number of service providers to that give e-services. In this section, we discuss the use of an ontology for the reputation and the quality of service (QoS).

9.6.1 Ontology for Reputation of a Service

Ontology for Reputation of a Service: In the service-oriented environment, the ontology for reputation of a service is defined as *the conceptualization of the reputation of the quality of service (QoS) from a service provider.*

9.6.2 Conceptual View of the Ontology for Reputation of a Service

The graphical view of the reputation of service ontology is shown in Figure 9.12 through the use of ontology presentation notation.

9.7 Ontology for Reputation of a Product

9.7.1 Ontology for Reputation of Product

Ontology for Reputation of Product: In the service-oriented environment, the ontology for reputation of a product is defined as the conceptualization that represents the reputation of a quality of product (QoP).

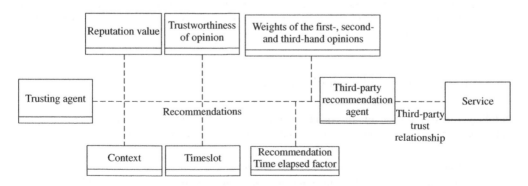

Figure 9.12   Ontology for reputation of service

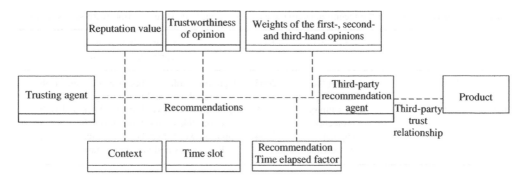

Figure 9.13 Ontology for reputation of product

9.7.2 Conceptual View of the Ontology for Reputation of a Product

The graphical view of the reputation of product ontology is shown in Figure 9.13 through the use of ontology presentation notation.

9.8 Reputation Databases

A reputation database is a useful repository of information that allows for a combination of the trust and reputation to be matrixed together. This section elaborates on this combination.

9.8.1 Reputation Database

Note that the reputation matrix stores the trusting peer and the trusted peer, as well as the initial trustworthiness value (this value is derived from the reputation value) (Table 9.5).

The trust matrix and reputation matrix together form the trust database. The trust database can be used by a software agent for valuation and validation. Valuation and validation are useful for examining the trust and reputation of agents in combination and for automatically assigning trustworthiness values.

In the later chapters, we shall learn how to assign trustworthiness values using the above trust ontology.

Table 9.5 An example of a reputation matrix

Trusting peer	Trusted peer	Context	Start	End	Initial trustworthiness
Sonya	Bob	Teaching offshore	1/1/00	2/2/00	3
Maja	Sarah	Proofread English	1/1/00	3/3/00	5
Faro	Liz	Writing a book	4/4/05	4/5/05	5
Joen	Andrew	Mate model IT project	4/4/00	30/4/05	4
Vidy	Sonya	Travel arrangement	9/9/03	9/9/04	4
Vidy	Faro	Share computing power	9/9/03	1/1/05	4.5

Table 9.6   Trustworthiness of the recommendation agent – an example if Alice and Bob were sports mates

Trusting agent	Recommendation agent	Context	Time	Trustworthiness of the recommendation/opinion
Alice	Bob	Winning the 400-m sprint	2005	Opinion worth five, because Alice and Bob were athletes and trained together for 5 years
Alice	Bob	Sports injury protection	2005	Opinion worth five, because Alice and Bob were athletes and trained together for 5 years
Alice	Bob	Computer conference	2004	Opinion worth five, because Alice knows that Bob knows nothing about music

To illustrate the recommendation relationship further, we look at the example given in Table 9.6.

In this relationship, the important thing from the trusting agent's perspective is the trustworthiness of the third-party recommendation agent in giving a correct recommendation in a particular context and time. Note that this implies that the expertise of the third-party recommendation agent is being judged. If Bob was an athlete, Alice would trust him to give the correct recommendation about who could represent our city to win a 400-m race. However, Alice may not trust Bob to organize a computer conference.

In view of Table 9.5, we could convert the reputation relationship to tabular form by using entity–relationship concepts (high-level only), as shown below in Table 9.7 (Sample of a reputation query), Table 9.8 (Sample of a recommendation), Table 9.10 (Sample of a recommendation agent's database), and Table 9.9 (Sample of recommendation agents).

9.8.2 Reputation Query Database

Table 9.7   Sample of a reputation query

Reputation query ID #	Context	Context time	Trusting agent
100	About Joe's mathematics ability	2004–2005	Mark
101	About Sonya's English skills	2004–2005	Mark
102	About HP Compaq desktop PC	2004–2005	Toby

9.8.3 Recommendation Database

Table 9.8   Sample of a recommendation

Reputation query ID #	Recommendation agent or ID #	Recommendation from recommendation agent	Time slot	First, second, third opinion	Trustworthiness of recommendation agent
100	Faro	5	Oct 2004	First	4
100	Liz	5	Jan 2005	First	5
100	Leo	3	Feb 2005	Second	2
102	Liz	4	Mar 2005	First	5

9.8.4 Recommendation Trust Database

Table 9.9 Sample of recommendation agents

Recommendation agent: Patricia
Trustworthiness in giving correct recommendation: 4.5 (5 is the maximum) ★★★★⯪

Recommendation context (from a reputation query)	Opinion or recommendation (from Patricia, the recommendation agent)	Trustworthiness of opinion Evaluation by trusting agent

---

Fishing	April and May fishing is good.	★★★
XYZ outsourcing	XYZ, productivity up 70%, recommend 4.5 out of 5.	★★★★⯪
Digital camera model ABC	The battery life is very short; recommend 2 out of 5.	★★★★
Human rights	The updated legislation is available in URL xxxx.	★★★★★

9.8.5 Recommendation Agent's Data

Table 9.10 Sample of recommendation agent's database

Recommendation agent or ID #	Trusted agents ID #	Recommendation context	Recommendation time	Trustworthiness of opinion
Faro	A1000	Computing	2004	5
Liz	A1000	IT	2003	4
Liz	A1000	Project management	2004	5
Leo	B1020	Logistics	2005	2

9.9 Seven Levels of Reputation Measurement

9.9.1 Trustworthiness of the Recommendation Agents in Giving the Correct Opinion

The trustworthiness of the recommendation agent is defined as '*a trustworthiness level of the third-party recommendation agent's willingness and capability in giving the correct recommendation to the trusting agent about a trusted agent*'. The trusting agent can reassign a trustworthiness level to the third-party recommendation agent on the basis of the agent's honesty in giving the correct recommendation in the previous response. The trustworthiness levels of the recommendation agents and their semantics are given in Table 9.11.

9.9.2 Seven Levels of Reputation for Trusted Agents

We define the domain of $[-1]$ to $[5]$ or range of $[-1]$ to $[5]$ as repute values used to represent the reputation of the reputation queried agent.

Since trust is personal (every agent may have his/her own perceived trustworthiness of the trusted agent for the same QoS based on its psychological type) and has an element of uncertainty in it, we propose a method by which unintentional interference in assigning trustworthiness to the reputation queried agent may be avoided from being propagated to other agents as its

Table 9.11 Trustworthiness of the third-party recommendation

Third-party recommendation agent's trustworthiness level	Trustworthiness of the recommender in giving a correct recommendation (semantics)
−1	Unknown
0	Not trustworthy at all
1	Minimal trustworthy
2	Little trustworthy
3	Partially trustworthy
4	Trustworthy
5	Very trustworthy

reputation. Additionally, this approach also takes into account the personal issues and uncertainty in reputation that is associated with trust. However, the following factors cannot be taken into consideration:

- the uncertainty in the trustworthiness as assigned by the trusting agent to the trusted agent;
- the personal nature that trust can have on the trusting agent while assigning a trustworthiness value to the trusted agent;
- unintentional minor errors by the trusting agent in assigning trustworthiness to the trusted agent.

Table 9.12 best illustrates the method proposed for a numerical representation of the reputation of a trusted agent.

In this section, we propose seven different repute levels. We also define each of the repute level semantics, so if a reputation querying agent is provided with a reputation level about the reputation queried agent from a recommendation agent, the meaning of the same can be understood (Table 9.13).

Table 9.12 Reputation of a trusted agent

Reputation level	Reputation of trusted agent (semantics) (linguistic definition)	Aggregated reputation value (user defined)	Visual representation (star rating system)
−1	Unknown	$x = -1$	Not displayed
0	Very bad reputation	$x = 0$	Not displayed
1	No reputation	$0 < x \leq 1$	From ☆ to ★
2	Little reputation	$1 < x \leq 2$	From ★☆ to ★★
3	Some reputation	$2 < x \leq 3$	From ★★☆ to ★★★
4	Good reputation	$3 < x \leq 4$	From ★★★☆ to ★★★★
5	Very good reputation	$4 < x \leq 5$	From ★★★★☆ to ★★★★★

Table 9.13 Repute levels along with their semantics

Value of repute	Meaning
−1	Unknown
0	The recommendation agent recommends that the reputation queried agent has a very well-known poor reputation.
1	The recommendation agent recommends that the reputation queried agent should not be trusted.
2	The recommendation agent recommends that the reputation queried agent has only little reputation.
3	The recommendation agent recommends that the reputation queried agent has some reputation.
4	The recommendation agent recommends that the reputation queried agent has good reputation and can be trusted.
5	The recommendation agent recommends that the reputation queried agent be trusted.

9.9.3 The Semantics of Reputation Levels

In this section, we explain the semantics of each of the above-mentioned repute levels. We propose that a repute value of '−1' means that reputation is unknown, and '0' means that for a given context and at a given point of time, the reputation queried agent has a very bad reputation. The recommendation agent discourages the reputation querying agent from entering into an interaction with the reputation queried agent for that given context and at that given point of time.

We propose that the following reasons may have prompted the recommendation agent to assign a repute value of '0' or '1' to the reputation queried agent:

- The reputation queried agent was involved in a very dishonest dealing with the recommendation agent and the recommendation agent is sure that there is no other malicious agent who was responsible for the undesirable outcome of its interaction with the reputation queried agent.
- The recommendation agent had a very clear idea of what it expected from the reputation queried agent from the interaction, and this expectation is realistic.
- The expectation of the recommendation agent was communicated to the reputation queried agent, in clear terms.

We propose that a repute value of '2' and '3' means that, for a given context and at a given point of time, the reputation queried agent has little or some reputation. The recommendation agent neither recommends the reputation querying agent to enter into an interaction with the reputation queried agent, nor recommends against it.

We propose that the following reasons may have prompted the recommendation agent to assign a reputation value of '2' or '3' to the reputation queried agent:

- Agents vary their behaviour according to the value of the interaction that they are involved in. In some conditions, considering the low value of the interaction, they may act in an honest way, while in some other situations, when the value of the interaction is not too high, they may act in a very dishonest or malicious way, without any regard for the consequences. These agents let the value of the interaction they are involved in decide their response. The value can be the financial value of the interaction.

- The recommendation agent does not have a very clear idea of what it expected from the reputation queried agent from its past interaction.
- The expectation the recommendation agent had from the reputation queried agent was not realistic.
- The expectation of the recommendation agent was not communicated to the reputation queried agent in clear terms.

We propose that a repute value of '4' or '5' means that, for a given context and at a given point of time, the reputation queried agent has a very good reputation. The recommendation agent recommends to the reputation querying agent to enter into an interaction with the reputation queried agent for that given context and at that given point of time. This is primarily based on past interaction/s for which the recommendation agent found the reputation queried agent to be very trustworthy.

Tables 9.14, 9.15 and 9.16 in the following sections apply the concept of repute values in the context of the seven levels of reputation measurement for QoSs, QoPs and levels of trustworthiness for third-party recommendation agents.

9.9.4 Seven Levels of Reputation for Quality of Services and Product

Table 9.14 Reputation for the quality of services

Reputation level	Reputation of the quality of the services or product (semantics) (linguistic definition)	Aggregated reputation value (user defined)	Visual representation (star rating system)
−1	Unknown	$x = -1$	Not displayed
0	Very bad reputation	$x = 0$	Not displayed
1	No reputation	$0 < x \leq 1$	From ☆ to ★
2	Little reputation	$1 < x \leq 2$	From ★☆ to ★★
3	Some reputation	$2 < x \leq 3$	From ★★☆ to ★★★
4	Good reputation	$3 < x \leq 4$	From ★★★☆ to ★★★★
5	Very good reputation	$4 < x \leq 5$	From ★★★★☆ to ★★★★★

9.9.5 Seven Levels of Trustworthiness for Recommendation Agents

Table 9.15   Seven Levels of trustworthiness for recommendation agents

Trustworthiness level of the third-party recommendation agents	Trustworthiness of the agent in giving a correct recommendation (semantics) (linguistic definition)	Aggregated reputation value (user defined)	Visual representation (star rating system)
−1	Unknown	$x = -1$	Not displayed
0	Not trustworthy at all	$x = 0$	Not displayed
1	Minimal trustworthy	$0 < x \leq 1$	From ⯪ to ★
2	Little trustworthy	$1 < x \leq 2$	From ★⯪ to ★★
3	Partially trustworthy	$2 < x \leq 3$	From ★★⯪ to ★★★
4	Trustworthy	$3 < x \leq 4$	From ★★★⯪ to ★★★★
5	Very trustworthy	$4 < x \leq 5$	From ★★★★⯪ to ★★★★★

9.9.6 Seven Levels of Trustworthiness of the Opinion

Trustworthiness of the Opinion (ToO) is defined as *'the trustworthiness of the recommendation agent's expertise in the given context that is called for in the reputation query'*. To validate the ToO, we look at the context of the opinion or the recommendation for which we believe the recommendation agent is able to give correct recommendations.

Therefore, we assign one of the following levels to the ToO illustrated in Table 9.16.

We can validate the ToO using the seven-scale system in Table 9.16.

In Table 9.17, we illustrate these levels in relationship to a visual scale of rating. This visual representation, under a rating scale using stars, is important for ease of understanding in service-oriented e-commerce environments, where websites predominate for facilitating business transactions, and customers require an easily understandable facility in the trust and Reputation of services.

Table 9.16   Semantics for the seven levels of trustworthiness of the opinion

Level	Semantics of the opinion	Opinion from the recommender
−1	Unknown	Undefined
0	Very untrustworthy	The opinion has no relevance at all
1	Untrustworthy	Not relevant to the recommender's expertise
2	Partially trustworthy	Partially relevant to the recommender's expertise
3	Largely trustworthy	Largely relevant to the recommender's expertise
4	Trustworthy	Relevant to the recommender's expertise
5	Very trustworthy	Very relevant to the recommender's expertise

Table 9.17 Representation for the seven levels of the trustworthiness of opinion

Trustworthiness level of the opinion	Trustworthiness of the opinion (semantics) (linguistic definition)	Aggregated reputation value (user defined)	Visual representation (star rating system)
−1	Unknown	$x = -1$	Not displayed
0	Not trustworthy at all	$x = 0$	Not displayed
1	Minimal trustworthy	$0 < x \leq 1$	From ☆ to ★
2	Little trustworthy	$1 < x \leq 2$	From ★☆ to ★★
3	Partially trustworthy	$2 < x \leq 3$	From ★★☆ to ★★★
4	Trustworthy	$3 < x \leq 4$	From ★★★☆ to ★★★★
5	Very trustworthy	$4 < x \leq 5$	From ★★★★☆ to ★★★★★

Table 9.18 Comparison of the seven-level reputation and trustworthiness scale

Trustworthiness level	Semantics (linguistic definitions)	Trustworthiness value (user defined)	Reputation level	Reputation of trusted agent (semantics) (linguistic definition)	Aggregated reputation value (user defined)
Level −1	Unknown agent	$x = -1$	−1	Unknown	$x = -1$
Level 0	Untrustworthy	$x = 0$	0	Very bad reputation	$x = 0$
Level 1	Minimally trustworthy	$0 < x \leq 1$	1	No reputation	$0 < x \leq 1$
Level 2	Partially trustworthy	$1 < x \leq 2$	2	Little reputation	$1 < x \leq 2$
Level 3	Largely trustworthy	$2 < x \leq 3$	3	Some reputation	$2 < x \leq 3$
Level 4	Trustworthy	$3 < x \leq 4$	4	Good reputation	$3 < x \leq 4$
Level 5	Very trustworthy	$4 < x \leq 5$	5	Very good reputation	$4 < x \leq 5$

9.9.7 Reputation Levels and Trustworthiness Levels

Table 9.18 illustrates the concepts of reputation levels and trustworthiness levels on an elaborated scale. This elaborated scale includes a wider representation of both the trustworthiness levels and reputation levels. The reputation of trusted agents is also more broadly illustrated in this comparison.

9.10 The Fuzzy Nature of Reputation

The quality of reputation is affected by the following factors: doubtfulness, diversity in opinions, endogenous factors and malicious attacks. In this section, we will elaborate on these critical issues, some of which have been introduced to readers in previous chapters.

9.10.1 Doubtfulness

The reputation of a reputation queried agent is conveyed by the feedback agent to the reputation querying agent. The feedback agents who convey the reputation of the reputation queried agent may or may not convey the correct reputation to the reputation querying agent.

Consider the following example. Alice assigns a trustworthiness value of 5 to Bob after her interaction with him. If Sonya (the reputation querying agent) asks Alice about the perceived trustworthiness of Bob, Alice may have little motivation to respond correctly. She may or may not provide an accurate account of her perceived trustworthiness of Bob (in other words, the reputation of Bob) to Sonya.

9.10.2 Diversity in Opinions

Reputation, like trust, has endogenous factors attached to it. These are important considerations in evaluating the reliability of trust and trust measures. This is best illustrated by the following example. Alice has had a previous interaction with Bob. Additionally, Charlie has had a previous interaction with Bob. Let us assume that, in these interactions, Alice and Charlie were the trusting agents and Bob was the trusted agent in both the interactions. Let us further assume that the context in which Alice interacted with Bob and the context in which Charlie interacted with Bob were similar. After the interactions, Alice assigned a trustworthiness value of 5 to Bob and Charlie assigned a trustworthiness value of 3 to Bob, although Bob had given the same QoS to both Alice and Charlie. Liz, in order to decide if she should interact with Bob, asks Alice and Charlie about their perceived trustworthiness of Bob. Alice and Charlie give their perceived trustworthiness of Bob (in other words, the reputation of Bob) as 5 and 3 respectively.

Here, we note that Alice and Charlie did not convey similar reputation values to Liz, and Liz obtains two different views of the reputation of Bob. The reason is that trust is based on personal belief and reputation too has a personal component attached to it.

9.10.3 Endogenous Factors

Trust is a very personal concept. The task of deciding whether to trust another agent involves a decision process. In making a decision, agents who have a 'thinking' preference have a tendency to analyse things in an objective and logical fashion before they trust another agent. They place no importance on personal values that they share with the other agent while deciding whether to trust them. On the contrary, agents who have a 'feeling' preference will have a tendency to place more importance on personal values before making a decision of whether to trust another agent. Similarly, they place more importance on personal values before assigning a trustworthiness value to the trusted agent. Additionally, depending on the psychological type of the trusting agent, the trustworthiness value assigned by the trusting agent to the trusted agent for the same QoS from the trusted agent may not be equal. Agents who have a 'thinking' preference will analyse things in an objective and logical fashion and assign trustworthiness values to the trusted agent based on reasoned and objective factors.

In our opinion, agents with 'thinking' preferences, while assigning trustworthiness values to the trusted agent will not attach any importance to the personal rapport that they share with the trusted agent. In contrast, agents who have 'feeling' preference will not analyse things in an objective and logical fashion and place more weight on the personal rapport that they share with the trusted agent, while assigning a trustworthiness value to the trusted agent.

Let us consider an example to explain our argument. Assume that Alice, Liz and Bob are owners of three logistic companies. Also assume that these companies are all located in distinct places. Furthermore, let us assume that Bob and Liz want to use the warehouse space of Alice. After Bob has used the warehouse space of Alice, he will assign a trustworthiness value to Alice. Similarly, Liz, after her interaction with Alice, will assign a trustworthiness value to Alice. Let us assume that Alice provided the same QoS to both the Agents (Liz and Bob).

It is possible that for the same QoS provided by Alice, Bob and Liz have two different perceptions and hence assign different trustworthiness values to Alice. Similarly, it is possible that Bob and Liz evaluate the same QoS from Alice in two distinct ways. It is possible that Liz assigns a high trustworthiness value to Alice, as she perceives the QoS provided by Alice as good. In contrast, Bob might assign an average trustworthiness value as he perceives the QoS provided by Alice as mediocre.

Additionally, let us assume that both Liz and Bob have a personal rapport with Alice. Let us further assume that Liz has a 'feeling' preference while Bob has a 'thinking' preference. It is possible that Liz, owing to the personal rapport that she has with Alice, assigns a high trustworthiness value to Alice. In contrast, as Bob has a 'thinking' preference, he analyses the behaviour of Alice in an objective way with no regard for the personal rapport that he has with her. Hence, the trustworthiness value assigned by Bob and Liz to Alice may or may not be the same, for the QoS provided by Alice. Thus, Alice (the trusted agent) receives two distinct trustworthiness values. These trustworthiness values, when propagated by either Bob or Liz to other reputation querying agents, become the reputation for Alice.

9.10.4 Malicious

Intentional errors are sometimes made by feedback agents in communicating reputation. This is a major problem with the current trust models [2–6]. However, these models have not yet included a method by which an agent can discern between correct and incorrect recommendations. These models also do not suggest a method by which an agent can collect recommendations. A malicious agent may communicate an incorrect trustworthiness value to a reputation querying agent. None of the existing trust models address this problem.

Continuing with the above example of the interactions between three logistic companies, let us assume that Bob after interaction with Alice had assigned a trustworthiness value of '4' (denoting the highest possible trustworthiness value) and yet is a malicious agent in providing referrals.

Let us furthermore assume that Sarah is another agent in the logistic network. She asks Bob about the reputation of Alice. Bob, instead of communicating a value of '4', communicates a value of '2'. This value '2', when propagated by Bob, becomes the reputation of Alice.

9.11 The Dynamic Nature of Reputation

Reputation is built as an extension of trust that is generated by a trusted agent. It allows, therefore, apart from the inheritance of all of the dynamism of the trust, the introduction of a new set of dynamic characteristics of reputation, which are as follows:

- Longevity (rapid development and diminishing characteristics)
- Stability (incentives to maintain the reputation)
- Irregularity (changes in reputation context, quality criteria and identity)
- Abruptness (long-term vs short-term effects)
- Anomalies (evolution and change in domain expertise).

9.11.1 Longevity

Reputation manifests over a short period of time, yet may not last forever as an agent may not exist forever. In such a dynamic environment, the reputation of an agent may appear suddenly, reappear and disappear suddenly. If an agent suffers some loss to his/her reputation by any of the aspects of its fuzzy nature outlined above (doubtfulness, diversity in opinions, endogenous factors and malicious attacks) it is difficult for the agent to counter this loss, forcing changes in the business or social environment and impacting on future business opportunities.

The reputation of an agent may be gained or improved by continuing and sustaining the effort for the enhancement of the quality of the service provided by the business, countering the malicious attacks by achievement and being interested in developing a service profile in the community.

The collapse of the reputation of an agent by an instance of error or mistake in business and trading environments, particularly in technologically enabled circumstances, often leads to irreparable damage to the trust relationship, and therefore, business relationships are irretrievable. However, a well-established reputation may insulate organizations and their agents from dishonest behaviour or possible threats by malicious agents.

9.11.2 Stability

Maintaining stability in the reputation of an agent or organization depends on the sustained effort of a particular agent over the longer term in building up reputation and trust. If an agent's reputation has decreased over time through various circumstances, he/she needs to exert much more effort in re-establishing the reputation. Incentives may help maintain the reputation effort in the business or trading situation. Other support in the forms of moral and financial support for the agent, and long-term strategic support such as promotion, power of control and responsibility or pay-offs in economic and social terms may enhance the sustained effort of an agent in re-establishing the reputation and trustworthiness value.

9.11.3 Irregularity

Agents may vary their QoS from time to time, and thus there is an aspect of irregularity in the reputation of the business or trading context. This implies that the trustworthiness value is dynamic as the reputation context may fluctuate between stable and unstable trading or business environments, often dependent on exogenous factors such as economic cycles that are strongly influenced by consumer demand. Therefore, initiating and maintaining timely updates to the reputation of the agents (and/or their business or organization) is very important for maintaining consumer confidence in the provision of service, which ultimately affects trust and reputation.

9.11.4 Abruptness

An agent's/organization's reputation may take a long time to build and establish. However, reputation, and particularly trust, may be damaged by an instance, a single event or behaviour, mainly exogenous to the business itself, and lead to a slow process of recovery. It is a long-term effort to build a reputation, but it could be destroyed in an instance. This is particularly evident in malicious attacks in the electronic trading environments, which have been to date characterized by viruses, hackers and security threats (as malicious agents) that have posed enormous challenges, for example, to Internet service providers. Reputation is not reliable in the sense that people should not depend on it. Unlike a trustworthiness value that trusting agents assign themselves that can be updated frequently, reputation is somebody else's opinion and may not be updated easily.

9.11.5 Anomalies

Reputation can be either positive or negative. It may score highly, positively or negatively in an evaluation of trust. Malicious agents particularly may manipulate the reputation of another agent's trust. In electronic trading environments, it is easy for agents to hide their true identities from each other for competition purposes. In some cases, an agent ID may change over time. This may cause difficulties or anomalies in avoiding agents who have had very bad reputations and come back with different identities.

Thus, the anomalous or potentially anomalous nature of reputation is inherent in the context of electronic trading where dynamism is found in both time and context, in agent identity, in time slots and in service provision.

9.11.6 Managing the Dynamism of Reputation

In managing the dynamism of trust and reputation, therefore, it is important to have a well-established history of reputation and sophisticated trust and reputation assessment methods. For both the efficacy and integrity of the measurement of trust and reputation, we must maintain a system that ameliorates the effects of instant reputation values and unfair ratings that may mislead trusting agents. This should enable the provision of complete and adequate information about the reliability and trustworthiness of individual agents, business entities, products and services (these services also may include government, public and private agencies, services and events). It is also important to ensure the veracity of truthful feedback and recommendations, and therefore constant updating of the trustworthiness value of third-party recommendation agents is imperative.

The steady improvement of trust and reputation technologies within the organization for the coverage of the depth of trust and reputation evaluation is very important for business intelligence and consumer confidence. The provision of adequate and accurate documentation for the reputation domains (context) is also important so that the concept of reputation is explicitly defined, the context, the time dependency, quality criteria and assurance are made explicit, rather than a general implicit notion of reputation that claims to cover all these.

The provision of automated systems that facilitate the growing sophistication of trust and reputation systems is an increasing necessity in our automated and electronic trading environments where generally concerns about trust, reputation and reliability need to be addressed.

These general concerns lead on to our next chapter about the current state-of-the-art reputation methods and the utilization of multi-agent systems for automated reputation data collection, aggregation and evaluation processes.

9.12 Conclusion

In this chapter, we presented a significant discussion on the concept of the reputation ontology by elaborating on specific components in relationship to recommendations, trust, third-party issues and reputation queries. We also elaborated on a generic ontology and examined this in the context of agent, service and product reputation.

Importantly, we presented a seven-level scale for reputation measures. Finally, we related these ontological concepts back to previous discussion on the fuzzy and dynamic characteristics of trust, by extrapolating this to the concept of reputation. In summary, we specifically discussed the following:

- Reputation ontology
- Recommendation trust ontology
- Third-party trust ontology
- Reputation query ontology
- Generic ontology
- Specific ontology
- Agent reputation ontology
- Service reputation ontology
- Product reputation ontology
- The seven-level scale of reputation measure
- The fuzzy and dynamic nature of reputation.

It is our vision that trust and reputation technologies will eventually transform the electronic business environment from what is normally believed to be a very risky environment to a disciplined and trusted business one [7, 8], because they help change organizational and individual behaviour in a positive and professional way.

These technologies should have the potential to ameliorate the impact of dishonest services, unfair trading, biased recommendations, discriminatory actions, fraudulent behaviours and misleading advertising. They should help build business intelligence through the dissemination of customers' feedback, buyers' recommendations, third-party opinions, interference-free data and information sources, thus providing a guide for enhancing customer relationships, a method for learning consumer behaviour and a tool for capturing market reaction on products and services. In essence, they should encourage virtual collaboration and competition.

References

[1] Ba S. & Pavlou P. (2002) 'Evidence of the Effect of Trust Building Technology in Electronic Markets: Price Premiums and Buyer Behavior'. *MIS Quarterly*, **26**(3): 243–268.

[2] Abdul-Rahman A. & Hailes S. (2003) *Relying On Trust To Find Reliable Information*, Available: http://www.cs.ucl.ac.uk/staff/F.AbdulRahman/docs/dwacos99.pdf (7/08/2003).

[3] Yu B. & Singh M.P. (2002) 'Distributed Reputation Management for Electronic Commerce'. *Computation Intelligence*, **18**(4): 535–549.

[4] Cornelli F., Damiani E., Vimercati S., di Vimercati D.C., Paraboschi S. & Samarati P. (2003) *Choosing Reputable Servants in a P2P Network*, Available: http://citeseer.nj.nec.com/cache/papers/cs/26951/http:zSzzSzseclab.crema.unimi.itzSzPaperszSzwww02.pdf/choosing-reputable-servents-in.pdf (20/9/2003).

[5] Aberer K. & Despotovic Z. (2003) *Managing Trust in a Peer-2-Peer Information System*, Available: http://citeseer.nj.nec.com/aberer01managing.html (11/9/2003).

[6] Xiong L. & Liu L. (2003) A *Reputation-Based Trust Model for Peer-to-Peer eCommerce Communities*, Available: http://citeseer.nj.nec.com/xiong03Reputationbased.html (9/10/2003).

[7] Bakos Y. & Dellarocas C. (2002) "Cooperation without Enforcement?-A comparative Analysis of Litigation and Online Reputation as Quality Assurance Mechanism". *Proceedings of the 23rd International Conference on Information Systems (ICIS 2002)*, Barcelona, Spain.

[8] Dellarocas C. (2003) *"The Digitization of Word-of-mouth: Promise and Challenges of Online Feedback Mechanisms" Working Paper*, March 2003, Massachusetts Institute of Technology, http://ssrm.com.

10

Reputation Calculation Methodologies

10.1 Introduction

In this chapter, we introduce several different techniques for synthesizing the trustworthiness value from the recommendation agents' opinions. We give a brief introduction to the advanced techniques that can be used for calculating the reputation of a trusted entity. These techniques provide a method for calculating the trustworthiness value for use in the current time slot (not for the past or for the future). We introduce the following approaches:

(a) Methods that employ a deterministic approach
(b) Methods that utilize the Bayesian approach
(c) Methods that use a fuzzy systems approach.

We noted earlier that reputation is widely used in both the real and virtual world to

(a) determine the trustworthiness of a trusted peer, a service provider or a service when we neither have a direct experience nor an interaction with it in a specified context;
(b) refine the trustworthiness value even in the situation where we have had some previous interaction with the trusted entity.

In the previous chapters, we defined the notion of reputation and its semantics and developed an approach for modelling the reputation ontology. In this chapter, we will explore methods for collecting reputation values from various third parties. We will then aggregate these quantities in a suitable fashion and synthesize them in order to determine a trustworthiness value for the trusted peer.

There are many issues that need to be addressed before one can synthesize the value of reputation. In Chapter 8, we discussed the fact that the reputation opinion could be a first-, second- or third-hand opinion depending on whether the third party was directly known or referred to by a known entity or an unknown entity. It will be important for any reputation system to suitably weight these opinions to reflect their relative credibility. In addition, even within any one of these categories, the agents within these may have different credibilities. What exactly should be used to provide an assessment of this credibility is an important factor. As noted in Chapter 8, the credibility of a third-party agent (or trustworthiness of the recommending agent to give an opinion) is the likelihood that it will give an accurate and truthful opinion of the trusted agent in the given context and time slot.

Trust and Reputation for Service-Oriented Environments   Elizabeth Chang, Tharam Dillon and Farookh Hussain
© 2006 John Wiley & Sons, Ltd

This notion of third-party credibility contains the intrinsic notion that this credibility will vary with time; therefore, the reputation determination methodologies must enable the optimum approach to capture this factor. Specifically, they need to provide a mechanism for updating the credibility of the third-party agent after each interaction, where the agent has provided feedback based upon a quality assessment of the received opinion. Lastly, in the case of web services, as mentioned in earlier chapters, a web service can consist of several composite services, and we may be able to only obtain trustworthiness values for each of these components from a reputation methodology that synthesizes opinions. In this situation, we require an approach that facilitates ascertainment of trustworthiness values. This relates to composite service, which is based on the trustworthiness values of its components. We address all of these issues within this chapter.

10.2 Methods for Synthesising the Reputation from Recommendations

Several different techniques for synthesising the trustworthiness value from reputation opinions have been put forward in the literature [1–9, 11]. These techniques provide a method for calculating the trustworthiness value for use in the current time slot, given the recommendations from other agents. In this chapter, we will discuss the following approaches:

(a) Methods that employ a deterministic approach
(b) Methods that utilize the Bayesian approach
(c) Method that use a fuzzy systems approach.

10.3 Factors and Features that need to be Considered

In Chapter 8, we explained the importance of taking into account the credibility or trustworthiness of the opinion from a third party. In determining a measure, we could consider using the following:

 (i) The trustworthiness of the third-party agent.
(ii) A specific measure honed to assess the reliability of the third party in providing such opinions.

10.3.1 Trustworthiness of Agent: Intentional Error

The problem with using (i) is that the third-party agent could provide a misleading or false opinion about the trusted agent owing to competitive reasons or jealousy, even though in their own dealings with the trusting agent they might have always satisfactorily fulfilled the commitments in any agreement.

Thus, in the example in Figure 10.1, we note that a trusting agent that wants goods transported and warehoused in a logistics transaction might query Agent 2 about Agent 3, which is Warehouse provider 1. Agent 1 has previously interacted with Agent 2, for goods transportation. Note that since Agent 2 has always satisfactorily fulfilled the commitments under any service agreement, Agent 2 has a high trustworthiness value assigned by Agent 1 (the trusting agent) for goods transportation.

However, Agent 2 may also receive a commission from agent 4 (Warehouse provider 2) for any business it directs to Agent 4. Hence, Agent 2 may have a vested interest in giving a falsely negative opinion about Agent 3 (Warehouse provider 1), when Agent 1 is seeking an opinion. This would indicate that in many circumstances, using (i) the trustworthiness value of Agent 2 may not be a suitable measure for the credibility of Agent 2 or the trustworthiness of the opinion given by Agent 2.

10.3.2 Trustworthiness of Opinion

Hence, it is more suitable to use (ii), that is, a measure specifically honed to assess the credibility of Agent 2 or the trustworthiness of the reputation opinions given by Agent 2. Such a measure

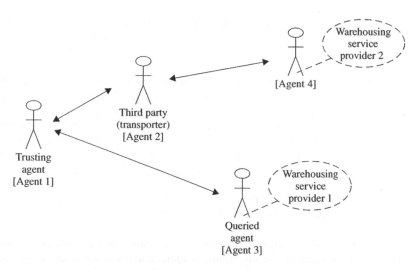

Figure 10.1 Third-party agents giving reputation recommendations

could be based on and reflect the reliability of the earlier opinions given by Agent 2. If we trust this opinion as a service provided by Agent 2 within the context of an implicit agreement to provide such an opinion, we could use a similar approach to assess this, but we would use a separate context and criteria.

The second issue that is of some importance is the basis of the opinion given by the third-party agent. The opinion may be an honest opinion. However because of poor judgement on the part of the third party agent or the lack of sufficient interactions with the queried agent, the opinion might still be wrong. This can be considered as the capability of the third-party agent to provide an opinion. Generally, the number of previous interactions can be directly obtained, but the quality of judgement is difficult to ascertain. However, the measure (ii) above would automatically take this into account and provide an indirect assessment of it.

10.3.3 Time Variability and Context

The next two issues, as mentioned in Chapter 8, are the time variability of the credibility and trustworthiness of opinion from a particular third-party agent. The time variability could have several sources and these are as follows:

(a) The third-party agent (say Agent 2) may have had more interactions in the intervening period with the reputation queried agent (Agent 3) about whom the reputation query is made, and, therefore can give a more accurate picture or alternatively might have had a more recent interaction with the reputation queried agent and, therefore, have a more up-to-date opinion of the agent.
(b) The third-party agent (say the transporter agent, Agent 2), might have had a change of ownership and/or management leading to different internal policies about giving reputation opinions.
(c) The third-party agent could be indulging in a fraudulent approach of giving some correct opinions where the impact or profit is small and giving a false opinion where the impact or profit is large in order to gain an advantage in the second situation.

This time variability from all these sources will have to be appropriately modelled. Thus, for (c), there might need to be a mechanism for reducing the credibility more in the event of an incorrect opinion than the rate at which it is built up in the case of a correct opinion.

Furthermore, as mentioned in Chapters 8 and 9, the context must be taken into account when providing the opinion. For example, where Agent 1 in Figure 10.1 asks Agent 2 for the opinion, it may be about storage of goods, such as wine, which requires a temperature and humidity-controlled environment.

It is no good, in such a situation, for Agent 2 to provide an opinion about Agent 3, the reputation queried agent, and merely assert that it is trustworthy in storing goods. Here, sometimes the Agent 2, which is answering the reputation query, may not be able to provide an opinion for the exact context about which the query was intended, but could do so for a context that is close [1].

For instance, Agent 2 might provide an opinion that Agent 3, in Figure 10.1, has got good refrigeration storage facilities. This could require a mechanism on the part of Agent 1 (the querying agent) to determine as to whether this is to be considered by the aggregation method, and if so how.

In addition, the opinion in a given context may involve several criteria and each of these criteria may have a different weighting, which would have to be taken into account in obtaining the aggregate reputation opinion.

Another factor that may be of interest is, one may need a measure of the variability of reputation opinions obtained from different third-party agents, in addition to determining say an average. This may give an indication of the variability of the quality of service provided by the reputation queried agent in its interactions.

There may also be the possibility that a small subgroup of agents may collude in giving a false opinion about the reputation queried agent and a mechanism to take this into account may be needed.

In addition to the reputation opinions from individual agents, it may sometimes be useful to have a group opinion. Such an opinion would be of considerable significance in virtual communities or trusted networks of agents. It would also be useful in the case of a large company that has several agents.

10.3.4 Aggregation of the Reputation for Component Services

Lastly, in the case of web services as mentioned in the last section, a method for determining reputation opinions for the composite web service will be necessary if the trusted agent provides opinions only about the component services.

This is the case of a logistics transaction involving a transporter and a warehouse, where we may have an individual reputation opinion for each of them. However, to obtain the opinion about whether the logistics provider would meet the total logistics transaction and deliver the services in a trustworthy fashion, the individual opinions need to be aggregated in some manner.

10.4 Deterministic Approach to Reputation Calculation

In the previous sections, we have noted several factors that need to be included in the calculation of trustworthiness of an agent based on the reputation opinions of third-party agents. We will initially start by considering the reputation recommendations of different individual agents or services providers.

10.4.1 Factors used for the Reputation Measure

The factors that need to be considered here are as follows:

(a) the credibility or trustworthiness of the recommendation agent to provide a reputation opinion (or the trustworthiness of the opinion);
(b) the time variability of the reputation opinion;
(c) the fact that the agents are first-, second- or third-hand third-party agents. [11]

10.4.2 Aggregation

Thus, the trustworthiness of the reputation queried agent k due to individual agent's opinion i, j, l in time slot n is given by Equation 10.1:

$$T_{RA}(k, n) = \alpha_1 \sum_i^I TC_A(i, n, c) * R_A(i, k, n, c) * TF_A(i, n, c)$$

$$+ \alpha_2 \sum_j^J TC_A(j, n, c) * R_A(j, k, n, c) * TF_A(j, n, c) \qquad (10.1)$$

$$+ \alpha_3 \sum_l^L TC_A(l, n, c) * R_A(l, k, n, c) * TF_A(l, n, c)$$

Here, α_1, α_2 and α_3 refer to a weighting factor for first-, second- and third-hand opinions of third-party agents.

$T_{RA}(k, n)$ is the aggregated reputation obtained from individual agents for agent k in time slot n and context c.

$TC_A(i, n, c)$ is the credibility or trustworthiness of reputation opinion for recommending agent i, in time slot n for context c.

$R_A(i, k, n, c)$ is the reputation opinion given by recommending agent i as a first-hand opinion about reputation queried agent k, in time slot n and context c.

$R_A(j, k, n, c)$ is the reputation opinion given by the recommending agent j as a second-hand opinion about reputation queried agent k, in time slot n and context c.

$R_A(l, k, n, c)$ is the reputation opinion given by the recommending agent l as a third-hand opinion about reputation queried agent k, in time slot n and context c.

$TF(n)$ is the time weighting factor, which provides a time-delay factor to the importance of the opinion, depending on the time since the last interaction between the recommending agent and the reputation queried agent. A function that would represent this effect would be

$$TF(i, n) = e^{\frac{-(n-m)}{N}} \qquad (10.2)$$

where, n is the current time slot; m is the last interaction time slot; and N is a term that characterizes the rate of decay.

Note that the larger the value of N, the smaller the rate of the decay, as seen in Figure 10.2. Other forms of time-related decay have also been considered [1].

10.4.3 Including Group Values

One can have a similar expression for the group reputation value, except that in this the credibility or trustworthiness of opinion will be a group credibility and the time-factor term will involve the time elapsed since the last interaction. Further, the reputation will be the synthesized reputation R_G of the group as a whole [1].

This gives the following expression for trustworthiness:

$$T_R(k, n, c) = \alpha_1 \frac{\left(\sum_{i=1}^{I} TC_A(i, n, c) * R_A(i, k, n, c) * TF_A(i, n, c) \right)}{\sum_{i=1}^{I} TC_A(i, n, c) * TF_A(i, n, c)}$$

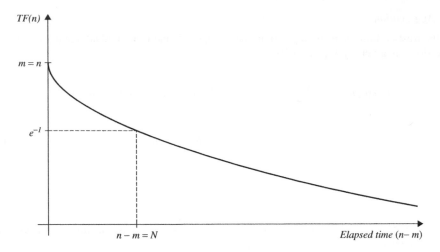

Figure 10.2   Rate of decay of reputation value

$$
+ \alpha_2 \frac{\left(\sum_{j=1}^{J} TC_A(j,n,c) * R_A(j,k,n,c) * TF_A(j,n,c) \right)}{\sum_{j=1}^{l} TC_A(j,n,c) * TF_A(j,n,c)}
$$

$$
+ \alpha_2 \frac{\left(\sum_{l=1}^{L} TC_A(l,n,c) * R_A(l,k,n,c) * TF_A(l,n,c) \right)}{\sum_{l=1}^{l} TC_A(l,n,c) * TF_A(l,n,c)}
$$

$$
+ \beta \frac{\left(\sum_{g=1}^{G} TC_G(g,n,c) * R_G(g,k,n,c) * TF_G(g,n,c) \right)}{\sum_{g=1}^{l} TC_G(g,n,c) * TF_G(g,n,c)}
\tag{10.3}
$$

where $0 \le TF \le 1$.

We allow for G groups.

$R_G(g,k,n,c)$ is the reputation of the gth group of the reputation queried agent k for the time slot n and context c.

$TC_G(g,n,c)$ is the group credibility or trustworthiness of opinion for time slot n and context c.

This aggregated reputation $T_R(k,n,c)$ reflects the aggregated reputation taking into account the reputation opinions of first-, second- and third-hand, third-party recommending agents as well as any group opinions.

10.5 Adjusting the Trustworthiness of Opinions

In addition to aggregating the reputation opinions of the recommendation agents and their associated groups, it is also necessary to have an approach for progressively modifying the credibility term in Equation 10.3, namely, TC (i.e. for modifying the trustworthiness of the opinion of the third-party recommending agent).

One can obtain the trustworthiness value of the reputation queried agent, if one follows up the reputation query with an actual interaction. This value can be termed $T_{actual}(k, n, c)$. T_{actual} can be used to determine the error in the reputation value provided by a particular third-party agent.

Let us call this error E_i for the current transactions. Then, E_i is given by

$$E_i = T_{actual}(k, n, c) - R_A(i, k, n, c) \tag{10.4}$$

for the ith third-party recommendation agent.

We note that if the detailed trustworthiness value was close to within a pre-specified tolerance ε to the actual trustworthiness we would expect to positively reinforce or increase the trustworthiness of the recommendation opinion of Agent i, $TC(i, n, c)$ (or credibility of Agent i). On the other hand, if the error E_i exceeded this pre-specified tolerance, we would expect to negatively reinforce or decrease the trustworthiness of the recommendation opinion. In order to prevent cyclic dishonesty, that is, giving the occasional dishonest opinion while generally providing correct opinions, it is necessary to decrease TC for $|E_f| > \varepsilon$ by a much larger value than the increase when $|E_f| \leq \varepsilon$, particularly where TC has values close to the maximum.

An updating mechanism that generally achieves this would be

$$TC'(i, n, c) = \eta * TC(i, n, c) + (1 - \eta) * (adj) * 5 \tag{10.5}$$

where $(adj) = +1$ for $|E_i| \leq \varepsilon$
$(adj) = -1$ for $|E_i| > \varepsilon$
$TC(i, n, c) = \max(0, TC'(i, n, c))$

Note that the value of trustworthiness of opinion would need to be modified for each recommending agent after each interaction.

If no actual interaction takes place, we could utilize this aggregated value $T_R(k, n, c)$ in place of the actual to calibrate the adjustment using Equation 10.5 in a similar fashion except that E is now given by

$$E_i = T_R(k, n, c) - R_A(i, k, n, c) \tag{10.6}$$

In this case, we assume that the collective opinions when aggregated over all the agents and groups can be used as an estimate for trustworthiness of Agent k in the time slot n and context c. Hence, the error E_i (given by Equation 10.6) in the reputation opinion offered by the third-party agent i is then used in place of Equation 10.4 in Equation 10.5 to progressively update the value of the credibility or the trustworthiness of the recommendation opinion.

10.5.1 Example

This example is used to illustrate the calculation of reputation using options from a third-party agent.

On June 15, 2001, Mr A had a consignment that needed to be transferred from Melbourne to Western Australia. This consignment consists of 10 boxes of fragile items, including computers, vases and glasses that needed to be delivered to Mr B in Perth. Mr A is looking for a company that can provide a packing service for the fragile items, and door-to-door pickup/delivery service for those boxes. He expects this consignment to arrive in Perth before 10:00 on February 18, 2002. In addition, Mr A prefers to deal with a potential service provider that offers online tracking and tracing, and signature basis proof of delivery service, which would enable him to know the status of this consignment at any time. After searching the market, Mr A considers having Fast Delivery Pty Ltd (FD) as his trusted agent to handle his consignment. As Mr A has had no previous interactions with FD in this context, he seeks the recommendation opinion from three agents X, Y and Z about the reputation of FD. The interactions and the trustworthiness values obtained by X, Y and Z for

Table 10.1 Trustworthiness values (reputation) provided by the third party agents about logistics providers

Agent	Transaction	Trustworthiness (reputation)	Time slot	n/m	Trusted agent
X	Perth	4.59	Jan–Jun 2000	1	
Y	Sydney	3.56	Jul–Dec 2000	2	FD
Z	Adelaide	3.07	Jan–Jun 2001	3	
X	Hong Kong	4.85	Jul–Dec 2001	4	
Z	Tokyo	4.30	Jan–Jun 2002	5	BA
Y	Beijing	4.56	Jul–Dec 2002	6	
Y	Singapore	4.33	Jan–Jun 2003	7	
Z	Bangkok	5.00	Jul–Dec 2003	8	CK
X	New York	5.00	Jan–Jun 2004	9	

the trusted agent FD are given in Table 10.1. The trustworthiness of opinion (credibility) of Agents X, Y and Z by Agent A at the date of the use of services (e.g. June 2001) are given in Table 10.2. Noting that since they are first-hand opinions, without any group opinions $\alpha_1 = 1$ and $\alpha_2, \alpha_3, \beta = 0$. Using these and Equation 10.1, with $N = 2$, we have obtained the calculated reputation value of 3.59 as shown in Table 10.2.

The trustworthiness of the reputation queried agent k can be found using Equation 10.1. In this case, the reputation of trustworthiness can be calculated on the basis of information that can be obtained from historic data in terms of trustworthiness. The values obtained by Mr A from Agents X, Y, and Z regarding the delivery service of FD are shown in Table 10.1. In Equation 10.1, opinion can be first-, second- or third-hand of third-party agents, and since all opinions in this case are first-hand opinions, we only need to consider for the value of $\alpha_1 = 1$. Assuming the historical data of past trustworthiness of the opinion of Agents X, Y and Z is shown in Table 10.2.

By assigning the value of $(n - m)/N$ to each interaction to indicate the time slot that we are interested in, the value of the time weighting factor $TF(n)$ can then be calculated; in this case, the value of $N = 2$ would be used. By using the value of $TF(n)$, combined with the reputation value given by recommending agent $R_A(i, k, n, c)$ and our own trustworthiness of opinion of the recommending agent $TC_A(i, n, c)$, the value for $T_{RA}(k, n)$, the aggregated reputation, can be calculated. The result of this is shown in the first row of Table 10.2 under the 'calculated reputation' column. Note that in this case we only have first-hand opinions, and that the value of trustworthiness of opinion would need to be modified for each recommending agent after each interaction. Therefore, by following up for the trustworthiness value of the reputation queried agent, we need to calculate the error E_i (Equation 10.4) for the current transactions. We note that the measured trustworthiness of the trusted agent after Agent A had the interaction was 4.5 using $\eta = 0.9$, $\varepsilon = 0.5$ and the measured and recommended values. We can calculate the adjusted values for trustworthiness of opinion as follows:

For X,

$$E = 4.50 - 4.59 = -0.09$$

$$|-0.09| < \varepsilon = 0.5$$

$$\therefore adj = +1$$

$$\therefore TC'(X) = \eta * TC(X) + (1 - \eta) * adj * 5$$

$$= 0.9 * 4.5 + 0.1 * 1 * 5$$

$$= 4.55$$

$$TC(X) = 4.55$$

Table 10.2 Credibility calculated and measured reputation
$N = 2$
$\eta = 0.9$
$\varepsilon = 0.5$

| Agent | Credibility | | | |
	Trustworthiness of opinion	Calculated reputation	Measured trustworthiness of trusted agent	Trusted agent	
X	4.50				
Y	4.00	3.59	4.50	FD	Jan 2000–Jun 2001
Z	3.00				
X	4.55				
Y	3.10	4.58	4.60	BA	Jul 2001–Dec 2002
Z	2.20				
X	4.60				
Y	3.29	4.89	4.90	CK	Jan 2003–Jun 2004
Z	2.48				
X	4.64				
Y	2.46				
Z	2.73				

For Y,

$$E = 4.50 - 3.56 = +0.94$$

$$|+0.94| > \varepsilon = 0.5$$

$$\therefore adj = -1$$

$$\therefore TC'(Y) = \eta * TC(Y) + (1 - \eta) * adj * 5$$

$$= 0.9 * 4.0 + 0.1 * (-1) * 5$$

$$= 3.10$$

$$TC(Y) = 3.10$$

For Z,

$$E = 4.50 - 3.07 = +1.43$$

$$|1.43| > \varepsilon = 0.5$$

$$\therefore adj = -1$$

$$\therefore TC'(Z) = \eta * TC(Z) + (1 - \eta) * adj * 5$$

$$= 0.9 * 3.0 + 0.1 * (-1) * 5$$

$$= 2.2$$

$$TC(Z) = 2.2$$

Thus, we have obtained the adjusted values of trustworthiness of opinion X(4.55), Y(3.10), Z(2.20). Note that the value for the trustworthiness of opinion of X has increased while those of Y and Z have decreased as their recommendations were more than 0.5 away from the measured value. This adjustment of the trustworthiness of opinion (Equation 10.5) of each agent in the group will then influence the outcome of the next reputation calculation. For the case of the dealing between Mr A and FD, the results are shown in the Table 10.2.

Similarly, when Agent A wanted to use logistics provider BA in Dec 2002, it sought the opinion of X, Y and Z using a similar approach and it got a calculated reputation of 4.58 versus a measured value of 4.6. Note that the values of X, Y and Z we obtained were X(4.6), Y(3.29) and Z(2.48). Hence they have all increased after this interaction. Similarly, for an interaction with CK by A in June 2004, we note that the calculated value of reputation based on the opinions of X, Y and Z was 4.89 and the measured value was 4.9. Hence, the new values of credibility of X, Y and Z have all increased.

It is interesting to observe that even though there have been two positive results for the opinions of Agent Z and only one negative result (the first one) (2.73), the value for trustworthiness of opinion of Z has not yet recovered to its values before the first interaction (3.0). This is because the approach weights a negative result more than a positive one as desired.

10.6 Bayesian Approach

10.6.1 Bayesian Model

The Bayesian approach uses a probabilistic approach to the determination of the reputation. It uses the well-known Bayes formula as the basis of the approach. In order to appreciate this approach, we will briefly discuss Bayes formula.

We first introduce the unconditional probability of occurrence of an event H as $p(H)$. Note that this probability is not dependent on the occurrence of any other event or any other evidence.

We next consider the notion of the probability of occurrence of an event H given some evidence E, $p(H|E)$. Here, the event H and E could be as follows:

(i) H is the occurrence of a defect in a product and E is the symptom such as a squeak, noise or rattle or overheating.
(ii) H is a satisfactory occurrence of a service and E is previous history of delivery of that particular service.

The conditional probability of H given E is given by the expression

$$p(H|E) = \frac{p(H\&E)}{p(E)} \tag{10.7}$$

that is, it is given by the probability of the occurrence of both H and E divided by the probability of occurrence of E.

Thus, in the case of occurrence of defect, $p(H|E)$ is the probability of occurrence of the defect given the occurrence of the symptom. Sometimes it is easier to know the probability $p(E|H)$ of having a symptom given that the product has the defect rather than of knowing $p(H|E)$ as we frequently record information by the defect and may indicate, in each instance of the product that has the defect, the symptoms that occurred. However, to compute $p(H|E)$ directly, we would need to store all cases of the symptom due to all causes and defects and determine the cases where those particular defects and the symptom occurred together, that is, $p(H|E)$.

Here, the Bayes formula, which is a well-known formula in probability theory, comes to the assistance. Bayes theorem states that

$$p(H|E) = \frac{p(H)p(E|H)}{p(E)} \tag{10.8}$$

$p(H)$ is referred to as the prior distribution and probability $p(H|E)$ is referred to as the posterior distribution or probability given evidence E.

10.6.2 Bayes Model for Interacting Agents

In applying this to two agents a and b interacting with each other, we can score each satisfactory interaction between them as '1' and each unsatisfactory interaction as '0'. Thus, a successful provision of a service by b to a would be scored as '1' and unsuccessful provision of services as '0'.

Alternately, a satisfactory interaction with a product b by a user a would be scored as '1' and an unsatisfactory interaction as '0'.

The outcome of these interactions between Agents a and b can be represented by the sequence

$$D = (X_{ab}(1), X_{ab}(2) \ldots X_{ab}(n)) \tag{10.9}$$

These outcomes of the previous n interactions can be treated as the evidence E. The event H can now be regarded as satisfactory outcome in the case of the $(n + 1)$th interaction, that is, H is $X_{ab}(n + 1) = 1$.

10.6.3 Bayes Trustworthiness Measure

In order to calculate the probability $(X_{ab}(n + 1) = 1|D)$ using Bayes formula, we need an estimate for H. Let us denote this by θ, that is, θ is the estimate for the proportion of successful delivery of services by Agent b to Agent a within the context c [2].

The prior probability distribution for this can be taken to be a Beta distribution (with parameters $\theta, \alpha_s, \alpha_f$) [12], that is,

$$p(\theta) = Beta(\theta, \alpha_s, \alpha_f)$$

$$= \frac{\Gamma(\alpha)}{\Gamma(\alpha_s)\Gamma(\alpha_f)} \theta^{\alpha_s - 1}(1 - \theta)^{\alpha_f - 1}$$

$\alpha_s > 0; \alpha_f > 0$ and $\alpha = \alpha_s + \alpha_f$ and Γ is the Gamma Function $\Gamma(x + 1) = x\Gamma(x)$ and $\Gamma(1) = 1$.

Let us assume each interaction $X_{ab}(i)$ is independent of all other interactions between a and b. Thus, the probability of approval in any observation is $\theta(1 - \theta)$.

If the number of successful interactions in D is s and unsuccessful (or failed) interactions is f, where $s + f = n$, then $p(D|\theta)$ can be modelled using the likelihood function for binomial sampling, that is,

$$L(D|\theta) = \theta^s (1 - \theta)^f$$

$L(D|\theta)$ is the likelihood of having s successful and $(n - s)$ unsuccessful interactions. Hence, we can use Bayes formula to obtain $p(D|\theta)$ as

$$p(\theta|D) = \frac{p(\theta)\theta^s (1 - \theta)^f}{p(D)} \tag{10.10}$$

where $p(\theta)$ is the Beta function.

Any standard statistics book shows that this leads to $p(\theta|D)$ being given by the Beta function [12]

$$p(\theta|D) = Beta(\alpha_s + s, \alpha_f + f) \tag{10.11}$$

The posterior estimate that the $X_{ab}(n + 1)$th encounter would be successful is given by the expected value of $p(\theta|D)$ and this is

$$E(\theta|D) = \frac{\alpha_s + s}{\alpha_s + \alpha_f + s + f} \tag{10.12}$$

If we assume that the first interaction corresponds to b being unknown to a and vice versa, we can consider b's trustworthiness estimated by a at this stage to be uniformly distributed. For the prior Beta if $\alpha_s = \alpha_f = 1$, we get a uniform distribution and henceforth in this section we use these values.

Hence, for subsequent interactions, leading to the $(n + 1)$th interaction, we get

$$E(\theta|D) = \frac{1 + s}{(1 + s) + (1 + f)} \tag{10.13}$$

This term $E(\theta|D)$ can thus be used as an estimate for the trustworthiness X_{ab} (i.e. b's trustworthiness estimate by a). In the earlier terminology this would be trustworthiness value of trusted agent b as determined by the trusting agent a, after direct interactions between Agent a and Agent b.

10.6.4 Bayes Reputation Measure

If Agent a has had no previous interactions with b but trusted third-party recommendation agents that are known to it have had interactions with b, it could use the recommendation opinion from these third-party recommendation agents to find the value of the reputation of b, r_{ab}. Consider that there are i such recommendations agents as shown in Figure 10.3 [2], [7].

In this case, it needs an assessment of the trustworthiness of the opinion of the intermediate (third-party) recommendation agents. We write w_{ai} as the trustworthiness of the opinion of agent i. (Note, w_{ai} is equivalent to $TC_A(i, n, c)$ in Section 10.4.2; however, its mode of determination and value could be different).

Note that r_{ib} would be determined by each i locally using the above expression for direct interaction between Agent i and Agent b. r_{ib} is equivalent to $R_A(i, k, n.c)$ in Section 10.4.2. r_{ib} would be determined for interactions between i and b in a similar manner to that indicated in Section 10.6.4. We assume that the opinion of each recommendation agent is independent of the opinion of the other agents.

Thus, the aggregated reputation is the sum of the recommendations each weighted by the normalized recommendation opinion weight.

$$r_{ab} = \frac{\sum_{i=1}^{l} w_{ai} r_{ib}}{\sum_{i=1}^{l} w_{ai}} \tag{10.14}$$

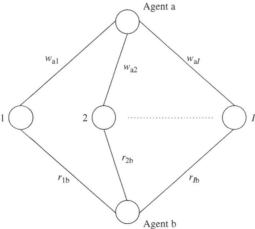

Figure 10.3   Network of recommendation agents

The question that still remains is how this recommendation opinion is arrived at. Some suggestions are to

(i) use the Chernoff bound [2];
(ii) to progressively update the value of w_{ai} using the expression (as explained in Section 10.5)

$$w_{ai}(n+1) = \eta w_{ai}(n) + (1 - \eta) \, adj$$

adj $= +1$ if the reputation value is in line with the assessment of satisfaction by Agent a with respect to Agent b.
adj $= -1$ otherwise
η is a learning parameter, $\eta > 0$

(i) above only estimates the number of interactions n necessary without noting the quality of the interactions and hence we prefer to use approach (ii).

10.6.5 Bayes Approach for Trustworthiness Calculation with Multiple Criteria

In the case of a service s_w that includes several criteria say c_1, c_2 we can determine the value of the reputation to provide the service satisfactorily.

In this case, we can consider satisfactory provision of service s_w as the hypothesis s_w, and c_1, c_2 as the evidence and can compute

$$p(s_w|c_1, c_2) = \frac{P(s_w)P(c_1, c_2|s_w)}{P(c_1, c_2)} \tag{10.15}$$

if c_1, c_2 are independent the joint probability $p(c_1, c_2) = p(c_1)p(c_2)$;
if c_1, c_2 are independent within the subset of cases with satisfactory service, $p(c_1, c_2|s_w) = P(c_1|s_w)$ $P(c_2|s_w)$.
Hence, we can write

$$p(s_w|c_1, c_2) = \frac{P(s_w)P(c_1|s_w)P(c_2|s_w)}{P(c_1)P(c_2)} \tag{10.16}$$

This expression provides a straightforward way for determining the conditional probability for s_w, given c_1 and c_2. Note, however, the assumptions are rather strong.

10.6.6 Evaluation of the Bayesian Approach

Several researchers have used the Bayesian approach for modelling trust and reputation. The most notable of these are as follows:

(a) L. Mui [2] who used an approach similar to that in Section 10.6.3 to develop an expression for trustworthiness.
(b) Wang and Vassileva [3] who used an approach similar to Section 10.6.4 to determine reputation in peer-to-peer networks employing approach (ii) for adjusting the trustworthiness of opinion value w_{ai}.

There are a number of weaknesses in the Bayesian approach and these include the following:

(a) Satisfaction/non-satisfaction of an Agent a with Agent b is modelled only using a [0,1] approach, not allowing fine gradations in satisfaction.

(b) Very strong assumptions of independence are made.
(c) The criteria c_1 and c_2 are independent (Section 10.6.5) for multi-criteria systems.

10.7 Fuzzy System Approach

Sections 10.3 and 10.4 have identified a number of factors that contribute to the determination of reputation of a trusted agent using the opinions of third-party agents. These factors can be considered to be imprecisely defined and sometimes cannot be accurately quantified. A methodology for dealing with such systems is fuzzy systems methodology. We need a representation of characterizing these factors individually and a means of combining them.

The factors identified previously are as follows:

(i) The recommendation opinion of the third-party agent $R_A(i, k, n, c)$.
(ii) The credibility or trustworthiness of the opinion TC.
(iii) The decay with the passage of time after the last interaction of the third-party agent with the reputation queried agent.

10.7.1 Fuzzy Representation of Factors

These factors can be represented by fuzzy sets. We first consider the recommendation opinion The recommendation opinion can be characterized using linguistic terms such as 'poor', 'average' and 'good'. The fuzzy term set, therefore, is (poor, average and good). If we employ the generalized bell (G-bell) shape, the membership functions for these are shown in Figure 10.4.

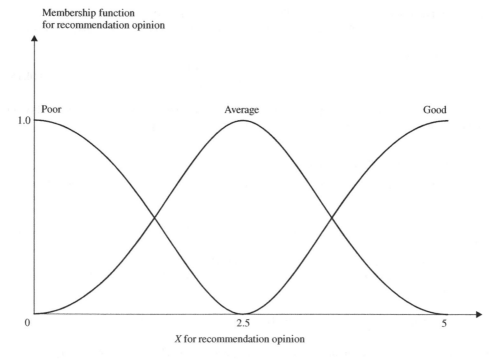

Figure 10.4   Membership function for recommendation opinion

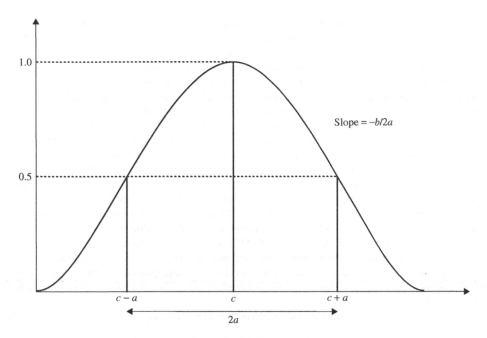

Figure 10.5 Generalized bell curve and parameters

The Membership function for recommendation opinion where the G-Bell is characterized by the parameter a, b, c is illustrated in Figure 10.5.

The credibility or trustworthiness of opinion has a similar representation. The time delay has a similar representation except that the fuzzy term set is now (long, average, short).

10.7.2 Fuzzy Inference System for Determination of Reputation Measure

In the last section, we developed a fuzzy model to express each of the reputation features or factors that we believed were important in characterising reputation. What we still need is a method of mapping the input space corresponding to these features or factors to the output space, which is a representation of reputation. The input space here will have several dimensions as explained earlier and will consist of the Recommendation Opinion (RO), Credibility (CR), and Time Delay (TD). The output space will have a single dimension, namely, the Reputation contribution of third-party agent i (R_i). The input features are all expressed using a fuzzy model, which transforms the variables of a particular input feature or factor into a linguistic variable. What we need is a method of fuzzy inference to perform the mapping from the input space to the output space. We will use fuzzy linguistic control rules for performing this mapping. The general form of a fuzzy linguistic control rule is as follows:

If A then B.

This fuzzy linguistic control rule states that if condition A is met, then carry out action B. Sometimes A is referred to as the antecedent and B as the consequent. A could be a composite condition or an antecedent consisting of premises joined together by connectives AND and OR.

There are two broad categories of fuzzy reasoning systems that need to be considered for our particular problem, and these are as follows:

(a) the Mamdani method [13] and
(b) the Takagi and Sugeno method [14] (T–S method).

In both the approaches, the input conditions are given in the form of fuzzy sets partitioned on the input universe of discourse. Further, in the Mamdani approach, the output is in the form of a fuzzy set partition on the output universe of discourse. In the case of the Takagi–Sugeno inference model, the output is a linear crisp function. Singletons represent a particular case of the Takagi–Sugeno model. This use of different forms for the output variable has considerable implications for the reasoning approach. In determining reputation employing the fuzzy systems approach, we have to choose between the two inference systems discussed above. What we would like to achieve is a value for the reputation for this inference system that will allow us to have some measure of this property. Each of the above input factors influences the value of the output reputation generated. We decided to choose the Takagi–Sugeno approach, as the single crisp value returned as a reputation is a more practical approach in that it would be more helpful to understand it as a measure of reputation. Further, there are several different approaches to aggregation and defuzzification, which could result in somewhat different values for the output in the Mamdani approach and this is likely to be confusing for the trust/reputation specialists. We can also obtain from previous interactions a number of examples that provide values of inputs and a target value of reputation, which can be used for training the Takagi–Sugeno model. Methods for directly tuning the parameters in the Takagi–Sugeno model have been widely researched and they provide a reliable approach.

In the reputation determination problem, it is, therefore, advantageous to use the Takagi–Sugeno inference system to allow this tuning of the parameters using the example data generated from previous interactions.

In separate studies, our group has also explored the Mamdani approach [15], but it is not discussed here.

10.7.3 Takagi–Sugeno Inference Approach

The Takagi–Sugeno (T–S) form of the rule is:

IF $(x_1$ is X_1 AND x_2 is $X_2 \ldots, x_n$ is $X_n)$

THEN $(y_q = a_{q0} + a_{q1}x_1 + \ldots a_{qn}x_n)$

WHERE: x_1, x_2 are scalar inputs; X_1, X_2 are fuzzy sets; $a_{q0}, a_{q1}, \ldots a_{qn}$, are real numbers; and y_q is the consequent of the rule.

This is the form of the fuzzy logic rule that was initially proposed by Takagi–Sugeno in their original work [14]. In the singleton case, the output y_q takes the form $y_q = a_{q0}$.

We consider a system with m fuzzy rules of the T–S form. The form of the crisp output is

$$y(\underline{x}) = \frac{\sum\limits_{q=1}^{m} \alpha_q \left(a_{q0} + \sum\limits_{s=1}^{n} a_{qs}x_s \right)}{\sum\limits_{q=1}^{m} \alpha_q} \tag{10.17}$$

Here, α_q is the firing strength of rule q. The actual approach to fuzzy reasoning in this case has the following steps:

(1) Fuzzify inputs.
(2) Obtain the firing strength α_q associated with each rule q.
(3) Obtain the output function of y_q associated with each rule q using the firing strength α_q.

Obtain the overall output $y(\underline{x})$ using the expressions (10.17) given above.

10.7.4 Fuzzy Rule Base for the Reputation Problem

As explained above, we intend to use the first-order Takagi–Sugeno approach for fuzzy inference with a linear function for the right-hand side. The inputs on the left-hand side of the fuzzy rule will consist of the factors or features that affect reputation.

These factors or features are RECOMMENDATION OPINION, CREDIBILITY, TIME DELAY. The input vector $\underline{x}$ is, therefore, defined to be $\underline{x} =$ [RECOMMENDATION OPINION, CREDIBILITY, TIME DELAY]. We will write this using short form notation as

$$\underline{x} = [\text{RO, CR, TD}]$$

A typical premise would be of the form 'Recommendation Opinion is *good*'. The structure of the left-hand side of the rule, therefore, is
IF ((RECOMMENDATION OPINION is X_1) AND
  (CREDIBILITY is X_2) AND
  (TIME DELAY is X_3))
$X_i i = 1, \ldots 3$ denotes in each case the fuzzy sets corresponding to the linguistic terms [POOR, AVERAGE, GOOD] for $i = 1,2$ and [LONG, AVERAGE, SHORT] for $i = 3$.

The actual form of the fuzzy sets and the associated membership function in each case were defined earlier.

In short form notation, the left-hand side of each of the rules would take the form
IF ((RO is X_1) AND (CR is X_2) AND (TD is X_3))
Since we are using the Takagi–Sugeno inference system, the right-hand side for the rule q has the form
Reputation $y_q = a_{qo} + a_{q1} x_1 + \cdots + a_{q3} x_3$, where $x_1, \ldots, x_3$ are the input variables.
that is, Reputation $y_q = a_{q0} + a_{q1} *$ (RECOMMENDATION OPINION) $+ a_{q2} *$ (CREDIBILITY) $+ a_{q3} *$ (TIME DELAY)
or in the short form notation
Reputation $y_q = a_{q0} + a_{q1} * \text{RO} + a_{q2} * \text{CR} + a_{q3} * \text{TD}$
Here $a_{qo}, a_{q1}, \ldots, a_{q3}$, are parameters. A typical example of the left-hand side in short form notation is:
IF (RO is GOOD) AND (CR is GOOD) AND (TD is SHORT)
The form of the qth rule in the rule base, in short form notation, therefore, is
IF (RO is X_1) AND (CR is X_2) AND (TD is X_3) THEN $a_{qo} + a_{q1} * \text{RO} + a_{q2} * \text{CR} + a_{q3} * \text{TD}$

The total number of possible fuzzy rules if we have n inputs and use K fuzzy sets to span the universe of discourse (UoD) for each input is equal to K^n. For the reputation problem, therefore, if we use 3 fuzzy sets to span each input factors UoD, the total number of rules that we can have is $3^3 = 27$. This takes into consideration all possible combinations of the inputs. These will be one such rule set for each third-party agent.

10.7.5 Calculation of Reputation

The reputation calculated by a particular third-party agent i is given by Equation 10.18, when the rule base is triggered by a particular set of value for $x = \{\text{RO, CR, TD}\}$, that is, $R_i = y(x)$. This value of R_i is weighted by the appropriate weighting factors for the first-, second- and third-hand opinion to obtain the aggregated reputation value for reputation queried agent k.

$$R_K = \alpha_1 \sum R_i + \alpha_2 \sum R_j + \alpha_3 \sum R_k \qquad \textbf{(10.18)}$$

10.7.6 Tuning Parameters and Trustworthiness of Opinion Adjustment

A set of data points was generated from third-party interactions and values of reputation opinion and trustworthiness of opinion. The form of G-Bell curves and the actual coefficients for the outputs of the T–S model were obtained using a fuzzy tuning algorithm [16]. The credibility or trustworthiness for the recommendation opinion for the ith agent can be calculated using the expression discussed in Section 10.5 for updating these, namely

$$TC(i, n, c) = \eta TC(i, n, c) + (1 - \eta)adj \qquad (10.19)$$

10.8 Summary

In this chapter, we addressed the situation where an agent needs to know the trustworthiness of another agent with whom it has had no previous interactions. The mechanism for doing this is to utilize the recommendation opinions of third-party agents. We discussed these methods of combining the recommendation opinions of third-party agents to obtain the aggregate reputation opinion of the collection of individual agents and group opinions. These included deterministic methods, Bayes method and a fuzzy system approach.

In addition to calculating the aggregate reputation opinion, we also discussed a method for updating the trustworthiness of the opinion of the third-party recommendation agent (or its credibility). These approaches provide methods of reputation assessment or reputation measurement.

References

[1] Sabater, J. and Sierra, C., (2002), "Reputation and social network analysis in multi-agent systems", in *Proceedings of the First International Joint Conference on Autonomous Agents and Multiagent Systems*, Bologna, Italy.

[2] Mui, L., (2002), *"Computational Models of Trust and Reputation: Agents"*, Evolutionary Games, and Social Networks", PhD thesis, Massachusetts Institute of Technology.

[3] Wang, Y. and Vassileva, J., (2003), "Bayesian network-based trust model", in *Proceedings of the IEEE/WIC International Conference on Web Intelligence*, Melbourne, Australia.

[4] Yu, B. and Singh, M.P., (2002), "Distributed reputation management for electronic commerce", *Comput. Intell.*, vol 18, no. 4, pp. 535–549.

[5] Abdul-Rahman and Hailes, (2000), "Supporting trust in virtual communities", *Proceedings of the 33rd Hawaii International Conference on System Sciences (HICSS)*, IEEE CS Press.

[6] Kamvar, S.D., Schlosser, M.T, and Garcia-Molina, H., (2003), "The eigentrust algorithm for reputation management in P2P networks", in *Proceedings of the 12th International WWW Conference*.

[7] Wang, Y. and Vassileva, J., (2003), "Bayesian network-based trust model in peer-to-peer networks" workshop on "Deception, Fraud and Trust in Agent Societies" at the autonomous agents and multi agent systems, 2003, *Conference (AAMAS-03)*, Australia.

[8] Damiani, E., Vimercati, S.D.C, Paraboschi, S., and Samarati, P., (2003), "Managing and Sharing Servents' reputation in P2P systems", *IEEE Trans. Knowl. Data Eng.*, vol. 15, no. 4, 840–854.

[9] Xiong, L. and Liu, L., (2004), "PeerTrust: Supporting reputation-based trust in peer-to-peer communities", *IEEE Transactions on Knowledge Data Engineering (TKDE)*, Special Issue on Peer-to-Peer Based Data Management, **16**(7), July.

[10] Hussain, F.K., Chang, E., and Dillon, T.S., (2004), "Classification of trust in Peer-to-Peer (P2P) communication", *Intl. J. Comput. Syst. Sci. Eng.*, vol. 19, no. 2, pp. 59–72.

[11] Dillon, T.S., Chang, E., and Hussain, F.K., (2004), "Managing the dynamic nature of trust", *IEEE Trans. Intell. Syst.*, vol. 19, no. 5, pp. 77–88.

[12] David, H., (1995), "A tutorial on learning with Bayesian networks", *Technical Report MSR-TR-95-06*, Microsoft Research.

[13] Mamdani, E. and Assilian, S., (1975), "An experiment linguistic synthesis with a fuzzy logic controller", *Int. J. Man Mach. Stud.*, vol. 7, pp. 1–13.

[14] Takagi, T. and Sugeno, M., (1985), "Fuzzy identification of systems and its applications to modeling and control", *IEEE Trans. Syst. Man. Cybernet.*, vol. 15, SMC-IS, no. 1, pp. 116–131.

[15] Schmidt, S., Steele, R., Dillon, T.S., and Chang, E., (2005), "Building a fuzzy trust network in unsupervised multi-agent environments," presented at *First IFIP WG 2.12 & WG 12.4 International Workshop on Web Semantics (SWWS '05), In conjunction with on the Move Federated Conferences (OTM '05)*, Agia Napa, Cyprus.

[16] Jang, R.J.S., Gupta, (1993), "ANFIS: Adoptive-networked-based fuzzy inference system" *IEEE Trans. SMC*, vol. 23, no. 3, pp. 665–685.

11

Reputation Systems

11.1 Introduction

Communications and transactions are often 'short-cut' over the Internet. A remote user or provider often receives brief information about another entity. The information might be incomplete. This inadequate information situation is mitigated through *trust and reputation* technologies. These provide an enhanced online shopping experience by allowing the customer to simulate the 'squeeze the tomato before you buy' principle. Decisions are made on the basis of inferences drawn from previous users' likes and dislikes. Furthermore, if the products' sellers could increase their appeal by communicating a 'virtually real' street presence, which would enable prospective buyers to try the product before making a decision to purchase, it would be to the vendor's benefit. In this chapter, we give an overview of the processes and basic techniques used in existing real-world business transactions. The Internet businesses that will be examined are as follows:

- BizRate
- Elance
- Alibris
- Money Control
- Yahoo!
- Epinions
- eBay
- CNET
- MovieLens.

Finally, we provide a general overview of their rating mechanisms and provide new guidelines for the development of a new generation of recommendation systems using the trust and reputation approach.

11.2 Reshaping e-Business with Reputation Technology

11.2.1 The Issues

The major issue in the distributed service-oriented network environment is that information exchanges between buyers and sellers are often asymmetrical. Problems include insufficient information about the merchant and goods and services offered. However, consumers often have to accept the fact that there is an element of risk involved in any transaction, and customers are often left in a very vulnerable position. Every consumer would like the opportunity to see a product three-dimensionally, experience its texture and try it to see if it meets his/her needs. A good analogy is

Trust and Reputation for Service-Oriented Environments   Elizabeth Chang, Tharam Dillon and Farookh Hussain
© 2006 John Wiley & Sons, Ltd

where the shopper visits a fruit market, and he/she applies the traditional quality test of 'squeezing the tomatoes before buying them'. This can be effectively done when the shopper is physically in the presence of the vendor, but the situation is different in the electronic marketplace. For example, the Internet seller might feel uncomfortable about sending a sample to a prospective buyer, as the former could not be certain of the latter's honesty. However, these e-business inefficiencies are now mitigated, and in many cases eliminated, through reputation technology.

11.2.2 Reputation Technology

Reputation technology is a computer-mediated communication tool for all agents in the distributed service-oriented environment. It collects information about products and the quality of services delivered by anonymous transactors or unknown business providers. *Reputation technology* provides, in real terms, the following benefits:

- It is a facilitator of quality control for businesses and their processes.
- It assists in the verification of sellers and fair trading.
- It provides insurance on transactions and real-world business situations, including assessment of a range of cues and recommendations relating to trustworthiness; this is normally not possible in the e-commerce or virtual world.
- It provides a technological platform for social recommendation and trustworthiness measures.
- It helps value-added relationships by improving the loyalty between the site and the customer, buyers and sellers, providers and end-users.
- It provides a business intelligence tool that helps business to learn from end customers and captures market needs, as well as monitoring competitors' performance and so forth.
- It has nothing to do with building a good-looking or sexy website, which gives little evidence of business solidity or information about the providers of the service.

Reputation technology is a computer-mediated communication tool, which collects information about anonymous transactions or unknown business providers in relation to the quality level of their products and services. The tool is useful to all agents in the distributed service-oriented environment, and it provides the following specific benefits:

(1) Quality control of business processes, seller verification, fair trading and insurance on transactions.
(2) Familiar real-world business situations and interaction, such as assessing a range of physical cues ('see and try' brick and mortar street presence).
(3) Ability to assimilate recommendations relating to trustworthiness from former users who can share their experiences. This is usually not possible in the e-commerce or virtual world.
(4) A social recommendation process and trustworthiness measure technological platform.
(5) Value-added relationship management that helps improve loyalty between the site and the customer, between buyers and sellers and between providers and end-users.
(6) A business intelligence tool that learns from end customers, captures market needs and watches competitor performance.
(7) A business analyser that accepts qualitative and quantitative feedback. Users can enter their profiles, opinions and comments. These data are converted to a statistical format and a list of recommendations is generated for the customer's information.
(8) A transparent assessment whiteboard with open system logic.
(9) 'At-one-glance' trustworthiness rating information, which generates simple, yet powerful, messages to be sent to buyers and sellers.

Trust and reputation technologies have nothing to do with building a good-looking or sexy website, which gives little evidence of a business' solidity or the providers behind it. Companies investing in these technologies will improve their business value, and customers reward these efforts by returning to the business that best matches their needs [1].

11.3 Trust and Reputation Systems versus Recommendation Systems

In this section, we discuss the difference between trust and reputation systems (TRS) and recommendation systems.

11.3.1 Reputation and Recommendation Systems for Business Intelligence

TRS and recommendation systems are two very important aspects of business intelligence. Both classes of systems are becoming increasingly popular with e-commerce site operators and service providers in service-oriented environments. We will discuss the specific features of both systems in the following text. This will help us to distinguish between the two classes of systems and their roles and dimensions within the context of the business intelligence environment.

Recommendation systems (RS) help to determine customer user profiles in relation to their goods and services usage record. Once RS has assigned a customer to a profile slot, based on data from his/her previous consumer behaviour patterns, the system is able to identify other goods and services that might be of interest. For example, if a customer has previously purchased optical cameras, laptops and other digital devices, then the idea of purchasing a digital camera might be appealing. However, RS do not make statements about product quality nor do they indicate or predict trustworthiness when the purchase is made under certain conditions.

TRS, in contrast to RS, address situations where the customer has decided to buy a certain product or service. TRS enables the shopper to evaluate brand quality by presenting a choice of products measured against certain quality assessment criteria. A decision can be made about which product to purchase, on the basis of the information provided to the purchaser. For example, a customer may want to buy a digital camera, but there are many brands and models available, such as Nikon and Pentax, from which to choose. The purchasing decision problem is compounded by the fact that there are thousands of websites selling digital cameras: which supplier should he/she buy from on the basis of assessment of supplier quality, and where does he/she find this information? It is necessary to obtain information on supplier reputation via TRS, which play a crucial role. It is clear that TRS as well as RS form two important dimensions of business intelligence.

11.3.2 Recommendation Systems

RS learns from a customer and recommends products that they would find most valuable from among the available options. Schafer *et al.* [1] present a detailed explanation of how RS help e-commerce portals increase sales, and they analyse six sites that use RS, including several sites that use more than one recommendation system. On the basis of the examples, they create a taxonomy of RS including the interfaces they present to customers, the technologies used to create the recommendations and the input they need from customers. RS have gained increasing popularity on the web, both in research systems (e.g. GroupLens and MovieLens) and online commerce sites (e.g. Amazon.com and CDNow.com), that offer RS as one way for consumers to find products they might like to purchase. RS are changing from novelties used by a few e-commerce sites, to serious business tools that are reshaping the world of e-commerce. Many of the largest commerce websites are already using RS to help their customers find products to purchase [1].

However, Swearingen and Sinha [2] argue that the effectiveness of any recommendation system is dependent on many factors and not just on the quality of the algorithm. They suggest that the goal of a recommendation system is to introduce users to items that might interest them and convince users to sample those items. They tried to answer the following question: 'what design elements of RS enable the system to achieve this goal'? They analysed the quality of recommendations and the usability of three RS for books (Amazon, RatingZone and Sleeper) and movies (Amazon.com, MovieCritic.com and Reel.com). Their research indicated that from a user's perspective an effective recommendation system would point users towards new items; furthermore, it provides details about recommended items, including pictures and community ratings. Finally, it provides ways to refine recommendations by including or excluding particular genres. Swearingen and Sinha [3] present design guidelines for a RS on the basis of the user's perspective towards the RS. They studied 11 online RS for designing these guidelines, which suggested which types of users are satisfied by interacting with the system and what specific system features lead to the satisfaction of those needs. The accuracy of recommendations made by any online RS is mostly dependent on the underlying collaborative filtering algorithm. Normally, the effectiveness of a recommendation system is measured by statistical accuracy metrics such as mean absolute error (MAE) [4]. However, satisfaction with a RS is only partially determined by the accuracy of the algorithm behind it [5].

RS rely on a series of techniques such as social filtering and matching, collaborative filtering, data mining and clustering. These are generally quite different from the class of the techniques that are described in the rest of this book for determining trust and reputation. For these reasons, we will not consider recommendation system further in this book. There is one exception, however, where they seem to be much closer to one another. This is where we have expert evaluation of a service, such as recommendation of movies. Here, it is important to know the reputation of the expert evaluator and, hence, there is a close link between these two classes of systems.

11.3.3 Reputation Systems

Reputation systems address the quality of goods and services, sellers or service providers, network agents or reviewers, which is based on a number of criteria. In the next few sections, we will mainly study the reputation systems of selected e-commerce portals. We will show examples of 10 websites that are adopting reputation systems. These sites are BizRate, Slashdot, Elance, BBC, Alibris, MoneyControl, Yahoo, Epinions, eBay and CNET. Other popular websites such as KuroHin.org, Reel.com, Amazon, CDNow.com, GroupLens and MovieLens and CitySearch, to name a few, have also adopted reputation systems. At the end, we briefly introduce a couple of RS. Owing to the space constraints of this book, we will not introduce all of them; however, readers are encouraged to visit websites mentioned above.

11.4 BizRate.com

BizRate.com is a shopping search engine, which lists stores and products from around the world. By listing price, availability and ratings information, shoppers can buy anything that is available for sale anywhere. BizRate.com indicates that they have an index of over 30 million product offers from more than 40 000 stores.

BizRate.com uses a proprietary shopping search and rating algorithm termed *ShopRank* that weighs price, popularity and availability of products against the reputations of merchants that sell them. BizRate indicates that they collect feedback about products and merchants from more than one million online buyers and sellers each month.

11.4.1 Reputation Rating of Products

All registered users can rate and review a product. Their ratings collectively form the *Overall Product Rating* shown as 1 to 5 stars, with 5 stars representing the highest rating. These stars are classified as in Figure 11.1.

Figure 11.1   Product rating scale. Reproduced by permission of Bizrate

Figure 11.2   Overall product rating with breakdown. Reproduced by permission of Bizrate

Many products have a further breakdown of the *overall product rating* based on the product, which further details the pros and cons of the product (Figure 11.2). Also, the reviews and comments of all the users are easily accessible. On the basis of this information, a user can decide if the product is suitable for his needs, which then facilitates his actual search for a suitable merchant.

11.4.2 Reputation Rating Merchants

Merchants can be found in the *Compare Prices and Stores* section of any product. This section lists the various merchants who carry the product, along with their price, availability and merchant rating. This rating is done on a four-level scale (Figure 11.3).

The smiley scale indicates the quality level of the store at the right price. BizRate requires a minimum of twenty surveys in the last ninety days to ensure it has a statistically significant rating

Figure 11.3   Merchant rating scale. Reproduced by permission of Bizrate

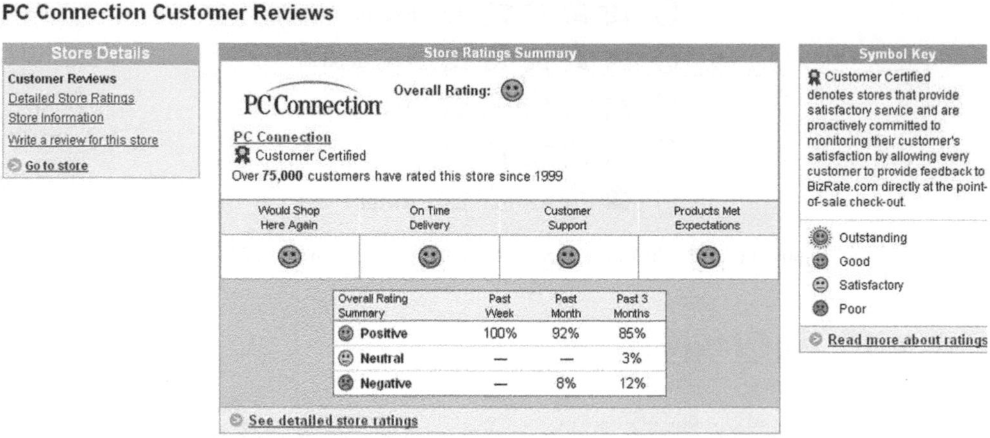

PC Connection Customer Reviews

Figure 11.4   Overall merchant rating. Reproduced by permission of Bizrate

$$\frac{\left(\begin{array}{c}\text{Average survey}\\\text{scores}\end{array} \times \begin{array}{c}\text{Number of}\\\text{surveys}\end{array}\right) + \left(\begin{array}{c}\text{Average member}\\\text{scores}\end{array} \times \begin{array}{c}\text{Number of}\\\text{member reviews}\end{array}\right)}{\left(\begin{array}{c}\text{Number of}\\\text{surveys}\end{array} + \begin{array}{c}\text{Number of}\\\text{member reviews}\end{array}\right)}$$

Figure 11.5   Merchant rating formula. Reproduced by permission of Bizrate

of a store's performance. The ratings once received are not permanent, but they can be changed for a valid reason, for example, if there is some evidence of suspicious review activity or if the store has changed its business model by introducing many modifications.

A *'customer certified' logo* provided by BizRate increases trust in that merchant (Figure.11.4). To gain this certificate, the store has to participate in a two-dimensional process. Firstly, the store should support BizRate to collect feedback and surveys from customers on completion of each individual deal and, secondly, customers should give at least a rating of 'satisfactory' or better on the 12 quality rating dimensions of service. If the store satisfies these criteria, it gets this certificate, which is displayed along with its rating by icon.

11.4.3 Reputation Algorithm

BizRate.com's rating system tries to ensure that it has reliable and up-to-date ratings. It utilizes a combination of information from (i) point-of-sale survey networks with (ii) member's feedback to calculate these ratings as shown in the equation (Figure 11.5).

To account for time variations, it utilizes only 90 days data in the calculations of the rating (Figure 11.6).

11.5 Elance.com

Elance's mission is to improve the way companies buy and manage services.

The Elance Online services marketplace is a web-based project marketplace that offers small businesses a quick, easy way to outsource projects, including web development, graphic design,

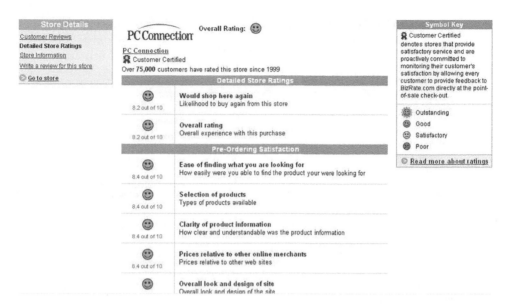

Figure 11.6   Detailed merchant overall rating. Reproduced by permission of Bizrate

software, engineering, administrative services, business strategy, writing and more. To do this effectively, it rates both service providers and buyers of the service.

11.5.1 Reputation Rating of Service Provider and Buyer

Feedback is provided on the basis of the quality and professionalism of the service provider. It is calculated on a five-point scale.

5.0 – Extremely satisfied
4.0 – Very satisfied
3.0 – Satisfied
2.0 – Not satisfied
1.0 – Extremely unsatisfied

In addition to leaving a feedback rating, Elance Online encourages all users to include specific comments with feedback whenever possible (Figure 11.7). In many cases, comments provide a more powerful indication of the service provider's performance. This provides a history of how an Elance Online service provider or an Elance Online buyer has been evaluated by others. Buyers and service providers show overall statistics along with individual feedback reviews for specific projects. Buyers rate and leave comments about service providers, while service providers leave only comments for buyers. Only transaction partners may leave feedback about each other.

11.5.2 Reputation Algorithm

Elance Online first calculates an overall score for each project. This per-project score is based on each criteria rating and the criterion's relative importance with respect to the provider's overall performance for the project.

- Responsiveness – 20 %
- Professionalism – 15 %

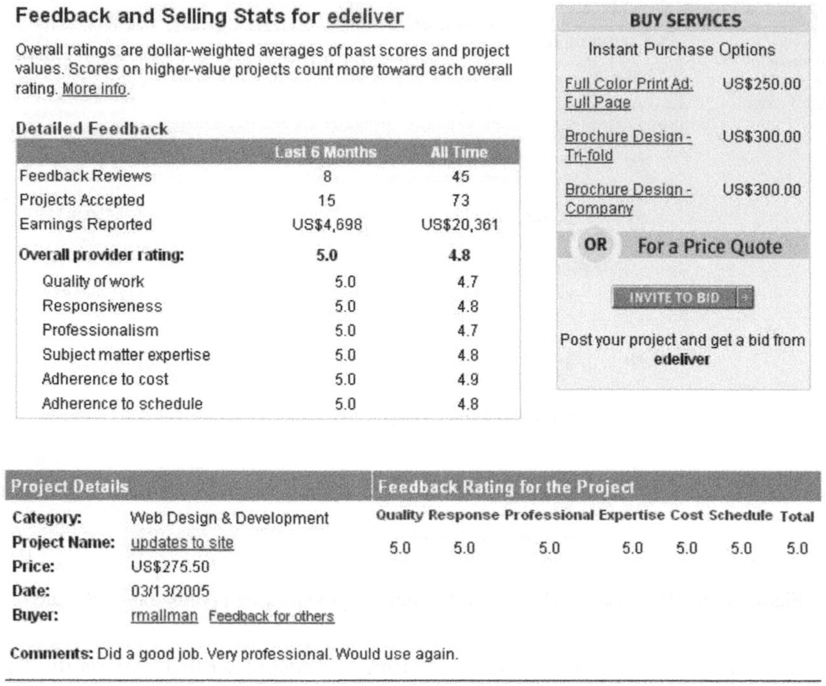

Feedback and Selling Stats for <u>edeliver</u>

Overall ratings are dollar-weighted averages of past scores and project values. Scores on higher-value projects count more toward each overall rating. <u>More info</u>.

Detailed Feedback

	Last 6 Months	All Time
Feedback Reviews	8	45
Projects Accepted	15	73
Earnings Reported	US$4,698	US$20,361
Overall provider rating:	**5.0**	**4.8**
Quality of work	5.0	4.7
Responsiveness	5.0	4.8
Professionalism	5.0	4.7
Subject matter expertise	5.0	4.8
Adherence to cost	5.0	4.9
Adherence to schedule	5.0	4.8

BUY SERVICES

Instant Purchase Options

<u>Full Color Print Ad:</u> <u>Full Page</u>	US$250.00
<u>Brochure Design -</u> <u>Tri-fold</u>	US$300.00
<u>Brochure Design -</u> <u>Company</u>	US$300.00

OR   For a Price Quote

[INVITE TO BID]

Post your project and get a bid from **edeliver**

Project Details		Feedback Rating for the Project						
		Quality	Response	Professional	Expertise	Cost	Schedule	Total
Category:	Web Design & Development							
Project Name:	<u>updates to site</u>	5.0	5.0	5.0	5.0	5.0	5.0	5.0
Price:	US$275.50							
Date:	03/13/2005							
Buyer:	<u>rmallman</u> <u>Feedback for others</u>							

Comments: Did a good job. Very professional. Would use again.

Figure 11.7   Elance's e-Delivers reputation rating. Reproduced with permission of Elance

- Subject matter expertise – 15 %
- Adherence to schedule – 10 %
- Adherence to cost – 10 %

Elance Online uses the per-project scores to calculate an overall numerical rating for each criterion, as well as an overall rating for the service provider. Scores on higher-value projects count more towards the overall ratings.

This way, when buyers rate service providers and vice versa, they are helping others make informed decisions about awarding and accepting projects in the future.

11.6 Alibris.com

Alibris provides information about independent sellers of books, music and movies. It claims to offer access to over 50 million used, new and out-of-print books.

11.6.1 Reputation Rating Merchants

Alibris uses seller ratings based on seller fill rate for the previous 31 to 120 days (Figure 11.8). The rating system and the associated fill rates are as follows:

- 95–100 % = 5 stars
- 85–94.99 % = 4 stars (new sellers are put in this category for the initial 90 days)
- 70–84.99 % = 3 stars
- 60–69.99 % = 2 stars
- 1–59.99 % = 1 star

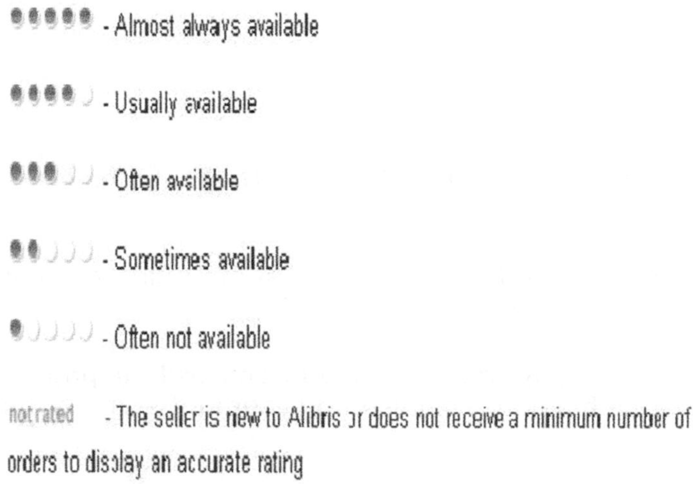

● ● ● ● ● - Almost always available

● ● ● ● ◡ - Usually available

● ● ● ◡ ◡ - Often available

● ● ◡ ◡ ◡ - Sometimes available

● ◡ ◡ ◡ ◡ - Often not available

not rated - The seller is new to Alibris or does not receive a minimum number of orders to display an accurate rating

Figure 11.8 Alibris' merchant rating scale. Reproduced by permission of Alibris. Copyright (c) 1998–2005 Alibris. All rights reserved

These ratings are updated regularly to allow for time variations. Sellers are also treated as 'Not Rated' if they have few sales (< 3) for the measurement period. The fill rate represents the number of Alibris orders that sellers fill/deliver. A minimum of 85 % is required for listings.

11.6.2 Reputation Rating of Products

Since Alibris is a second-hand store, it does not rate the quality of the product on the basis of user or expert reviews; instead, it rates the products on the basis of their condition (Figure 11.10). This is important because the quality of a second-hand product depends largely on its condition. A product's condition is rated on a six-scale basis, which is as follows: five red dots denote a new item, while one to five blue dots denote the condition of a used item, with five dots denoting the best product. This rating is illustrated in the Figure 11.9.

11.7 MoneyControl.com

MoneyControl is a personal finance portal built to track the Indian stock market. People can read articles and views from various people about their prediction of the stock market and rate these opinions. On the basis of all these views and information, they can present their own insights or make decisions about the stocks they want to invest in.

11.7.1 Reputation Rating of Opinions

Opinions given by various users can be rated by peers. This rating is on a 5-star scale:

- 1 – Miserable
- 2 – Poor
- 3 – Average
- 4 – Good
- 5 – Excellent.

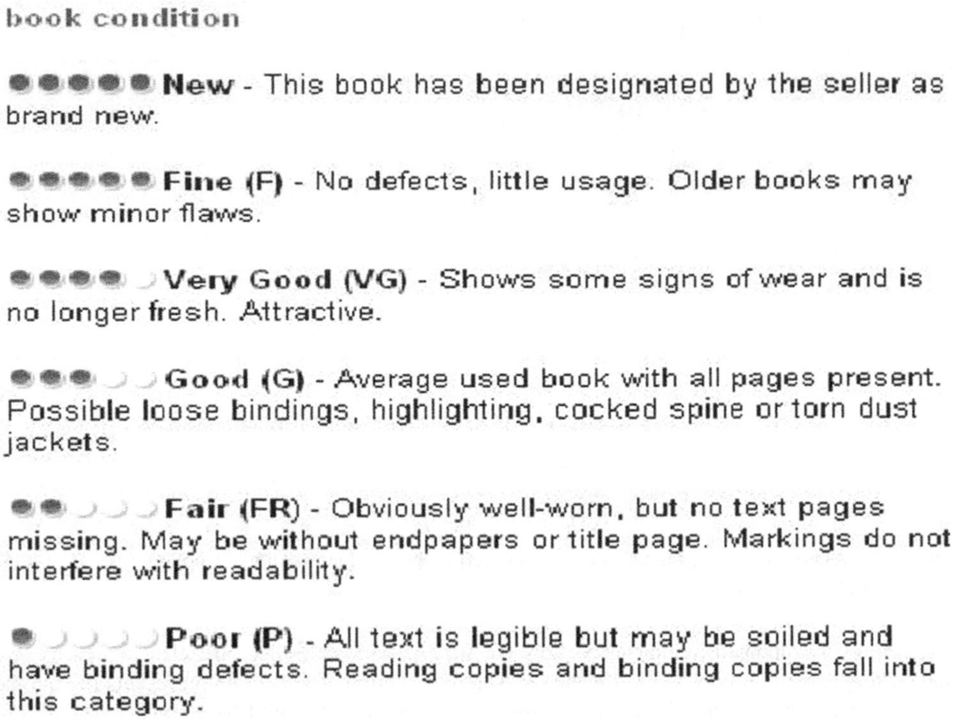

book condition

●●●●● **New** - This book has been designated by the seller as brand new.

●●●●● **Fine (F)** - No defects, little usage. Older books may show minor flaws.

●●●● **Very Good (VG)** - Shows some signs of wear and is no longer fresh. Attractive.

●●● **Good (G)** - Average used book with all pages present. Possible loose bindings, highlighting, cocked spine or torn dust jackets.

●● **Fair (FR)** - Obviously well-worn, but no text pages missing. May be without endpapers or title page. Markings do not interfere with readability.

● **Poor (P)** - All text is legible but may be soiled and have binding defects. Reading copies and binding copies fall into this category.

Figure 11.9   Rating the condition of books. Reproduced by permission of Alibris. Copyright (c) 1998–2005 Alibris. All rights reserved

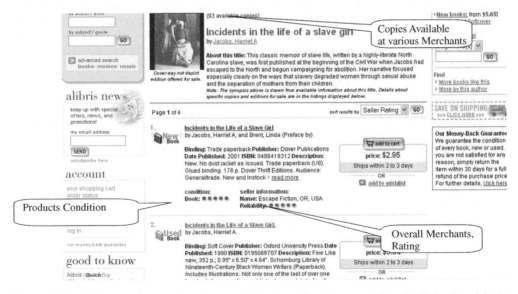

Figure 11.10   Rating merchants and products available. Reproduced by permission of Alibris. Copyright (c) 1998–2005 Alibris. All rights reserved

Having read the opinion and the peers' rating of the opinion, a member can decide whether he wants to accept the opinion, or else, he may decide that he has an opinion already about something and can rate the opinion of others. It is possible to track a particular sector of stocks, which will present you with the views of various people about that sector.

Another very interesting feature is the tracking of the stock price at the time the message was posted and the current stock price. This way it can be easily seen if the opinion was in the correct direction or if the stock movement is in the opposite direction to that of the prediction. Please refer to http://moneycontrol.com/for details.

11.7.2 Reputation Algorithm

The overall rating of an article or opinion is a simple average of all the ratings by the various users. So, this overall rating is not an integer but is expressed as a number between 0 and 5. All the ratings from different users are stored on the system and can be viewed easily, if required. On the basis of the views and ratings, a user can decide if he wants to track the particular member's views or not.

11.7.3 Reputation Rating of Reviewers

Members can track other members who they find consistently give good opinions or have knowledge of a sector of stocks. The number of members tracking your opinions can be viewed as your trusting agents, and the number of members tracked by you forms your trusted agents network. Members can also choose to ignore certain members whose opinions they do not trust, and these can be viewed as malicious agents. All this information about the trusted agents, trusting agents and malicious agents can be seen in the profile of a member along with the messages that have been rated by peers and the messages rated by the member.

An example of tracking a given member is illustrated below. The MoneyControl's tracking members page includes information such as:

— The date since tracking took place
— Number of messages received and posted
— How many members have rated you
— The rating level you are at the moment
— The tracker details
— How many ratings you should ignore
— Details of ratings and messages.

Additional information is available at http://moneycontrol.com.

11.8 Yahoo.com

Yahoo is one of the most popular portals in the world which offers a variety of customer services, including search engine services, sales and products, personalized services, small business services, free email services, chat rooms, news, sports, investment advice, health information and entertainment services.

11.8.1 Reputation Rating of Products

Normally, the overall product ratings are displayed in search results. The average product rating is a simple average of the total number of posted ratings for that product. All *registered users* are allowed to rate the products, which forms the *user rating* (Figure 11.11).

Along with this there is another type of rating called the *expert rating*, which is the rating given to this product by known experts or trusted sources. This inspires higher confidence among users than user ratings (Figure 11.12).

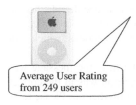

Average User Rating: ★★★★★ (249 Ratings) Read reviews
Average Expert Rating: ★★★★★ (1 Rating) Read 1 review
Hard Drive Size: 20 GB
Memory Type: Hard Disk Drive
Supports: MP3, WAV, AAC, AIFF
Battery Life: 12 hour(s)
Save this product | Set an alert | Email to a friend

Average User Rating
from 249 users

Figure 11.11   Average user rating. Reproduced with permission of Yahoo! Inc. © 2005 by Yahoo! Inc.
YAHOO! and the YAHOO! logo are trademarks of Yahoo! Inc

Expert Reviews (BETA)

Trusted Source

Average Expert Rating: ★★★★★ (1 Rating, 1 Review)

Source	Brief Review	Rating
CNET	Apple's fourth-generation iPod delivers notable improvements to an already excellent product--and at a lower price. more...	★★★★★

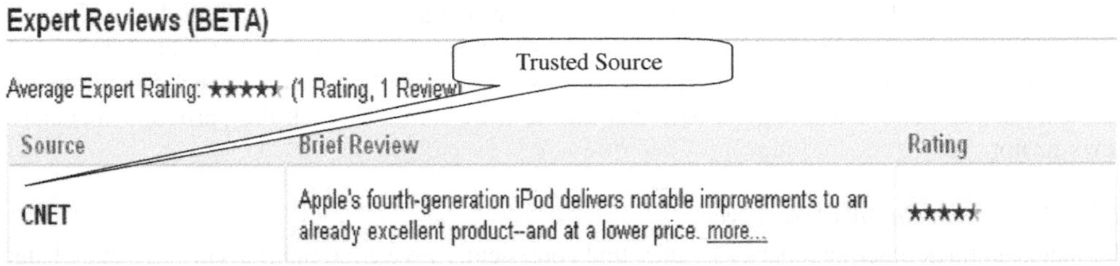

Figure 11.12   Average expert rating. Reproduced with permission of Yahoo! Inc. © 2005 by Yahoo! Inc.
YAHOO! and the YAHOO! logo are trademarks of Yahoo! Inc

Excellent	☆☆☆☆☆
Very good	☆☆☆☆☆
Good	☆☆☆☆☆
Fair	☆☆☆☆☆
Poor	☆☆☆☆☆
No rating submitted	N/A

Figure 11.13   Merchant rating scale. Reproduced with permission of Yahoo! Inc. © 2005 by Yahoo! Inc.
YAHOO! and the YAHOO! logo are trademarks of Yahoo! Inc

11.8.2 Reputation Rating for Merchants

The rating system collects feedback from shoppers about merchants and then uses the feedback to
determine an overall rating for that merchant (Figure 11.13).

11.8.3 Reputation Algorithms

The *average merchant rating* is an average of total users' response (Figure 11.14). Ratings by
customers who have made an online purchase and submitted a rating based on a post-transaction
survey are given *twice* the weight of ratings submitted by others who did not make the purchase.
From the ratings of a number of users, the overall average rating is calculated.

Figure 11.15 is an example of how a merchant rating appears on Yahoo! Shopping.

Against some merchants, there may be a 'Not Yet Rated' icon because no rating is displayed
until a merchant has at least five posted ratings. This is to improve the statistical relevancy of the
displayed merchant ratings.

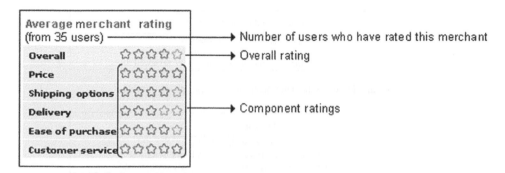

Figure 11.14   Overall rating and five-criterion rating. Reproduced with permission of Yahoo! Inc. © 2005 by Yahoo! Inc. YAHOO! and the YAHOO! logo are trademarks of Yahoo! Inc

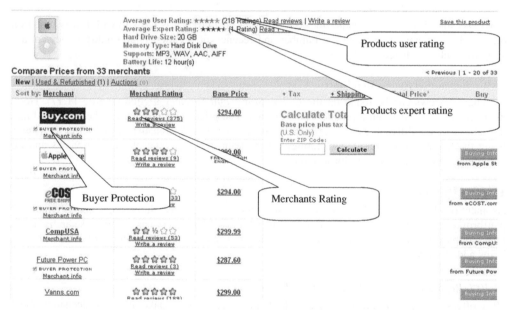

Figure 11.15   Yahoo's overall reputation system of products and merchants. Reproduced with permission of Yahoo! Inc. © 2005 by Yahoo! Inc. YAHOO! and the YAHOO! logo are trademarks of Yahoo! Inc

11.9 Epinions.com

Epinions.com is a consumer reviews portal on the Internet. It offers product ratings, merchant ratings, review ratings as well as reviewer ratings. The objective is to provide reliable sources that provide valuable consumer insight, unbiased advice, in-depth product evaluations and personalized recommendations to the public and help customers make informed buying decisions. It tries to capture customers' needs by learning what people like and dislike about shopping, how they use the Internet in their everyday lives, and what would make them buy online rather than off-line.

11.9.1 Reputation Rating of Products

Epinions' *product ratings* ranks the products from 1 to 5 stars, with 5 stars denoting the best products (Figure 11.16). All of the products in a category or subcategory are ranked on the basis of their overall star rating.

Product name and product image
Overall product rating with respect to ease of ordering, customer service, on-time delivery, selection along with the total number of epinion users who have reviewed it.

Figure 11.16   Schematic diagram of overall product rating

Store name
Overall store rating with respect to ease of ordering, customer service, on-time delivery, selection along with the total number of epinion users who have reviewed it.

Figure 11.17   Schematic diagram of overall merchant rating

11.9.2 Reputation Algorithms

The exact algorithm for calculating the overall rating of a product is not known but it is based on the weighted average rating and the number and recency of reviews. High-quality reviewers are assigned high weightage.

11.9.3 Reputation Rating Merchants

Merchant reviews can be found on the *Online Stores and Services* section (Figure 11.17). Here, the different stores that sell a product can be found. The prices of a product can be compared and the ratings of the different stores can be seen. Users rate these merchants from 1–5 ticks, with 5 being the highest. The overall store ratings can be seen as an average of all the reviews and all individual reviews can be seen for any merchant.

11.10 eBay.com

As mentioned in Chapter 7, eBay is a virtual online marketplace. It provides an international platform where anyone can trade anything. It sells goods and services for diverse individuals, business players and service providers. So far, it has handled 25 million items of sale with new items added daily.

11.10.1 Reputation Rating Members (Both Buyers and Sellers)

eBay offers *reputation rating systems* to help its prospective users make informed decisions about which members to buy from. It only gathers feedback from *eBay members* (eBay sellers or eBay buyers) and calculates the reputation of target eBay members. The rating system considers eBay's members as most important; hence, all positive and negative ratings of members stay with them for life.

In eBay, members only provide feedback about their interaction with another member. eBay determines a member reputation rating based on feedback that buyers and sellers provide on each other. They use a three-point rating feedback scale namely (positive, negative, neutral). Information on this is given on http://pages.ebay.com.au/help/feedback/evaluating-feedback.html. Note that unless you are actually making a transaction you are not permitted to provide feedback.

11.10.2 Reputation Algorithms

It utilizes a differencing mechanism by calculating the difference between positive ratings and negative ratings to give the feedback score.

In the example shown in Figure 11.18, the feedback score is $1715 - 3 = 1712$. Positive feedback is shown as a percentage of the positive over a combined total, which is 1715/1718, that is, 99.8 %.

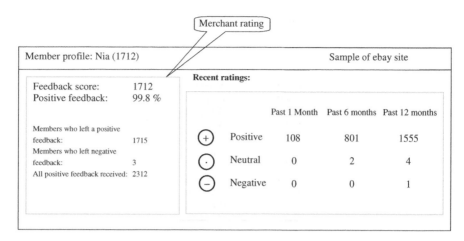

Figure 11.18   Schematic diagram of a member rating in eBay

'Members who left a positive feedback' is the number of unique members who have given this member a positive rating. If a member has had several transactions with another member and leaves more than one positive rating, they will still count only once in this number.

On the basis of the points scored, an eBay member is assigned stars. As the number of positive feedback score increases, the higher is the star level assigned, from yellow star, which is between 1 to 49 points, to red shooting star, which is over 100 000 points.

The portal also defines Power Sellers who have achieved 98 % positive feedback ratings (http://pages.ebay.com.au/services/buyandsell/powersellers.html)

11.11   CNET.com

CNET attempts to provide expert and unbiased advice on technology products and services to inform users and expedite purchases. By integrating an extensive directory of more than 200 000 computer, technology and consumer electronic products with editorial content, downloads, trends, reviews and price comparisons, CNET attempts to give users the most up-to-date and efficient shopping resources on the Web.

11.11.1 Reputation Rating of Products

CNET *Product rating* is done by an expert panel of editors. CNET editors are special employees of CNET with expertise in providing reviews of products. CNET indicates that their editors offer hands-on experience. Thus, they get the actual products i.e. they see them, touch them and test them. Most of the products they review are provided to them by manufacturers for trial. Once the review is complete, the product is returned and with the consent of the seller the review is published.

There are five categories of generic rating criteria for all products:

1. Set up
2. Design
3. Features
4. Performance
5. Service and support.

These categories change for different products. For example, laptops may have battery life taken into consideration while cameras may have image quality as the criterion. In each of the above categories,

Table 11.1 Product quality
rating

Rating	Explanation
1–1.9	Abysmal
2–2.9	Terrible
3–3.9	Very poor
4–4.9	Poor
5–5.9	Mediocre
6–6.9	Fair
7–7.9	Good
8–8.9	Very good
9–9.9	Excellent
10	Perfect

there are further criteria to be measured. These criteria could be the same for all categories or they might vary, depending on the type of product. For example, for a camera, the criteria for each category above is as follows:

- Point-and-shoot
- Midrange
- Semipro.

11.11.2 Reputation Algorithm

In order to calculate the quality of the product, CNET has set coefficients that weight each criterion differently as a percentage of the total. The overall scale for quality is between 0 and 10. On the basis of the input from the above criteria, the rating is calculated using the coefficients set for the particular product. For example, for a *mainstream Notebook* the rating is calculated as follows:

$$Rating = ((design * 0.2) + (features * 0.3) + (performance * 0.1) + (batterylife * 0.2)$$

$$+ (service\ and\ support * 0.2))/10$$

The outcome of this equation is a number in the range from 1 to 10 (Table 11.1).

Along with editor ratings, CNET also lets all its registered members express their personal views on any article or product (Figure 11.19). *User ratings* allow the users to rate a product and express their views in the form of *pros* and *cons*. These ratings collectively form the *average user rating*.

11.11.3 Reputation Rating Merchants

CNET merchant rating is termed *certified store rating* and a store can get this rating if it participates in the CNET certified store programme (Figure 11.20). All the stores in this network have to adhere to the CNET code of conduct and a strict set of quality service guidelines and CNET certified rating criteria. These ratings are indicated as stars next to the stores' sites. Stores can receive a rating from half a star to 5 stars, depending on compliance. The total number of stars received for each of the four categories below will equal the store's overall rating (new stores display the status of *Not Yet Rated* during their initial 30-day evaluation period). The star legend is shown in Figure 11.21 for reference. These ratings are updated weekly and registered users can rate a store on the basis of their experience and write a review for the store.

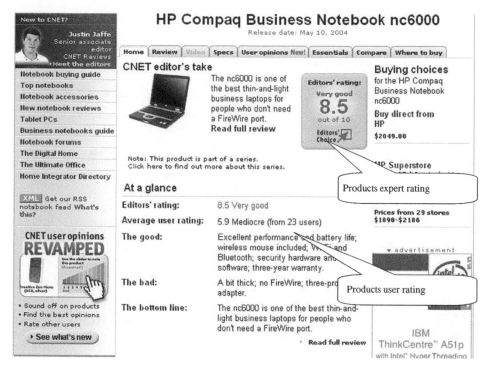

Figure 11.19 CNET product rating. Used with permission from CNET Networks, Inc. Copyright © 1995–2005 CNET Networks, Inc. All Rights Reserved

Rating description	Visual representation (Star rating)
Poor	From ☆ to ★
Fair	From ★☆ to ★★
Good	From ★★☆ to ★★★
Very good	From ★★★☆ to ★★★★
Excellent	From ★★★★☆ to ★★★★★

Figure 11.20 CNET merchant rating scale. Used with permission from CNET Networks, Inc. Copyright © 1995–2005 CNET Networks, Inc. All Rights Reserved

11.11.4 Reputation Algorithm

There are four categories that are involved in deciding the rating for a store.

- Site functionality – Maximum of 1 star
- Store standards – Maximum of 1 star
- Order fulfilment – Maximum of 1 star
- Customer feedback – Maximum of 2 stars.

Each category is further decomposed into several criteria that the store has to satisfy.

Site functionality: Stores are required to tick several boxes that represent criteria associated with each category. For site functionality, these include shopping carts, secure servers, credit card

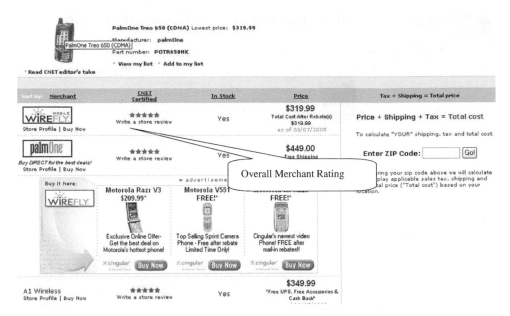

Figure 11.21 Certified merchant rating. Used with permission from CNET Networks, Inc. Copyright © 1995–2005 CNET Networks, Inc. All Rights Reserved

Figure 11.22 Detailed merchant rating. Used with permission from CNET Networks, Inc. Copyright © 1995–2005 CNET Networks, Inc. All Rights Reserved

acceptances, return policies, FAQ/help policies, credit card charges and product specific pages (Figure 11.22).

Store standards are judged on the basis of prompt email responses, privacy policies, store contact information and supply of up-to-date prices, shipping costs and in-stock information to CNET.

Order fulfilment is based on efficiency of resolving order problems, displaying of prices without hidden costs, e-commerce compatible site, efficiency of executing orders without major time delays and delivery of products in a professional manner.

Customer feedback is based on efficiency of customer service and staff, accurate on-site information (price, description, costs, etc.), timely delivery of products and positive feedback greater than 85 %.

11.12 MovieLens Recommendation Systems

Movielens.umn.edu utilizes a RS. However, it is discussed here since it involves evaluation of a service. The site uses 'collaborative filtering' technology to make recommendations of movies/videos that might be enjoyed by a user and helps him/her avoid the ones that would not be that enjoyable. On the basis of the ratings of the movies that have been seen by a user, MovieLens generates personalized predictions of the movies that would be enjoyed by the user. The more the user uses the system to rate the movies he/she has seen, the more accurate the predictions.

Figure 11.23   Movie rating scale. Reproduced with permission of Movielens

> You've searched for **all titles**.
> Found **30** movies, sorted by **Prediction**
> Genres: **All** | Exclude Genres: **None**
> Dates: **All** | Domain: **Ratings** | Format: **All** | Languages:
> Show **Printer-Friendly Page** | **Download Results** | **Suggest a Title**
>
> Page 1 of 2   page 2>

(hide) Predictions for you ⤳	Your Ratings	Movie Information	Wish List
★★★★★	5.0 stars ⌄	**Independence Day (ID4) (1996)** DVD info|imdb Action, Sci-Fi, War	⬜
★★★★★	5.0 stars ⌄	**Lawrence of Arabia (1962)** DVD info|imdb Adventure, War	⬜
★★★★★	5.0 stars ⌄	**Lord of the Rings: The Return of the King, The (2003)** DVD VHS info|imdb Action, Adventure, Fantasy	⬜
★★★★⭑	4.5 stars ⌄	**American History X (1998)** info|imdb Drama	⬜
★★★★⭑	4.5 stars ⌄	**Minority Report (2002)** DVD VHS info|imdb Action, Sci-Fi, Thriller	⬜
★★★★⭑	4.5 stars ⌄	**Raiders of the Lost Ark (1981)** DVD info|imdb Action, Adventure	⬜
★★★★⭑	4.5 stars ⌄	**Terminator 2: Judgment Day (1991)** DVD info|imdb Action, Sci-Fi, Thriller	⬜
★★★★	4.0 stars ⌄	**Bad Boys (1995)** DVD info|imdb Action	⬜

Figure 11.24   Movies rated for the recommendation system. Reproduced with permission of Movielens

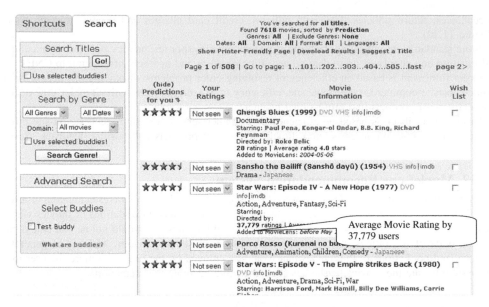

Figure 11.25   Recommended movies with ratings. Reproduced with permission of Movielens

11.12.1 Recommendation of Products

The above site is dedicated to movies and videos. All registered users can rate movies/videos on the basis of a 5-star scale, with 5 stars representing the best (Figure 11.23).

Before the *recommendation system* can *predict* the kind of movies enjoyed by a user, it must gather data about the user's likes/dislikes (Figure 11.24). To enable this, a user is first prompted to rate at least 15 movies. All the user ratings can be seen in his profile and contribute towards the overall rating for the movie.

All ratings given by users for a movie they have seen, collectively form the overall rating of the movie (Figure 11.25). This is calculated as a simple average of all the ratings and can be viewed when a movie is recommended by the system. The unseen movies can be added to the wish list if the user chooses and he can later printout a copy of his wish list for reference.

11.13 Review of Reputation Systems

Table 11.2 gives a high-level view of the technology adoption (see the vertical bar and the black dots) by the listed companies (see horizontal bar) for their business intelligence.

Table 11.2   Summaries of reputation systems

	BizRate	Elance	Alibris	Money control	Yahoo	Epinions	eBay	CNET	Movie lens
Rating merchants' reputation	●	●	●		●	●	●	●	
Rating products' reputation	●		●		●	●		●	●
Rating customers' reputation		●					●		
Rating reviews' reputation				●					
Rating reviewers' reputation				●					
Recommendation systems									●

11.14 Summary

In this chapter, we opened by discussing the difference between RS and trust/reputation systems. These two classes of systems address two very important dimensions of business intelligence. RS identify what other goods and products the user might purchase or consume, on the basis of information drawn from his/her previous consumption and purchasing behaviour patterns. The inferences are drawn without reference to either the quality of the goods or the effectiveness/efficiency of the supplier.

Reputation systems, on the other hand, help customers or businesses to evaluate the quality of goods and services that they wish to purchase or consume against certain important features of these goods and services.

As these reputation systems actually make recommendations to the consumer about the trustworthiness or reputation of goods and services, people sometimes tend to confuse them with RS. However, as explained above, they are two quite distinct classes of systems. This book is mainly concerned with TRS; therefore, in this chapter we have examined a number of commercial implementations for these systems on the Internet on several e-commerce sites.

The key concepts are the following:

- Trust and reputation systems
- Reputation values
- Reputation aggregation methods
- Recommendation systems
- The difference between reputation and recommendation systems
- Business intelligence with trust and reputation systems
- Business intelligence with recommendation systems.

References

[1] Schafer J.Ben., Konstan J. & Riedl J. (1999) 'Recommender systems in e-commerce' In *Proceedings of the First ACM Conference on Electronic Commerce*, pp. 158–166, ACM Press, 1999.
[2] Swearingen K. & Sinha R. (2001) 'Beyond algorithms: An HCI perspective on recommender systems'. *Recommender Workshop. SIGIR 2001*, UC, Berkeley.
[3] Swearingen K. & Sinha R. (2002) 'Interaction design for recommender systems'. In *DIS 2002*. ACM Press 2002.
[4] Breese J., Heckerman D. & Kadie C. (1998) 'Empirical analysis of predictive algorithms for collaborative filtering', *Proceedings of the 14th Conference on Uncertainty in Artificial Intelligence*, pp. 43–52, Madison, Wisconsin, July 1998.
[5] Herlocker J., Konstan J.A. & Riedl J. (2000) 'Explaining collaborative filtering recommendations', *Proceedings of the ACM 2000 Conference on Computer-Supported Collaborative Work, Philadelphia, Pennsylvania, United States*.

WEBSITES

eBay 'Star Signs' Available online: http://pages.ebay.com/help/feedback/reputation-stars.html Bottom of Form.

http://www.amazon.com
http://www.amazon.com/gp/help/seller/feedback-popup.html, accessed 2005.
http://www.amazon.com/exec/obidos/subst/community/reviewers-faq.html, accessed 2005.
http://www.amazon.com/exec/obidos/tg/browse/-/537868/pop-up/ref=olp_wa_1/, accessed 2005.
http://www.yahoo.com, accessed 2005.
http://help.yahoo.com/help/us/shop/shop-77.html, accessed 2005.
http://help.yahoo.com/help/us/shop/shop-06.html, accessed 2005.
http://help.yahoo.com/shop/shop-68.html, accessed 2005.

http://www.Epinions.com, accessed 2005.
http://www.Epinions.com/help/, accessed 2005.
http://www.ebay.com, accessed 2005.
http://pages.ebay.com.au/help/feedback/evaluating-feedback.html, accessed 2005.
http://pages.ebay.com.au/services/buyandsell/powersellers.html, accessed 2005.
http://www.bizrate.com, accessed 2005.
http://www.bizrate.com/content/ratings_guide.html, accessed 2005.
http://reviews.CNET.com, accessed 2005.
http://reviews.CNET.com/4002-5_7-5100969.html, accessed 2005.
http://www.brainyencyclopedia.com/encyclopedia/m/mo/movielens.html, accessed 2005.
http://moneycontrol.com/stocks/index.php, accessed 2005.
htpp://sellers.alibris.com/policies.cfm#4, accessed 2005.

12

Trust and Reputation Prediction

12.1 Introduction

In the previous chapters, we presented a technique for determining the trust that an agent 'a' has in another agent 'b' or a service provider 'b' for a particular context in time slot N. In the case of a service provider, it was noted that the service agreement between an agent and the service provider may involve several criteria $C_1, C_2, \ldots, C_e$. With respect to each of these criteria, it was emphasized that it was important for three factors to be taken into account in determining the value of trust. These factors were the clarity, the commitment and the influence of each criterion. The calculation of trustworthiness by an agent for a service provider in the specific context and time slot N can then be determined as explained in Chapters 6 and 10. The important point to realize is that this value of trustworthiness represents a snapshot for the interaction that took place in the time slot N. After n interactions, the agent **a** could have stored in the database trustworthiness values corresponding to the n interactions, where $n \leq N$. This is illustrated in Figure 12.1, where we have assumed one interaction per time slot. (Note $X = -1$ is not included for this time slot as the agent is known after the interaction).

Using the stored information from a previous interaction enables calculation of the trustworthiness value for time slot $N + 1, N + 2, \ldots$; this is the major consideration at this stage. Note that instead of relying on the outcome of a previous interaction, although this information is very important, to assess future trustworthiness, it is important to recognize that it is only one source of data. One could be faced with the situation where a dishonest service provider lulls the agent into a false sense of security about its trustworthiness by producing a satisfactory interaction, which might result in the unsuspecting customer becoming a victim of fraud or poor service.

Furthermore, we do not appear to be making use of all available information, namely, the n trustworthiness values stored in the database. In this chapter, we will explain methods used to track the history of previous interactions, which will enable us to more accurately determine future trustworthiness levels.

12.2 Considerations in Trustworthiness Prediction

As noted in Figure 12.1, it is possible to have n previous values of trustworthiness stored in the database after n previous interactions. The question that arises here is, 'is it useful to utilize this information to predict future trust?' An examination of Figure 12.1 shows that the trustworthiness value may vary over time. This variation could be due to several sources:

(a) The service provider may have different capabilities to provide the service depending on the load being experienced at a particular time.

Trust and Reputation for Service-Oriented Environments   Elizabeth Chang, Tharam Dillon and Farookh Hussain
© 2006 John Wiley & Sons, Ltd

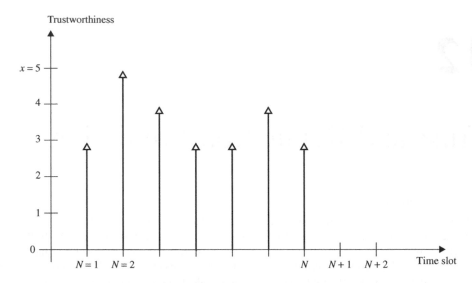

Figure 12.1 Trustworthiness values over service time intervals

(b) There could be a change of management or a change of policy in the provision of the service by the service provider.
(c) There could be the possibility that the service provided is slowly deteriorating over time.
(d) There could be the possibility of cyclical fraud.

This variability over time needs to be modeled in any prediction method.

12.2.1 Recency is Important

Generally speaking, given the fact that the service provider's service level could be varying over time, we would like to utilize the historical sequence of trustworthiness values generated to predict future trustworthiness values. However, it is also important to weight the trustworthiness values obtained in recent interactions more heavily, so as to avoid modeling behaviors that may no longer be relevant to the fulfilment of the service agreement by the service provider in the future. Hence, any prediction method should weight more recent trustworthiness values more heavily, progressively reducing the effect of older values. At the same time, it should be able to filter out any spike or sudden high trustworthiness value as not providing a good basis for the future prediction.

12.2.2 Understanding Trend is Important

The service provider could be providing a progressively deteriorating level of service as shown in Figure 12.2.

The trustworthiness values indicate a general but progressive decline in trustworthiness value. Thus, at the $(N + 1)$th interval it may be more appropriate to use the dotted value indicating the $(N + 1)$th trustworthiness value rather than trustworthiness indicated at time slot N. Hence, it is important to know not only the snapshot value at the time slot N but also the trend line over the last few time slots. In fact, sometimes this trend could be superimposed in other variations as can be seen in Figure 12.3. Here, the dotted envelop indicates there is a progressive trend downwards in trustworthiness values even though they may be moving up and down within the

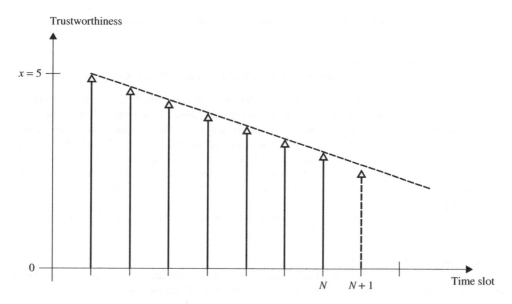

Figure 12.2   Progressive decline in trustworthiness values

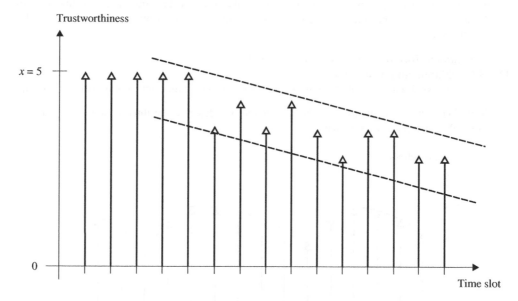

Figure 12.3   Decline in trustworthiness value with time

downtrending envelop. It is important that any prediction technique is able to pick up these trends in trustworthiness variation if they exist.

The need for modeling the trend component can be ascertained by using a statistical test [1] that fits a linear regression line to the reputation or trust series and determines if its gradient is zero by using a t-distribution.

12.2.3 Detecting Cyclical Fraud or Cyclical Poor Performance is Important
Sometimes, a service provider or an agent can either:

(i) perpetrate cyclical fraud or
(ii) periodically provide poor performance in a cyclical fashion owing to other circumstances.

Cyclical fraud can occur where a service provider or agent can provide good service levels or good-quality goods (in the case of a seller) on several occasions. Every now and then, the seller may provide poor-quality service or sell poor-quality goods. If the trusting agent is averaging the trustworthiness over a number of transactions, this occasional fraud is likely to be masked. Frequently, the dishonest service provider might link the poor service to items where it makes a large gain, whereas he/she provides good service for items where there is a lower gain or cost. The periodical deterioration in trustworthiness could also be associated with periods of high demand on the service provider.

Such cyclical variations are sometimes referred to as *seasonality*. The prediction should be capable of detecting such seasonality (Figure 12.4) [7].

12.2.4 Variability in Trustworthiness
In some cases, there may be considerable variability in the quality of service delivered by a service provider. This variability could be caused by different subcontractors or different component sub-services that a service provider may use. It may also be determined by the work load the service provider is experiencing at a particular time. It could also be due to a change of management or a different management policy. Some of this variability could occur within a given time slot. In which case, it is useful not only to know the mean value of trustworthiness within the time slot but also the variance.

12.2.5 Initiation, Trustworthiness Measure and Prediction
12.2.5.1 Initiation and Association
In this section, we discuss the following two important concepts in the context of a trust relationship:

(i) *Initiation of the trust relationship*: In this section, we describe the three different ways that a trust relationship can be initiated.

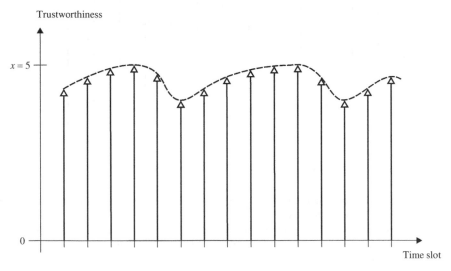

Figure 12.4   Trustworthiness prediction

(ii) *Association*: The initiation of the trust relationship results in the actual association or involvement of the trusting agent in an interaction with the trusted agent.

The *association* between the trusting agent and the trusted agent is defined as the actual interaction between two agents with a common need or that share a common interest in a particular time slot.

There are three scenarios that form the association:

- Association through direct personal contact or interaction
- Association through third-party introduction
- Association through review of one's historical record.

These three scenarios are called the *initiation of the relationship*. In other words, they describe how the association is formed.

We define the *initiation of the trust relationship* as a type of initial introduction that results in an association or a relationship and that *provides methodologies* for *calculating or deriving* the trust value. There are *mainly three* different types of initiation (Table 12.1):

(a) Direct interaction
(b) Reputation
(c) History review.

Each of these plays an important role in defining the trustworthiness by measurement or prediction.

Direct interaction
Initiation of the relationship by direct interaction: This is defined as a relationship begun by direct contact between the agents without any mediator or without the parties knowing each other upfront or through any recommendation. The relationship generally begins from a mutual sense of requirement, for example, 'she became his new boss', or a random meeting, 'they met in the train', or direct agent to agent interaction, 'she was buying a footy ticket from him'. The *direct interaction* can result in trust and assigning of a trust value by the trusting agent to the trusted agent.

Recommendation
Initiation of the relationship by recommendation (also known as *introduction or obtaining reputation*): This is defined as a relationship begun by a third-party mediator who provides an introduction or recommendation. For example, Alice knows Liz through Budi. However, this relationship may

Table 12.1 Trustworthiness prediction with initiation

Initiation of the trust relationships	Trustworthiness measure	Trustworthiness prediction
Direct interaction	Yes	No (It only gives a trustworthiness value after a direct interaction)
Reputation	Yes (known as *reputation measure*)	No (Can use reputation value to predict future trustworthiness value)
History review	Yes	No (Can use history data to predict or forecast the future trustworthiness value)

not start with mutual trust. It depends on what Budi tells Alice about Liz and how much Alice believes Budi. This initiation method is known as *recommendation*. *Recommendation* is a method that helps to form a trust relationship by deriving the initial trust value, also known as *reputation*, based on references or recommendations collected from other parties. The *aggregated* recommendation value is called the *reputation of the agent*.

History review
Historical (or past knowledge) review or a look at past records may result in a new or renewed trust relationship. Historical data could be obtained from the trusting agent's own history repository (past personal interaction data). Therefore, historical review can be part of 'personal interaction'. Historical review only gives you trust values of the past. This relationship may or may not start from mutual respect if one party knows the history of the other party upfront. Over the service-oriented network, sometimes we interact with a service provider only a few times, and the more we interact with this service provider, the more precise is our determination of the trustworthiness of the service provider.

In this chapter, we shall focus on the trustworthiness prediction with the reputation values and historical data.

Trustworthiness measurements result in trustworthiness values. A series of interactions from the past results in a series of trustworthiness values. Only when we have a set of previous trustworthiness values are we able to carry out trustworthiness prediction.

Trustworthiness measurement is carried out by a trusting agent after a direct interaction or direct experience with the trusted agent, and it is a method of deriving a trustworthiness value. This trustworthiness value could be for a trusted agent or a service or a product.

The result of trustworthiness measurement gives a better idea of whether one should go ahead in dealing again with the trusted agent.

Before we carry out any prediction, we must have some historical data, and we use these data to predict the future trustworthiness, even though, in the real world, this method has not been used often in relation to trust, trustworthiness and reputation. However, in the computer and networked world, since agents are becoming autonomous, and malicious agents exist in the network, if there is historical data collected, one should be able to carry out trustworthiness prediction to help filter this data.

The historical data can be collected from past experiences or from third-party opinion. Note that, similar to the trustworthiness measure, the trust value or trustworthiness value is used only by the trusting agent and may not necessarily be shared with any agents on the networks. If the measure or prediction is wrong, the trusting agents have to take sole responsibility for this error.

12.3 Example – Logistics Service

In the following subsections of this Section, we give a logistics example, which will be used for studies of trustworthiness prediction using the methods developed in this chapter. In the first part (single interaction scenario), we illustrate how the trustworthiness criteria are calculated. In the second part, multiple interaction scenarios, we use a similar approach to find several data points for a sequence of time slots. This sequence of data points will be used to carry out the trustworthiness prediction.

12.3.1 Single Interaction Scenario, Calculation of a Trustworthiness Measure

On February 12, 2000, Mr A had a consignment that needed to be transferred from Melbourne to West Australia. This consignment consisted of 10 boxes containing fragile items such as computers, vases and glasses that needed to be delivered to Mr B in Perth. Another 10 boxes were mainly cloth to be delivered to Mr C in Albany. Mr A is looking for a company that can provide a packing service for the fragile items, and door-to-door pickup/delivery service for those 20 boxes of items, and he expects this consignment will arrive in Perth and Albany before 10:00 and 14:00 on February 18, 2000 respectively. In addition, Mr A prefers to deal with a potential service provider that offers online tracking and tracing, and signature basis proof of delivery service, which will enable Mr A to know the status of his consignment any time.

After searching the market, Mr A chooses Fast Delivery (FD) Pty Ltd as his trusted agent to handle his consignment. The following summary highlights the key points in the contract between Mr A and FD.

- Total cost was AU$1500 (AU$600 for the delivery service from Melbourne to Perth and AU$900 for the delivery service from Melbourne to Albany), which includes GST, packing, door-to-door delivery service from Melbourne to Perth and to Albany, and transit insurance.
- Packing and picking time is on the same day, which is on February 15, 2000. FD would pack the fragile items and pick consignment at 9:00 at Mr A's office in Melbourne.
- Ten boxes fragile items would be delivered to Mr B in Perth at 9:00, February 18, 2000 in an intact status.
- Ten boxes of cloth would be delivered to Mr C at 14:00, February 18, 2000.
- Mr A should be able to track and trace his consignment by using Easytrack system offered by FD any time.
- FD should offer signature basis proof of delivery service once Mr B and Mr C received their shipment, respectively.
- Money Back Guarantee: if consignment is not delivered on time, FD would, at Mr A's request, refund freight costs or give a replacement product. A refund would not be payable if on-time delivery is not made owing to circumstances such as inclement weather, industrial disputes, or traffic control.

On the day when the contract between Mr A and FD was signed, Mr A made promises to Mr B and Mr C regarding the respective delivery times on the basis of the above agreements.

However, FD did not fully fulfill its commitment. Mr A was told that Mr C did not receive the 10 boxes of cloth until 18:00, February 19, owing to FD's partner in Perth being unable to dispatch its vehicle to deliver consignment from Perth to Albany. Moreover, FD failed to offer online track and trace service because of system breakdown when Mr A needed to know where his consignment was. Therefore, Mr A got AU$900 back according to the 'money back guarantee' agreement.

12.3.2 Trustworthiness Measure Formulae

Tables 12.2 and 12.3 give a summary of the different terms of the CCCI (Correlation of delivery quality against defined quality, Quality Commitment to each of the defined Quality Assessment Criteria, Clarity of each criterion from both parties' views, and Influence of each criterion on the overall quality assessment) metric developed in Chapter 3.

$$\text{Trustworthiness} = 5 * \left(\frac{\sum_{C=1}^{N} \text{Commit}_{\text{criterion}\,c} * \text{Clear}_{\text{criterion}\,c} * \text{Inf}_{\text{criterion}\,c}}{\sum_{C=1}^{N} 5 * \text{Clear}_{\text{criterion}\,c} * \text{Inf}_{\text{criterion}\,c}} \right) \tag{12.1}$$

Equation 12.1 is the expression developed in Chapter 3 for trustworthiness. This is used in the next section to calculate the trustworthiness for the single interaction scenario (given in Section 12.3.1) and the multiple interaction scenarios (given in Section 12.3.4).

Table 12.2   Criteria for service provision

1	Commit$_{\text{criterion } c}$	The *commitment to the criterion* represents the fulfilment of each commitment (each criterion) of the trusted agent.
2	Clear$_{\text{criterion } c}$	The clarity of the criterion represents the clarity of each commitment (each criterion) in the service agreement (contract), and whether it is understood in the same way by both parties.
3	Inf$_{\text{criterion } c}$	The influence of the criterion represents the importance of each commitment (each criterion) that affects the trustworthiness determination.
4	Corr$_{\text{quality}}$	To measure the fulfilment of the mutually agreed service, we carry out the *correlation* of the *quality of the service defined in the contract or mutually agreed* against the *actual delivered services* of the trusted agent.

Table 12.3   Semantics, levels, values of commit criteria for service provision of a service

Seven-level scale	Semantics (deliver the service)	Description	Values of Commit$_{\text{criterion } c}$	Visual representation (star rating system)
-1	Ignore	No agreement was drawn up.	$x = -1$	Not displayed
0	Nothing is delivered	The provider (trusted agent) did not fulfill any of the commitments.	$x = 0$	Normally not displayed
1	Minimally delivered	The provider only delivered a little bit of what was committed.	$0 < x \leq 1$	From ★ to ★
2	Partially delivered	The provider only delivered half of what was committed.	$1 < x \leq 2$	From ★★ to ★★
3	Largely delivered	The provider delivered most of the service committed.	$2 < x \leq 3$	From ★★★ to ★★★
4	Delivered	The provider's service delivery is satisfactory.	$3 < x \leq 4$	From ★★★★ to ★★★★
5	Fully delivered	The provider has fully delivered what was committed.	$4 < x \leq 5$	From ★★★★★ to ★★★★★

12.3.3 Trustworthiness Measure for the Example

12.3.3.1 Criteria

From Mr A's (trusting peer) perspective, the criteria that can be utilized to measure the service quality of FD are as follows:

- Agreed price (AU\$600 + AU\$900)
- On-time pickup service
- On-time delivery service
- Intact delivery
- Tracking and tracing capability

- Signature basis proof of delivery
- Money back guarantee.

12.3.3.2 Importance of criterion (Inf$_{criterion}$)

On-time service

On-time service in this case includes on-time pickup and delivery. Nevertheless, it is important to distinguish between on-time delivery and on-time pickup as their impact on Mr A are different. In addition, this distinction enables Mr A to measure service more precisely.

In terms of on-time pickup, Mr A expected that the transport company should pick this consignment at the particular time that was agreed to by both parties. Obviously, any late pickup service would cause disruption to Mr A's daily schedule and cause inconvenience for Mr A.

Regarding on-time delivery, any late delivery to Perth and Albany would affect the business relationship between Mr A and his clients (Mr B and Mr C) negatively since Mr A promised the delivery time to them. In other words, Mr A is focused more on on-time delivery than on on-time pickup in the service delivered.

Accordingly, Mr A assigned the value of importance for on-time delivery and on-time pickup as 5 and 3, respectively.

Intact delivery

Owing to the fact that this consignment consists of ten boxes of fragile items, intact delivery becomes one of the key criteria to measure the performance of the service provider. In other words, receiving fragile items such as computers, vases and glasses in a broken condition does not make any business sense. Hence, Mr A assigned the value of importance for this criterion as 5.

Online tracking and tracing capability

Online tracking and tracing ability allows Mr A to gain instant consignment status information. Thus, the value of importance in terms of tracking and tracing capability was 4.

Agreed costs

An increment in the fixed costs with the same service requirement and standard agreed to by both sides of service would affect Mr A's perception of the service. Therefore, Mr A gave a value of 3 for this criterion.

Signature basis proof of delivery

Signature basis proof of delivery was one of the key factors for Mr A to select the service provider. Mr A considered this criterion to be important as he wants to know when the consignment was delivered and to whom. In the case that Mr B or Mr C was unable to receive the shipment by himself and appointed someone else on his behalf to receive the consignment, signature basis proof of delivery allows the consignment carrier and Mr A, B and C to know who received the cargo. Consequently, the value of importance for this criterion was 4.

Money back guarantee

Before interaction occurs, trustworthiness between trusting peer and trusted peer is zero since both parties in the service interaction are new to each other. Thus, money back guarantee item would make financial sense for Mr A in the case where FD failed to fulfill its commitment. Accordingly, money back guarantee was given a value of 4 in terms of its importance.

Table 12.4 Importance and clarity of each criterion

Criterions	Influence	Clear
Intact delivery	5	5
On-time delivery	5	5
Agreed costs	3	5
Tracking and tracing capability	4	5
Signature basis proof of delivery	4	5
On-time pickup service	3	5
Money back guarantee	3	5

12.3.3.3 Clarity of criterions (Clear$_{criterion}$)

In this case, all criteria were written very clearly in the contract; hence, Clear$_{criterion}$ for each criterion was 5. In other words, trusting peer (Mr A) and trusted peer (FD) understood their responsibilities for this business transaction without any ambiguity.

Table 12.4 shows the importance and clarity of each criterion

Based upon the actual performance (commit$_{criterion}$) of FD, together with the importance and clarity of the criteria mentioned previously, the corr$_{criterion}$ can be worked out straight away. Ultimately, the trustworthiness value can be obtained. The following Table 12.5 summarizes Commit$_{criterion}$, Inf$_{criterion}$, Clear$_{criterion}$ and Corr$_{criterion}$ for each criterion selected by the trusting peer.

$$RelCorr_{service} = Corr_{service}/Max_{service} = 512.5/675$$

$$Trustworthiness = 5 * RelCorr_{service} = 3.796$$

This trustworthiness value indicates that the trusted peer is largely trustworthy, which predicts a higher possibility for the next potential business transaction between the trusted peer and the trusting peer under the same circumstances. More precisely, the trusting peer (Mr A) would choose the trusted peer (FD) to handle another consignment if the criteria, namely Inf$_{criterion}$ and Clear$_{criterion}$ are the same as in the previous case.

12.3.4 Multiple Interaction History Scenario

Given the high value of trustworthiness (Trustworthiness = 3.796) based on the first interaction with FD, Mr A decides that he will continue his business relationship with FD. However, as mentioned

Table 12.5 Summary of Commit$_{criterion}$, Inf$_{criterion}$, Clear$_{criterion}$ and Corr$_{criterion}$ for selected criterions

Criterions	Commit$_{criterion}$		Inf$_{criterion}$	Clear$_{criterion}$	Corr$_{criterion}$	
	Actual	Maximum	Actual	Actual	Actual	Max
Intact delivery	5	5	5	5	125	125
On-time delivery	2.5	5	5	5	62.5	125
Agreed costs	5	5	3	5	75	75
Tracking and tracing capability	0	5	4	5	0	100
Signature basis proof of delivery	5	5	4	5	100	100
On-time pickup service	5	5	3	5	75	75
Money back guarantee	5	5	3	5	75	75
				Total	512.5	675

Table 12.6 Trustworthiness values corresponding to different times

Transaction	Trustworthiness	Time
Perth	4.59	Jan–Jun 2000
Sydney	3.56	Jun–Dec 2000
Adelaide	3.07	Jan–Jun 2001
Hong Kong	4.85	Jun–Dec 2001
Tokyo	4.30	Jan–Jun 2002
Beijing	4.56	Jun–Dec 2002
Singapore	4.33	Jan–Jun 2003
Bangkok	5.00	Jun–Dec 2003
New York	5.00	Jan–Jun 2004
LAX	5.00	Jun–Dec 2004
Average	4.43	Time slot 2000–2005

previously, Mr A merely trusts FD under the same criteria as was the case for the previous business interaction. More precisely, Mr A would ask FD to handle his consignments based on not only the same criteria but also the same values for $Inf_{criterion}$ as in the previous business interaction.

General speaking, the trusting peer should collect the historical data regarding the value of trustworthiness. Nevertheless, attention should be paid to the influence of $Inf_{criterion}$ over the trustworthiness value.

Example one illustrates how a varied $Inf_{criterion}$ can incorrectly affect trustworthiness.

Example one
Consider the example where Mr A asks FD to handle his second consignment. However, Mr A changes the value of $Inf_{criterion}$ over on-time delivery criterion from 5 to 4, and maintains the same value of $Inf_{criterion}$ over the remaining criteria. In addition to maintaining the same value of $Clear_{criterion}$, FD performs service of the same standard as in handling the first consignment given by Mr A. $Commit_{criterion}$, $Inf_{criterion}$, $Clear_{criterion}$ and $Corr_{criterion}$ for each criterion selected by trusting peer in the second business interaction, must be used in the calculation.

On the basis of these 10 historic data in terms of trustworthiness, Mr A got the average value of trustworthiness as shown in Table 12.6:

In the next few sections, we shall use this example to illustrate how the trustworthiness prediction formulae utilize the data to carry out trustworthiness predictions and forecasting.

12.4 Prediction Methods

In Section 12.2, we considered the different issues that must be taken into account by the different trustworthiness prediction techniques:

(a) Recency of information about trust
(b) Trends in trustworthiness
(c) Cyclical nature or seasonality in trustworthiness
(d) Variability in trustworthiness.

The collection of trustworthiness values belonging to different time slots for a given context can essentially be considered to be a time series. The area of prediction of future values given a time series has been widely studied by statisticians and has been applied to a range of problems other than trust. In the rest of this chapter, we present a few of these and their application to the trustworthiness time series.

Specifically, we will discuss the following approaches to trustworthiness and reputation prediction:

(a) Exponential smoothing
(b) Markov modeling.

We will term the difference between the forecast value for trustworthiness using the prediction and the actual trustworthiness value generated by the interaction as the *error term*; thus,

$$E = (T^F - T^A) \tag{12.2}$$

where, T^F is the predicted value of trustworthiness and T^A is the actual value of trustworthiness obtained after the interaction.

In order to get a measure of the goodness of the prediction approach, we will utilize the mean square error (MSE), over all the forecast points. Thus, if the forecast and actual values were available for f points, then the mean square error is

$$MSE = \frac{1}{f} \sum_{t=1}^{f} E_t^2 = \frac{1}{f} \sum_{t=1}^{f} (T_t^F - T_t^A) \tag{12.3}$$

The better the overall mean square error MSE, the better the forecast.

12.5 Exponential Smoothing

Let us rewrite the trustworthiness series as shown in Table 12.7.

12.5.1 No Trend, No Seasonality

An approach would be to examine a smoothing function that produces a smoothed valued that could be used as the forecast and utilize the error between the smoothed value T^S and the actual value T^A to update the smoothed value T^S for the next time step. The exponential smoothing method

Table 12.7 Time series for trustworthiness

Trustworthiness	Time slot
4.59	1
3.56	2
3.07	3
4.85	4
4.30	5
4.56	6
4.33	7
5.0	8
5.0	9
5.0	10
4.43	Average

does this. A good introduction to exponential smoothing is given in Reference [2]. Essentially, the smoothed value for time step t is obtained by modifying the $(t-1)$th smoothed value by a proportion of the error $(T_{t-1}^A - T_{t-1}^S)$ experienced at the tth step. Here T_{t-1}^A is the actual value obtained after interaction at time $t-1$ and T_{t-1}^S is the smoothed value at time step $t-1$.

Thus, the smoothed trustworthiness value at the tth step is given by the equation

$$T_t^S = T_{t-1}^S + \alpha(T_{t-1}^A - T_{t-1}^S) \tag{12.4}$$

where α is the smoothing constant, $0 < \alpha \le 1$.

This is normally rearranged to give the exponential smoothing formula

$$T_t^S = \alpha T_{t-1}^A + (1-\alpha)T_{t-1}^S \tag{12.5}$$

where $0 < \alpha \le 1$ for $t \ge 3$.

One could initialize this process by setting $T_2^S = T_1^A$ or $T_2^S = \frac{1}{3}\sum_{t=1}^{3} T_t^A$.

Also note we always have T_1^A as greater than or equal to zero, as there is an interaction and, therefore, the agent is known.

We note the following:

1. The larger the value of α approaching, $\alpha \approx 1$ the more rapid the change in the smoothed value T_t^S and the smaller the value of α the slower the change.
2. The weight for an older value of actual trustworthiness to obtain the smoothed value reduces rapidly.

Solving for, T_t^S we obtain

$$T_t^S = \alpha \sum_{j=1}^{t-1}(1-\alpha)^{j-1} T_{t-(j-1)}^A + (1-\alpha)^{t-2}T_2^S, \quad t \ge 2 \tag{12.6}$$

Let us write the weight for T_{t-1}^A term in the above equation as

$$W_{t-1} = \alpha(1-\alpha)^{t-1} \tag{12.7}$$

Then for, $\alpha = 0.5$ we get the values in Table 12.8.

$$T_5^S = 0.5T_4^A + 0.25T_3^A + 0.125T_2^A + 0.0625T_1^A \quad (if \ we \ set \ T_2^S = T_1^A)$$

Hence, we note that we are very rapidly reducing the weight of the actual value of trustworthiness that corresponds to an older time slot in the sequence.

Table 12.8 Weight values for $\alpha = 0.5$

t	W_{t-1}
1	0.5
2	0.25
3	0.125
4	0.0625
5	0.03125

3. The value of α can be chosen by trial and error or knowledge of the domain or be chosen to minimize the MSE.
4. In order to use it for forecasting past the point, $t - 1$ we can use the expression

$$T_{t+j}^S = \alpha T_{t-1}^A + (1 - \alpha) T_{t+j-1}^S \tag{12.8}$$

Note we use T_{t-1} as it is the last point for which we have an actual trustworthiness value.

12.5.2 With Trend Factor but no Seasonality

Note as discussed in Section 12.2.2 modeling the trend is important. Here we model the trend term but not seasonality.

We write the smoothed trend component of trustworthiness as T_t^T for the tth time slot.

We need to enhance the expressions given in Section 12.5.1 in order to effectively model the trend component. This is done by introducing a second smoothing expression that utilizes the difference in the first-order smoothing terms and involves a second smoothing constant β.

However as the value of trustworthiness is bounded in the range, $0 \leq T \leq 5$ we modify the formulae for exponential smoothing as shown below.

The modified formulae for exponential smoothing with trend (sometimes called *double exponential smoothing*) are;

(i) Add trend term to prediction equation of T_t^S

$$T_t^{S^1} = \alpha T_t^A + (1 - \alpha)(T_{t-1}^S + T_{t-1}^T) \tag{12.9}$$

$$T_t^S = \min\{5, T_t^{S^1}\} \text{ if } T_t^{S^1} > 0$$

OR

$$T_t^S = \max\{0, T_t^{S^1}\} \text{ if } T_t^{S^1} \leq 0$$

(ii) Update trend turn

$$T_t^T = \beta(T_t^S - T_{t-1}^S) + (1 - \beta)T_{t-1}^T \tag{12.10}$$

In order to initialize the trend turn T_1^T can be set to

$$\Delta T_1^A \text{ or } \frac{1}{3}\left(\sum_{i=1}^{3} \Delta T_i^A\right) \text{ where}$$

$$\Delta T_i^A = T_{i+1}^A - T_i^A \tag{12.11}$$

Again as before α, β can be chosen from experience, trial and error or so as to minimize MSE using a search algorithm.

If one has the actual value for trustworthiness for time, t, that is, T_t^A but no actual values for $t + 1$ to $t + j$, then the forecast for $t + j$ is

$$T_{t+j}^{F^1} = T_t^s + j T_t^T \tag{12.12}$$

$$T_{t+j}^F = \min\{5, T_t^{F^1}\} \text{ if } T_t^{F^1} > 0$$

OR

$$T_{t+j}^F = \max\{0, T_t^{F^1}\} \text{ if } T_t^{F^1} \leq 0$$

If we consider the data set in Table 12.8, and compute the smoothed values with and without trend factors we obtain the results in Table 12.9. This table includes forecast values for time slots 11 to 15.

Table 12.9 Predicted values with and without inclusion of trend component
The value below are calculated with $\alpha = 0.5$ and $\beta = 0.3$

Time slot	T_t^A	Without trend			With trend		
		T_t^S	E_t^2	T_t^T	T_t^S	E_t^2
1	4.59					
2	3.56	4.59	1.06	−0.49	4.59	1.06
3	3.07	4.08	1.01	−0.64	3.59	0.27
4	4.85	3.57	1.63	−0.36	3.90	0.91
5	4.30	4.21	0.01	−0.24	3.92	0.15
6	4.56	4.26	0.09	−0.11	4.12	0.20
7	4.33	4.41	0.01	−0.06	4.17	0.03
8	4.90	4.37	0.28	0.06	4.50	0.16
9	5.00	4.63	0.13	0.12	4.78	0.05
10	5.00	4.82	0.03	0.14	4.95	0.00
11		4.91		0.11	5.00	
12		4.95		0.08	5.00	
13		4.98		0.05	5.00	
14		4.99		0.04	5.00	
15		4.99		0.03	5.00	
		Sum Sqr	**4.26**		Sum Sqr	**2.81**
		$1/f$	**0.11**		$1/f$	**0.11**
		MSE	**0.47**		MSE	**0.31**

These, in turn, give for
without Trend: $\qquad$ MSE $= 0.47$
with Trend: $\qquad$ MSE $= 0.31$
The plot of these is given in Figure 12.5.

12.5.3 With Trend and Seasonality

If there are variations in the trustworthiness associated with the periodic or cyclic fraud, then the trustworthiness would show seasonality or a cyclical nature. Then, one needs to use the Holt–Winters method. The basic equations (after modification to account for the fixed range of T) are as follows:

$$T_t^S = \alpha \frac{T_t^A}{T_{t-t_c}^C} + (1 - \alpha)(T_{t-1}^S + T_{t-1}^T) \qquad (12.13)$$

$$T_t^S = \min\{5, T_t^{S^1}\} \text{ if } T_t^{S^1} > 0$$

OR

$$T_t^S = \max\{0, T_t^{S^1}\} \text{ if } T_t^{S^1} \leq 0$$

$$T_t = \beta(T_t^S - T_{t-1}^S) + (1 - \beta)T_{t-1}^T \qquad (12.14)$$

and a seasonal smoothing equation

$$T_t^C = \gamma \frac{T_t^A}{T_t^S} + (1 - \gamma)T_{t-1}^C \qquad (12.15)$$

where T_t^C is the seasonal term and t_c is the length of the cycle. The forecast for j time slots beyond the last time slot t for which we have an observed actual trustworthiness value is given by

$$T_{t+j}^F = (T_t^S + jT_t^T)T_{t-t_c+j}^C \qquad (12.16)$$

$$T_{t+j}^F = \min\{5, T_t^{F^1}\} \text{ if } T_t^{F^1} > 0$$

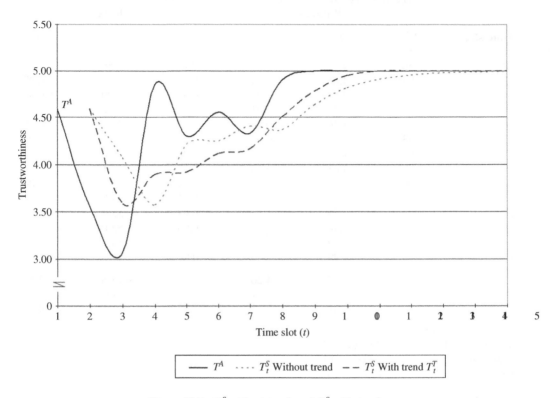

Figure 12.5 T_t^S without trend, and T_t^S with trend

OR

$$T_{t+j}^F = \max\{0, T_t^{F^1}\} \quad \text{if } T_t^{F^1} \leq 0$$

Initialization of the trend term and the seasonal term needs to take the seasonality cycle period t_c into account.

12.6 Markov Approach Plus Trend and Seasonality

We examine an alternative approach here for modeling the general nonstationary reputation services. The original nonstationary stochastic reputation series will be written as

$$R_{p,\tau}^A, \quad p = 1, \ldots M, \tau = 1, \ldots N \tag{12.17}$$

for the pth season (cycle) and the τth periodic point, where N is the number of sample points per season.

If the reputation series $\{R_{p,\tau}^A, \ p = 1, \ldots M, \ \tau = 1, \ldots N\}$ exhibits a trend, then we have to carry out trend component analysis as in Section 12.6.2 and it shows seasonality component analysis as in Section 12.6.3. If the reputation series does not show a significant trend or seasonal component, we would proceed directly to using the Markov chain in Section 12.6.5, 12.6.6 and 12.6.7.

12.6.1 Deterministic Components

The trend component and the seasonal components of the reputation series are the deterministic components. Once they are removed, we get the stochastic components discussed in Section 12.6.4 to 12.6.7.

12.6.2 Trend Component Analysis

An alternative approach to filling the whole series is to use a moving average approach over a subset of data [3]. In fact, a version of the experimental smoothing procedure can be used for this.

In fitting a functional form as trend, one fits the whole course of observation, using a function. Let $R^T_{mp,\tau}$ be the trend in the mean. Then, the mean detrended series is $R'_{p,\tau}$ where

$$R'_{p,\tau} = R_{p,\tau} - R_{mp,\tau} \quad p = 1, \ldots M, \tau = 1, \ldots N \tag{12.18}$$

Several functions are also useful for modeling reputation series. However, for convenience, we will use a polynomial equation of the form given in Equation 12.19 to model the trend in the mean.

$$R^T_{mp,\tau} = a_{m_0} + a_{m_1} t + a_{m_2} t^2 + \ldots \ldots \tag{12.19}$$

where $t = (p - 1)N + \tau$ over the whole range of data.

The parameter 'a' of the polynomial equation may be computed using least square techniques.

To determine the degree of the polynomial, one can use a statistical test of significance, say, F–significance test of the highest order term in its ability to represent the variance of the reputation series [6].

12.6.3 Seasonal Component Analysis

Seasonal components may be estimated using a nonparametric approach that estimates the periodic mean and variance by a simple averaging process, as shown below.

The detrended series may be written as

$$R''_{mp,\tau} = \mu_\tau + \sigma_\tau \varepsilon_p, \tag{12.20}$$

where μ_τ and σ_τ are the periodic mean and standard deviation respectively, and $\varepsilon_{p,\tau}$ is a stochastic component.

The mean and standard deviation are estimated by the following equations:

$$\mu_\tau = \frac{1}{M} \sum_{p=1}^{m} R_{p,\tau} \qquad \tau = 1, \ldots N \tag{12.21}$$

$$\sigma_\tau = \left[\frac{1}{M-1} \sum_{p=1}^{m} (R_{p,\tau} - \mu_\tau))^2 \right]^{1/2} \qquad \tau = 1, \ldots M \tag{12.22}$$

12.6.4 Stochastic Component Analysis

The stochastic components can be written as

$$\varepsilon_{p,\tau} = \frac{R''_{p,\tau} - \mu_\tau}{\sigma_\tau} \qquad p = 1, \ldots M, \quad \tau = 1, \ldots N \tag{12.23}$$

This stochastic process may or may not be time dependent, that is, persistent. The persistence of the reputation series may be modeled by a simple Markov process, as discussed in Section 12.6.5.

Note that if the initial time series for reputation does not have a significant trend or seasonal component, we could use the Markov chain directly.

12.6.5 Markov Chain of Finite States

A first-order Markov process is a special case of a stochastic process such that the next state $(t_N = N + 1)$ that the system is likely to be in is [5]:

(a) determined by its present state $(t_N = N)$
(b) not determined by its state at any prior time before time, t_N that is, its states for $t_N \leq N$
(c) not determined by the time that it reached its present state.

We will utilize a discrete state, discrete time version of a first-order Markov process for modeling the stochastic component of the reputation series.

A Markov chain of six states corresponding to the reputation levels (0 to 5) is postulated. The transition probability is denoted by

$$p_{ij}(t_N), \qquad i, j = 1, \ldots, 6 \qquad (12.24)$$

$p_{ij}(t_N)$, denotes the probability of the process in ith state transferring to the jth state at time t. Thus we modeled the process by a first-order Markov chain where the number of states is assumed to be six for a known agent, each corresponding to one level on the reputation scale given in Tables 9.11 and 9.12. The asymptotic transition probabilities [4] are estimated as

$$p_{ij}(t_N) = \frac{n_{ij}}{n_i}, \qquad (12.25)$$

where n_{ij} is the total number of transitions from ith state (i level) to the jth state (jth level) and

$$n_i = \sum_i n_{ij} \qquad (12.26)$$

This ensures that the normalizing equation $\sum_i p_{ij} = 1$ is fulfilled.
We assume that there are sufficient data points for estimating these transition probabilities p_{ij}.

12.6.6 Markov Model in Reputation Prediction

The trusting entity must make a trust decision within the current time slot N and does not know about the trusted entity's trustworthiness value in that time slot.

In this case, the trusting entity must issue a reputation query for the prospective trusted entity and specify the context in which it wants to interact with the other entity, along with the time space.

The reputation-querying entity first classifies the obtained reputation into the different time slots using the time spot when the interaction took place. It uses the witness entities' trustworthiness values to weed out reputation from the untrusted entities. It then combines the reputation obtained from trustworthy and unknown entities using the following expression to determine the reputation-queried entity's trustworthiness at each time slot:

$$\text{Repute value } (m, A, T_A) = \left(\sum_{i=1}^{K} TC(i)\langle\rangle(R[m, i, A, t]) \right) / K$$

$$(12.27)$$

$$+ \beta * \left(\sum_{j=1}^{L} \text{Trustworthiness } [j] \right) / L$$

where K and L are the number of trusted entities and number of unknown entities, respectively, in time slot T_A; A denotes the context; T_A denotes the time slot on the time space; and T denotes the time spot that the reputation-querying agent finds in time slot T_A and m denotes the identity of the reputation-queried agent. $R[m, i, A, t]$ is the trustworthiness value that witness entity i communicates about the reputation-queried entity m at time spot t, in context 'A'. $TC(i)$ represents the witness i's trustworthiness value, as perceived by the reputation-querying agent, in the context of communicating recommendations. $\diamond$ is an operator that adjusts the trustworthiness value communicated by the witness entity with the witness entity's witness trustworthiness value. β gives an appropriate weighting to the recommendations that the unknown entities communicated.

The first term in Equation 12.27 combines the reputation-queried entity's reputation values from trusted entity's (whom the reputation-querying entity trusts to communicate accurate recommendations) reputation values, and the second term combines the reputation values from the unknown agents (with whom the reputation-querying entity has no previous experience of soliciting recommendations).

12.6.7 Markov Chain for R

The trusting agent has to make a prediction for the reputation value at the time slot $N + 1$. In this case, the trusting agent has to have a model for the dynamic nature of reputation/trust, so that it can use it for predicting the reputation value at the time slot $N + 1$.

If the trust-based decision is to be taken at a time in the future, then the trusting agent can use the Markov method. It adopts the procedure, explained in Section 12.6.6, to gather the reputation-queried agent's repute values and subsequently obtain the corresponding level for each value for each time slot from 0 to N.

We define a reputation Markov chain as a given agent's sequence of aggregated repute values that correspond to a sequence of time slots.

12.6.8 Constructing Current State Vector C

Then, the trusting agent must construct the current state vector c and the Markov matrix M to make a trustworthiness prediction for the next time slot $N + 1$ using the Markov model *Constructing current state vector c*. The current state vector shows the reputation-queried agent's repute value at time slot N. It will be a 1×6 matrix because we use six reputation levels (excluding the unknown agent) that is, $C = [C_1, C_2, C_3, C_4, C_5, C_6]$. We determine the reputation-queried agent's repute value at time slot N using Equation 12.27, denoting it with a '1' corresponding to the repute level of the agent at time slot N, that is, $C_i, = 1, C_j = 0, i \neq j$. We denote all other reputation levels with '0'.

12.6.9 Constructing Markov Matrix

12.6.9.1 Constructing the Markov Matrix

A given agent's Markov matrix denotes the probability of its transiting from one reputation level to another on the basis of its past behavior, which we capture using the Markov chain. To determine the probability of an agent transiting from reputation Level 1 to reputation Level 2, we find the ratio between the number of times that agent has transited from Level 1 to Level 2 and the total number of times the agent has transited from Level 1 to any other level as shown in Equation 12.22. We denote an agent's Markov matrix as M. This would be a 6×6 matrix with rows corresponding to the trustworthiness level at time slot N and columns corresponding to the trustworthiness level at time slot $N + 1$ of a given agent. An element in the matrix denotes the probability of the agent transiting from the trustworthiness level corresponding to the row in which the element occurs at time slot N to the column in which the element occurs, at time slot $N + 1$.

12.6.9.2 Determining the Future Reputation Value at Time slot '$N + 1$'

Once we determine the Markov matrix and the current state, we determine the agent's future trust state vector by multiplying the current state vector with the Markov matrix. The future state vector denotes the probability that the agent will behave with a reputation level I at time slot $N + 1$. We denote an agent's future state vector as f, $f = [f_1, f_2, f_3 \ldots\ldots, f_6]$. Mathematically, we represent this as

$$f = c \times M. \tag{12.28}$$

From the future state vector, we choose the reputation-queried agent's future reputation level (at time slot '$N + 1$') as the level to which the agent has the highest probability of transiting to. A trusting agent decides to go ahead and interact with the trusted agent only if the trusted agent's reputation level is ≥ 5 because 5 and 6 denote positive trust. This is the same for all three cases we have described.

12.7 Rejustification of Third-Party Recommender's Trust Value

After interacting with the chosen agent, the trusting agent uses CCCI metrics to rate the trusted agent's behavior (see Chapter 6). On the basis of the trustworthiness value assigned to the trusted agent after interacting with it, the trusting agent modifies all the recommending agents' trustworthiness values for their credibility. These agents are the recommending agents from whom the trusting agent had solicited recommendations. The changes to the credibility of the recommending agent are made using the approach given in the rest of this section.

Trustworthiness value of the queried agent's reputation can be obtained if the query is followed up with an actual interaction. This value can be termed T_{actual}. T_{actual} can be used to determine the error of the reputation value provided by a particular third-party agent.

Let us call this error E_i for the current transactions. Then E_i is given by

$$E_i = T_{actual} - R(i) \tag{12.29}$$

for the ith third-party recommendation agent.

We note that if the predicted trustworthiness value was close to a prespecified tolerance ε to the actual trustworthiness level, we would expect to positively reenforce or increase the trustworthiness of the recommendation opinion of agent i, TC (or credibility of the agent i). On the other hand, if the error E_i, exceeded this prespecified tolerance, we would expect to negatively reenforce or decrease the trustworthiness of the recommendation opinion. In order to prevent cyclic dishonesty, that is, giving the occasional dishonest opinion, while generally providing correct opinions, it is necessary to make large decreases TC for $|E_f| > \varepsilon$ than the increase made to TC when $|E_f| \leq \varepsilon$. This is particularly the case where TC has values close to maximum.

An updating mechanism that generally achieves this would be

$$TC(i) = \eta TC(i) + (1 - \eta)(adj) \times 5 \tag{12.30}$$

where $(adj) = +1$ for $|E_i| \leq \varepsilon$
$\qquad (adj) = -1$ for $|E_i| > \varepsilon$

Note that the value of trustworthiness of opinion would need to be modified for each recommending agent after each interaction.

If no actual interaction takes place between the trusting agent and the reputation queried agent, we could utilize the aggregated repute value given by Equation 12.27 in place of the actual trust value T_{actual} obtained from the interaction. Equation 12.29 to calculate the error is then replaced by

Equation 12.31. The adjustment to the credibility is still done using Equation 12.30 except that E_i is now given by Equation 12.31.

$$E_i = \text{Repute Value}(m, A, t) - R(i) \tag{12.31}$$

In this case, we assume that the collective opinions when aggregated over all agents and groups can be used as an estimate for trustworthiness of agent R in the time slot n and context c. Here Repute Value(m, A, t) (the aggregate opinion) could be obtained using Equation 12.27.

12.8 Summary

In this chapter, we identified the dynamic character of trust and reputation. In particular, we noted that a trustworthiness series or reputation series over time can show both a trend component and a seasonal (or cyclic) component. The second is often related to seasonal conditions in delivery of a series or alternatively with cyclical fraud.

We presented two different approaches to modeling these series and these included:

(i) Exponential smoothing
(ii) Markov process with trend and seasonal modeling.

These models can be used for trustworthiness and reputation predications.

References

[1] Pleines, W.W. and Zajic, K.U. A method of estimating demand, *IEEE Trans. on Power Apparatus and Systems*, Vol. 88, pp. 375–384, 1969.
[2] Engineering Statistics Handbook, available at *http://www.itl.nist.gov/div898/handbook*, accessed 2005.
[3] Kendall, M.G. *Time Series*, Griffin, 1973.
[4] Walpole, R.E. and Myers, R.H. *Probability and statistics for engineers and scientists*, Collier MacMillan Int. 7th Edition, 1972.
[5] Anderson T.W and Goodman L.A. Statistical inference about Markov chain, *Annals of Math. Statistics*, Vol. 28, pp. 89–110, 1957.
[6] Hager H. and Antle C. The choice of the degree of a polynomial model, *Royal Stat. Society, B*, Vol. 30, pp. 469–471. 1968.
[7] Dillon T.S. Morsztyn K. and Phua K.S. Short-term load forecasting using adoptive pattern recognition and self-organizing techniques, *Proceedings of PSCC Conference*, Cambridge, 1975, paper 2.4/3.

13

Trust and Reputation Modeling

13.1 Introduction

In this chapter, we use a pictorial or graphical modeling language to model *trust* and the *trust relationship* between agents in the service-oriented network environment. This is the first attempt in the trust literature to develop a conceptual modeling language for these concepts. Our proposed language for modeling trust and trust relationships is based on extensions and semantic modifications to the existing notation systems of the Entity–relationship (E–R) diagram [1] and the Unified Modeling Language (UML) diagrams. These modified notation systems are tailored for trust relationship modeling, trust property diagram, trust context diagrams, trust transition diagrams and trustworthiness assessment diagrams in order to model trust at different levels of abstraction, specifically the dynamism of the trust. The aims of the trust modeling language are as follows:

(a) to help the conceptual visualization of trust relationships;
(b) to help software engineers communicate with business providers or end-users; and
(c) to help automate the generation of the trust matrix and the reputation matrix (the trust databases).

13.2 Significance of Pictorial Modeling

Modeling is a method of representing reality. A model is created to produce an abstraction of a system. Additionally, a model of a system is easier to understand than the actual system itself. Generally, it only represents those elements of reality that are relevant to the investigation.

In the field of software engineering, a model of the system is built before building the system. UML is used to create an initial model of the project before the project itself commences. This model represents the structure and behavior of the system and other details of the system in a simple pictorial language that is much easier to follow and comprehend compared to the project specification. In other words, UML is used to represent the model of the system. This model of the system represented in UML is a pictorial reflection of abstractions of the project specification. As it is a pictorial representation, it is a much easier way to understand the structure and behavior of the system than using the project specification itself. Additionally, this model hides many unwanted details from the user and depicts only the relevant details.

In a similar way, we need a language that is powerful and expressive enough to represent the trust relationships and the various properties of trust relationships, between two or more agents, in the appropriate level of detail. However, so far, there is no existing work in the literature for defining a pictorial language to model trust and trust relationships between two agents. Trust and trust relationships between two agents can be modeled in the following ways:

(1) Using English, or any other natural language, we can model the trust relationships and their various properties between agents.

Trust and Reputation for Service-Oriented Environments Elizabeth Chang, Tharam Dillon and Farookh Hussain
© 2006 John Wiley & Sons, Ltd

(2) Using a pictorial language, we can model or represent trust relationships and the characteristics of trust between agents.

The benefits of using a pictorial language over a natural language to represent a trust relationship and its attributes between two agents are as follows:

It allows clear representation of entities, attributes and designation of these to entities. In a trust relationship between two agents, we have the following features:

(1) The trusting agent in the trust relationship
(2) The trusted agent in the trust relationship
(3) The context of the trust relationship
(4) The initiation of the relationship
(5) The trustworthiness value
(6) Start time (the time the trust relationship was established)
(7) End time (the time the trust relationship will cease).

If we depict all of these attributes and entities of the trust relationship using a natural language, it is difficult to comprehend all attributes and properties. For a reader, it will be difficult to grasp all the properties or attributes of the trust relationship. We believe that by using a modeling language, we can easily represent the trust relationship between agents along with the trust relationship attributes:

- Between two agents, more than one trust relationship may exist, each of which may be in a different context. Each of the trust relationships between the agents, irrespective of whether they are in the same context or in a different context, will have all of the attributes mentioned above. By using a natural language such as English, it is difficult to represent the multiple trust relationships that exist between two agents along with their attributes in such a way that it is easily understood by readers.
- By using a pictorial language, one can represent trust and its attributes in a way that can be understood by people irrespective of the languages and vocabulary that they understand.
- An agent may have trust relationships with multiple agents. By using a modeling language, we can represent these multiple relationships in a very efficient and, importantly, a more comprehensible way than by using a natural language.
- Through the use of a pictorial language, trust and its attributes can be understood more quickly than through the use of natural language.
- By using a pictorial language, transitive trust relationships are implicitly evident. There is no need to model or state them explicitly.
- There are no efficient natural language processors that can convert a natural language representation of a trust relationship between two agents to the internal machine representation. Mechanisms do exist in taking graphical (or pictorial) models and converting them to code. Hence, the pictorial representation will form the basis of a Model-driven Approach (MDA) for trust and reputation.
- If trust and its attributes and methods are represented using a pictorial modeling language, this representation can easily be converted to another form of representation by the help of a 'compiler' (or a parser). If a trust repository or a trust database is developed, a user need not worry about designing the schema and maintaining the trust repository. A user just needs to specify the *trust relationships* with the others (along with the attributes) as explained later in the chapter. The 'compiler' (or a parser) will automatically convert this pictorial representation to the schema for the trust repository. Similarly, any changes made by the user to the attributes of an already existing trust relationship will be reflected in the trust repository.

However, it is not an easy task to convert a natural language representation of trust or trust relationships to a pictorial representation. There are no efficient natural language processors to do so. As can be inferred, the use of a pictorial modeling language would be preferred over natural language modeling. Hence, in this chapter, we propose a pictorial modeling language to model trust and trust relationships between two agents. We believe that a pictorial language for modeling trust and trust relationships along with their attributes is imperative for information systems research and development. Our proposed language involves semantic and syntactic extensions and modifications to UML.

13.3 Notation Systems

In this section, we introduce all the notations needed to model trust and trust relationships. Since trust relationships are between two entities, the parties involved and the relationships between them have to be modeled using some unique and clearly defined notations. At present, trust and trust relationships cannot be modeled adequately.

13.3.1 Notations for Representing Different Types of Agents

In our trust modeling *language,* we make use of agent notations to represent any entity that is involved in a trust relationship. Depending on the role being played by an entity in a trust relationship, it will be represented by a different modeling notation. In our modeling language, we will make use of the notation outlined in Figures 13.1 to 13.3 to represent a trusting agent and a trusted agent.

As can be seen from Figures 13.1 and 13.2, this notation for representing a trusting entity or agent has a straight solid hand or the trusted entity or agent has a straight clear hand.

Figure 13.3 shows other agent notations. An 'unknown' agent or an 'untrusted' agent has a *sad-hand* and a 'malicious' agent has a *twisted-hand* and these can be used as part of the pictorial representation for agents.

13.3.2 Relationship Notations

In order to model the fact that two entities/agents are involved in a trust relationship, we make use of the following notations. A *relationship symbol* is used to connect each agent involved in a trust relationship.

In the previous chapters, especially in Chapters 2, 5 and 8, the above notations were used quite frequently. For details, please refer to these chapters.

13.3.3 Initiation Relationship Notations

In Chapter 2, we introduced the concept of *initiation* of a relationship. It states that the *relationship* may be formed by '*direct interaction*' between two agents, or '*recommendation*' from a third-party

Trusting agent   Trusted agent   Unknown agent   Malicious agent

Figure 13.1   Notation for representing an agent

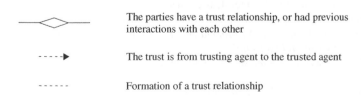

The parties have a trust relationship, or had previous interactions with each other

The trust is from trusting agent to the trusted agent

Formation of a trust relationship

Figure 13.2   Notation for relationships

Figure 13.3   Agent property notations based on E–R system

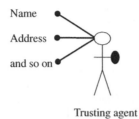

Trusting agent

Figure 13.4   Agent property notations based on the agent diagram developed in this book

agent, or *'history'* review from stored records or files and this historical data is used as the basis to form an initial trust.

13.3.4 Notation for Agent Properties

In representing agent properties, we can borrow E–R based notation [1] as well as UML-based notation. Both notations can be used in modeling the same domain of interest. In modeling the property of an agent, either notation can be selected and it does a good job for modeling the trust. We show the different alternatives in Figure 13.3 (based on E–R), Figure 13.4 (based on agent diagrams), and Figure 13.5 (based on UML). Note that a solid black dot represents the trusting agent, and a white dot represents the trusted agents.

Note that 'not known' in the above tuple field in Figure 13.5 indicates that the 'end time' of the trust is not known; the agent may only know the start time of the trust.

13.3.5 Tuple Notation and Presentation Rules

In order to model trust and the trust relationship between two agents, we introduce the notion of a *trust tuple*. The *trust tuple* consists of an ordered set of all of the attributes that completely describes a trust relationship in the appropriate level of detail. The idea behind introducing a trust tuple is to qualify each trust relationship, represented using the trust relationship notation as described in Section 13.3.2, by a trust tuple. The trust tuple contains all the attributes about a trust relationship that describe the trust relationship in detail.

The trust tuple, introduced in Chapter 4, is defined as:

[Trusting agent, trusted agent, context, start time, end time, trustworthiness value]

A trust tuple can be defined as *'An ordered set of attributes of a trust relationship that conveys information about the trust relationship'*.

We define the tuple as an *'ordered set'* because we want to have a standard order for representing elements in a trust tuple.

The elements in the trust tuple are the attributes and methods that will be present in any given trust relationship.

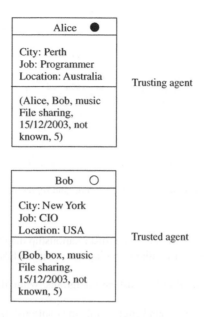

Figure 13.5 Agent property notations based on UML system

The proposed rules for representing the trust tuple are the following:

- The attributes of the trust relationship should be represented in the same order as specified in the above tuple.
- Any trust tuple should always begin with '['. In order words '[' denotes the beginning of the trust tuple.
- The trust tuple should always end with ']'. In order words ']' denotes the end of the trust tuple.
- All elements are necessary and should be present in all trust relationships, except ' "start time' and 'end time'. These two elements are optional.
- In situations where the value of the 'start time' and 'end time' elements are not known, they should be explicitly specified by 'not known'.
- The successive elements of the trust tuple, except for the first and last one, should be separated by a comma ',' (for representation purposes).
- Each trust relationship should be represented by no more than and no less than one trust tuple.
- The value of the date in the 'end time' field should be greater than the value of the date in 'start time'. If this is not satisfied, then this represents an invalid trust relationship.
- The fields HH, MM, SS in the 'time slot start' are not necessary (the date is the most important).

For example, let us assume that East Logistics Pty Ltd and West Warehouse Pty Ltd are two logistic companies, located in Sydney and Canberra, respectively. Let us further assume that they have their areas of operation specific to the area that they are located in. East Logistics wanted to store some of its consignment of goods in the warehouse belonging to West Warehouse. West Warehouse subsequently obliged and met the request from East Logistics to use the former's warehouse space. Let us assume that the interaction took place on 15/07/2003 and that East Logistics, as the trusting entity, assigned a trustworthiness value of '5' to West Warehouse.

The above trust relationship between East Logistics and West Warehouse can be abstracted and represented using the trust tuple in Figure 13.6.

Figure 13.6   Trust tuple notation and representation

13.3.6 The Trust Case Notation

Note (a) The trust case diagram is similar to the trust relationship diagram. The differences between the trust case diagram and the trust relationship diagram are as follows:

(1) the tuple representation is used in the trust case diagrams, while the relationship only indicates the key context;
(2) the trust case is represented by a round circle, while the relationship is represented by a diamond.

Note (b) The trust case can be used to represent the *trust relationship* by giving a detailed trust tuple. The advantage of using the trust case diagram is to see the context, start time, end time and trust value in one glance. The trust relationship diagram is good for a high-level visual representation and discussion, and the use of the diamond is a tradition for those from E–R background.
Note (c) The trust case notation (Figure 13.7) is based on the UML use case diagram, except that the 'case' is represented by the trust tuple.

13.4 Trust Relationship Diagrams

13.4.1 Trust Relationship Modeling

To model the trust or reputation relationships, we adopt two kinds of notation systems, one utilizing traditional modified E–R notation, and the other using the agent notation (Figure 13.8) developed in this book.

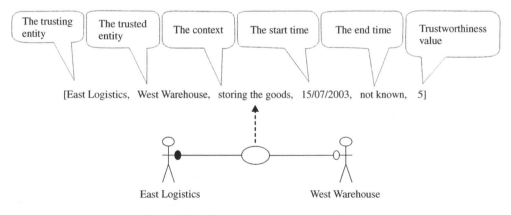

Figure 13.7   The trust case notation and diagram

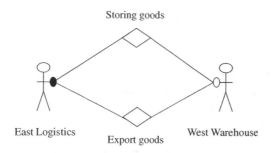

Figure 13.8   Trust relationship modeling using agent notation

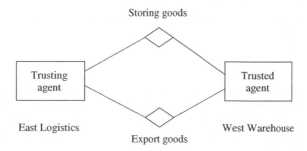

Figure 13.9   Trust relationship modeling using modified E–R notation

Figure 13.9 is an example of modeling trust relationships. In a real-world situation, we often encounter multiple relationships that coexist between two agents, such as a working relationship and a personal relationship that can coexist between co-workers, and so on. In the example given earlier, East Logistics Pty Ltd may trust West Warehouse Ltd for storing the goods, but not for exporting its goods. (Note that exporting and importing goods is one of the functions that many warehouses perform).

13.4.2 Higher-level and Lower-level Trust Relationship Modeling

In this book, we have frequently used trust relationship diagrams throughout every chapter. We also interchange notation using E–R based notation and agent-based notation. We found that for high-level modeling, the agent-based notation system is much more intuitive and understandable by nontechnical users. On the other hand, the E–R based notation can give more technical details, such as cardinalities and attributes.

In the previous chapters, especially in Chapters 2, 4, 5 and 8, we have used the trust relationship diagram to represent the following:

- Single trust relationships
- Multiple trust relationships
- Single agent with multiple relationships
- Multiple agents in multiple trust relationships.

Therefore, in this section, we will not repeat these diagrams.

In the next section, known as 'trust case diagrams', we shall revisit the above complex relationships. We shall look at it from a trust case perspective, and see how the above four categories of relationships are represented by trust case diagrams. The reader should pay particular attention to the difference between trust relationship diagrams and trust case diagrams. The preliminary distinction is already explained in Section 13.3.6.

13.5 Trust Case Diagrams

The concept of a *trust tuple,* introduced and explained earlier, is very important for modeling trust cases. In this section, we explain how we can use the *relationship notation* explained in Section 13.3.2, for trust cases:

- A trust relationship in a single case (single context) (the trust relationship that a trusting agent has with the trusted agent in a single context)
- A trust relationship in more than one context (multiple cases) (the trust relationship that a trusting agent has with a given trusted agent in multiple contexts or cases)
- Multiple trust cases (the trust relationships that an agent has with more than one trusted agent, possibly in a different context)
- Multiagent trust cases.

13.5.1 Single Trust Case Diagram

In this section, we show how, using the *trust case notation* representation of the agent introduced in Section 13.3 and the notation of representing the trust case, introduced in Section 13.3.6, we can model trust cases between a trusting agent and a trusted entity in a single context case (Figure 13.10).

For explaining the modeling, we continue with the example of East Logistics and West Warehouse explained in Section 13.3.5.

The steps involved are as follows:

(1) The trusting agent is drawn in a trust case diagram and labelled with the identity of the trusting agent.

   In our example, since East Logistics is the trusting entity, we draw the trust case diagram of the trusting agent and label it as East Logistics.
(2) The trusted agent is drawn in the trust case diagram and labelled with the identity of the trusted agent.

   In our example, since West Warehouse is the trusted agent, we draw the trust case diagram of the trusted agent and label it as West Warehouse.
(3) We connect the trust case diagrams of the trusted agent and the trusting agent with the trust case notation to show that trust relationships exist between them. We label and annotate the trust case with the trust tuple that conveys meaningful information about the trust relationship.

   In our example, since East Logistics and West Warehouse are involved in a trust relationship, we connect the trust case diagrams of both the entities by a trust case notation (circle) and annotate it with the trust tuple. The trust tuple has already been explained in Section 13.3.5. (See also Figure 13.13.)

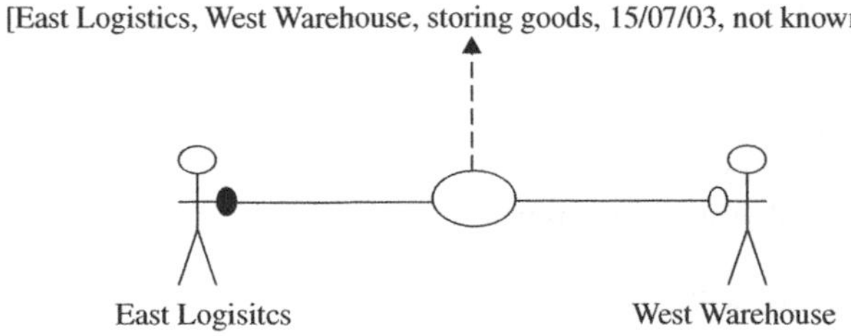

[East Logistics, West Warehouse, storing goods, 15/07/03, not known, 5]

East Logisitcs   West Warehouse

Figure 13.10   Single trust case diagram

Note that 'not known' indicates that the 'end time' of the trust is not known; the agent may only know the start time of the trust interaction.

13.5.2 Modeling a Trusting Agent that has Multiple Trust Cases with Multiple Trusted Agents

It is possible that an agent has trust relationships with one or more agents, each of which may or may not be in the same context. Our proposed modeling method allows for these trust cases (where an agent has trust relationships with more than one agent) to be modeled easily (Figure 13.11).

As an example for explaining how these multiple trust cases that an agent has with another agent may be modeled, let us assume that East Logistics has the following trust relationships with South Field, apart from the trust relationship that it has with West Warehouse.

East Logistics had made use of the services of South Field (subcontractor) in order to deliver goods to Kalbarri (a remote area in Western Australia). The date on which the interaction took place was 10/05/2000. East Logistics had assigned a *trustworthiness value* of '5' to South Field.

From the above description of the trust cases between East Logistics and West Warehouse as well as South Field, we can infer that the trust tuple for this trust relationship is as follows:

[East Logistics, West Warehouse, storing goods, 15/07/2003, not known, 5] and
[East Logistics, South Field, delivering the goods, 10/05/2000, not known, 5]

Note that 'not known' indicates that the 'end time' of the trust is not known; the agent may only know the start time of the trust.

The steps involved in modeling trust cases with multiple agents (irrespective of the context in which the trust case falls) are as follows:

(1) The trusting agent is drawn in the trust case diagram and labelled with the identity of the trusting agent.

In our example, since East Logistics is the trusting entity, we draw the trust case diagram of the trusting agent (Figure 13.1) and label it as East Logistics.
(2) Each agent with whom the trusting agent shares a trust case is drawn in the trust case diagram as a trusted agent and labelled with the identity of the trusted agent/s.

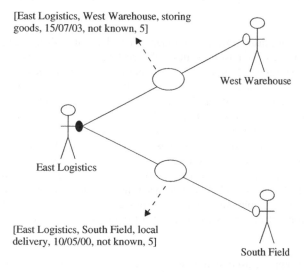

Figure 13.11   A multiple trust case diagram

In our example, since West Warehouse and South Field are the trusted agents with which East Logistics shares a trust case, we draw two trust case diagrams, one for South Field and one for West Warehouse (as in Figure 13.2) and label them with the identities of the respective trusted agents.

(3) We connect the trust case diagrams of the trusting agent with the trust case diagram of each of the trusted agents with whom the trusting agent has different trust cases.

In our example, since East Logistics has a trust relationship with (1) West Warehouse and (2) South Field, we connect the trust case diagrams of East Logistics with West Warehouse with a trust case notation and the trust case diagram of East Logistics with South Field with a separate trust case notation.

(4) Annotate each of the trust case notations between the trusting agent and the trusted agent/s with the corresponding trust tuple that conveys a more meaningful information about the trust case.

In our example, since East Logistics has a trust relationship with

(a) West Warehouse, we annotate the trust case notation between East Logistics and West Warehouse with the trust tuple for this case.

(b) South Field, we annotate the trust case notation between East Logistics and Southfield with the trust tuple for this case.

13.5.3 Modeling Multiple Trust Cases between Same Agents

It is possible that two agents share more than one trust relationship between them, each of which is in a different context or case. Our proposed methodology allows for *multiple* trust relationships between the same agents in different contexts to be modeled easily.

Let us assume that East Logistics, apart from the trust relationship outlined above with West Warehouse, has another trust relationship with West Warehouse in the context of 'delivering goods'. The trust relationship was established on 10/07/2004. East Logistics had assigned a *trustworthiness value* of '4' to West Warehouse at the time of the interaction, that is, 10/07/2004. The trust tuple for this trust relationship would be as follows:

[East Logistics, West Warehouse, delivering goods, 10/07/2004, not known, 4]

Note that 'not known' indicates that the 'end time' of the trust is 'not known'; the agent may only know the start time of the trust.

The steps involved in modeling trust cases between two agents in *multiple contexts or cases* are as follows:

(1) For the trusting agent, draw a trust case diagram and label it with the identity of the trusting agent. In our example, since East Logistics is the trusting entity, we draw the trust case diagram of the trusting agent (as per Figure 13.1) and label it as East Logistics.

(2) For the trusted agent, draw a trust case diagram and label it with the identity of the trusted agent. In our example, since West Warehouse is the trusted agent, we draw the trust case diagram of the trusted agent (as per Figure 13.2) and label it as West Warehouse.

(3) Connect the trust case diagrams of the trusted agent and the trusting agent with the trust case notation for each trust relationship that exists between them. As there can be several trust relationships between the two agents, there may be many trust cases between the two agents.

(4) Annotate each of the trust case notations between the trusting agent and the trusted agent with the corresponding trust tuple that conveys a more meaningful information about the trust relationship. In our example, since East Logistics and West Warehouse are involved in two trust cases, we connect the trust case diagrams of both the entities by two trust case notations (circles).

Let us assume that the upper trust case notation is used to model the trust relationship in the context of storing the goods and the lower trust case notation is used to model the trust relationship in the

[East Logistics, West Warehouse, storing goods,
15/07/2004, not known, 5]

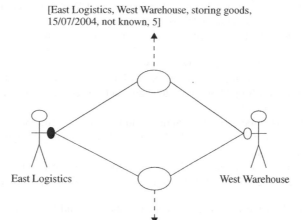

[East Logistics, West Warehouse, delivering goods, 10/07/2004, not known, 4]

Figure 13.12   Modeling multiple trust cases with the same agents

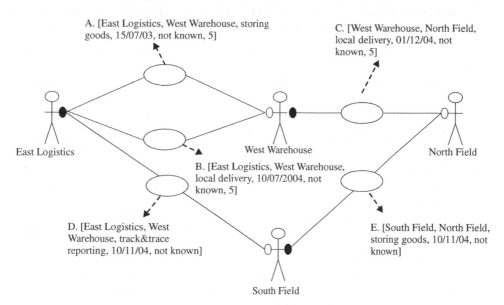

Figure 13.13   Modeling multiple relationships with multiple agents

context of delivering goods. We then annotate each of the trust cases with the corresponding tuples as shown in Figure 13.12.

Note that the trust relationship is represented by trust tuples, and there are multiple trust cases between the same agents.

13.5.4 Modeling Multiagent Trust Cases

Our proposed modeling method allows for multiagent trust cases between agents to be modeled and represented in an easy way (see Figure 13.13). Let us assume that the following trust relationships exist between four agents: East Logistics, West Warehouse, North Field and South Field.

(A) East Logistics had used the warehouse belonging to West Warehouse on 15/07/2003 and had assigned a trustworthiness value of '5' to West Warehouse.
(B) East Logistics had used the services of West Warehouse to deliver goods to a location. It had used the services of West Warehouse on 10/07/2004 and had subsequently assigned a trustworthiness value of '5' to West Warehouse.
(C) West Warehouse had used the services of North Field to deliver goods to a location. It had used the services of North Field on 01/12/2004 and had subsequently assigned a trustworthiness value of '5' to North Field.
(D) East Logistics had used the services of South Field to perform some *track and trace* operations on the goods due for delivery on 10/11/2004 and had subsequently assigned a trustworthiness value of '5' to South Field.
(E) South Field had used the services of North Field to store some of the goods on 10/11/2004 and had subsequently assigned a trustworthiness value of '5' to North Field.

The trust tuple for each of the five different scenarios above would be as follows:

For scenario A: [East Logistics, West Warehouse, storing the goods, 15/07/2003, not known, 5]
For scenario B: [East Logistics, West Warehouse, local delivery, 10/07/2004, not known, 5]
For scenario C: [West Warehouse, North Field, local delivery, 01/12/2004, not known, 5]
For scenario D: [East Logistics, South Field, track and trace reporting, 10/11/2004, not known, 5]
For scenario E: [South Field, North Field, storing the goods, 10/11/2004, not known, 5]

Note that 'not known' indicates that the 'end time' of the trust is not known; the agent may only know the start time of the trust.

The steps involved in modeling such multiagent trust cases between agents are as follows:

(1) Determine those agents that act as trusting agents *only*. Draw their trust case diagrams (as explained above in Sections 13.3.1 – 13.3.3) and label them.

From the above five distinct trust cases, we can see that East Logistics does not play the role of a trusted agent in any scenario. Hence, we draw a trust case diagram for the trusting agent, East Logistics in (Figure 13.13) and label it as East Logistics.

(2) Determine those agents that act as trusted agents *only*. Draw their trust case diagrams (as explained above in Sections 13.3.1 – 13.3.3) and label them.

From the earlier five distinct trust relationships, we can see that North Field does not play the role of a trusting agent in any scenario. Hence, we draw a trust case diagram for the trusted agent North Field (Figure 13.13) and label it as North Field.

(3) Determine those agents that act as both trusting agents as well as a trusted agent. Draw their trust case diagrams (as explained above in Section 13.3.1 – 13.3.3) and label them.

From the five distinct trust cases, we can see that West Warehouse and South Field play the role of a trusting agent as well as a trusted agent.

We can also see that West Warehouse and South Field play the role of a trusting agent as well as a trusted agent. In some cases, they play the role of a trusting agent and in some other cases they play the role of a trusted agent. For example, West Warehouse plays the role of a trusted agent in scenario A and B whereas it plays the role of a trusting agent in scenario C. Similarly, South Field plays the role of a trusted agent in scenario D while it plays the role of a trusting agent in scenario E. For West Warehouse, we draw the trust case diagram representing the trusting agent, and for North Field, it is represented by a trusted agent (Figure 13.13). We label them with their respective identities. Finally, we need to ensure that the notation for representing West Warehouse as a trusted agent (with a clear circle) faces those agents for whom it acts as a trusted agent (East Logistics in this case, can be seen in scenario A and B) and the notation for representing West Warehouse as a trusting agent (with a solid circle) faces those agents for whom it acts as a trusting agent (North Field in this case, can be seen in scenario C).

Similarly, we need to ensure that the notation for representing North Field as a trusted agent (with white hand) faces those agents for whom it acts as a trusted agent (West Warehouse and South Field in this case, as can be seen in scenario C and E) and the notation for representing South Field as a trusting agent (with a solid circle) faces those agents for whom it acts as a trusting agent (North Field in this case, as can be seen in scenario E).

For each of the trust relationships outlined in the above five scenarios, we connect the trust case diagram of the trusting agent with the trust case diagram of trusted agent with a trust relationship notation and annotate them with the trust tuple for that trust relationship.

The modeling of multiple trust cases using trust case diagrams is shown in Figure 13.13.

Note that 'not known' indicates that the 'end time' of the trust is not known; the agent may only know the start time of the trust.

13.6 Trust Class Diagrams

As discussed in the above section, we can use trust case diagrams to represent an agent involved in a trust relationship. We discussed how, coupled with the trust relationship symbol and the trust tuple, we could model the trust relationships between any number of agents. However, as can be observed from the above examples, trust case diagrams do not allow any attributes of the agent to be represented apart from the identity of the agent. Therefore, we introduce trust class diagrams.

In this section, we propose trust class diagrams that will permit us to represent all the attributes or properties of the agent. The trust class diagrams are based on the notion of class, which we borrowed from UML. However, we have made semantic and syntactic modifications to the notion of class in order to model and represent an agent in a trust relationship.

13.6.1 Trust Class Diagram

In order to model trust, we will make use of the class symbol to represent any agent involved in the trust relationship. Just as in UML, the class that we use for representing an agent involved in a trust relationship has three sections (Figure 13.14). The semantics and the purposes of each of these sections of the class are, however, different from UML.

Note that 'not known' in the tuple field indicates that the 'end time' of the trust is not known; the agent may only know the start time of the trust.

The first section in the trust class diagram contains the identity of the agent involved in a trust relationship. We propose the following rules for representing the name of the agent:

- This section can contain *no more than* and *no less than one* identity corresponding to an entity (corresponding to the identity of the trusting peer/trusted peer).

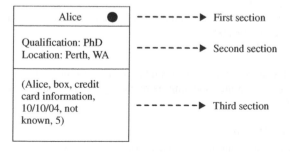

Figure 13.14 The trust class notation

- If the agent is recognized using a pseudo-anonymous identity, then this section should contain the pseudo identity of the trusting agent/trusted agent involved in the trust relationship.
- The trust class diagram for a trusting agent would have a solid black dot as shown in Figure 13.18 and the trust class diagram for a trusted agent would have the clear white dot. We propose to distinguish a trusting agent and a trusted agent by these dots.

We propose that the second section contain the attributes of the agent. This section can contain any attribute about the agent; however, we propose the following rules:

- Each new attribute of the agent should specifically mention the name or type of the attribute.
- Each new attribute should start on a new line immediately below the previous attribute.
- When an attribute is specified, its value cannot be left as undefined, that is, each attribute should have a value.
- The value of a given attribute should be from a valid domain.
- The trust tuple/s denoting the trust relationships should not appear in this section as attributes of the agent.
- The attributes and their value should be separated by a colon.
- If no attribute of the agent is being specified, then this section should contain the remark 'NIL'.
- All the attributes in this section should be of the agent whose identity is specified in the first section.

The third section specifies all of the trust relationships that need to be modeled for the agent whose identity is mentioned in the first section. For this section, we propose the following rules:

- The trust tuple/s representing the trust relationship/s that the agent whose identity is mentioned in the first section has with the other agents can only be contained in the third section of the trusting agents class.
- The format of representing the trust tuples in this section follows the rules that have been specified in Section 13.5.3.
- Each trust tuple should start on a new line immediately below the previous trust tuple.
- The section should contain only those trust relationships that are being modeled for the agent whose identity is mentioned in the first section. It is irrelevant to represent those trust relationships of the agent that are not being modeled.

Using the above notation of a trust class diagram, the trust relationship notation and the trust tuple, we will explain how we can model the following trust relationships:

- Trust relationships in a single context; (the trust relationship that a trusting agent has with the trusted agent in a single context);
- Trust relationships in more than one context (the trust relationship that a trusting agent has with a given trusted agent in different contexts);
- Multiple trust relationships (the trust relationships that an agent has with more than one trusted agent possibly in different contexts);
- Multiagent trust relationships.

We shall explain how the above relationships can be modeled using trust class notations introduced in Section 13.3. For elucidating the modeling more clearly, we continue with the example that we outlined in Section 13.5.

13.6.2 Single trust Class Diagram
In this section, we show how, using the trust class notation representing the agent introduced in the previous section and the notation of representing the trust relationship, introduced in Section 13.3,

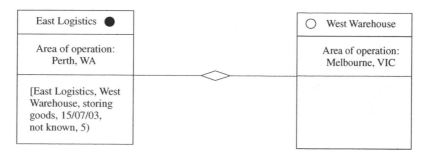

Figure 13.15   Single trust relationship diagram

we can model the trust relationship between a trusting agent and a trusted entity in a single context (see Figure 13.15).

For the purpose of elucidation, we use the trust relationship between East Logistics and West Warehouse outlined in the preceding text. The steps involved are as follows:

(1) For the trusting agent, draw a trust class diagram and label it with the identity of the trusting agent in the first section. The trust class diagram for a trusting agent has a solid black circle in the first section.

In our example, since East Logistics is the trusting entity, we draw a trust class diagram of the trusting agent and label the first section of the trust class diagram (see also Figure 13.15) with the identity of East Logistics.

(2) For the trusted agent, draw a trust class diagram for the trusted agent and label it with the identity of the trusted agent.

In our example, since West Warehouse is the trusted agent, we draw a trust class diagram for the trusted agent and label it with the identity of West Warehouse.

(3) Connect the trust class diagrams of the trusted agent and the trusting agent with the trust relationship notation to show that the trust relationships exists between them and label/annotate the trust relationship with the trust tuple that conveys meaningful information about the trust relationship.

In our example, since East Logistics and West Warehouse are involved in a trust relationship, we connect the trust class diagrams of both the entities by a trust relationship symbol and annotate it with the trust tuple. The trust tuple has already been explained in Section 13.3.5.

13.6.3 Multiple trust Class Diagram

It is possible to make use of the trust class diagrams to model and represent the trust relationships that an agent has with more than one agent, each of which may or may not be in the same context (Figure 13.16).

As an example, we model the two trust relationships that East Logistics has with West Warehouse and South Field to show how trust relationships between two agents in multiple contexts can be modeled.

The steps involved for modeling trust relationships that an agent has with multiple agents (irrespective of the context in which the trust relationship follows) using trust class diagrams are as follows:

(1) For the trusting agent, draw its trust class diagram and label it with the identity of the trusting agent.

In our example, since East Logistics is the trusting entity, we draw the trust class diagram of the trusting agent and label it with the identity of East Logistics.

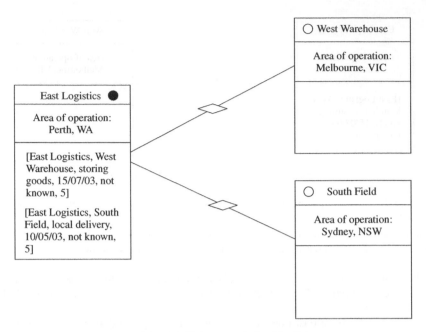

Figure 13.16   Multiple trust relationship diagram

(2) For each agent with whom the trusting agent shares a trust relationship, draw the trust class diagram of the trusted agent and label it with the identity of the trusted agent/s.

In our example, since West Warehouse and South Field are the trusted agents with which East Logistics shares a trust relationship, we draw the two trust class diagrams of the trusted agents, one for South Field and one for West Warehouse and label it with the identities of the trusted agents, respectively.

(3) Connect the trust class diagrams of the trusting agent with the trust class diagram of the each of the trusted agents with whom it shares a trust relationship by the trust relationship notation.

In our example, since East Logistics has a trust relationship with (1) West Warehouse and (2) South Field, we connect the trust class diagrams of East Logistics with West Warehouse with a trust relationship notation and the trust class diagram of East Logistics with South Field with a separate trust relationship notation.

(4) Annotate each of the trust relationship notations between the trusting agent and the trusted agent/s with the corresponding trust tuple that conveys more meaningful information about the trust relationship.

In our example, since East Logistics has a trust relationship with

(a) West Warehouse, we annotate the trust relationship notation between East Logistics and West Warehouse with the trust tuple for this relationship.

(b) South Field, we annotate the trust relationship notation between East Logistics and South Field with the trust tuple for this relationship.

13.6.4 Multiagents and Multitrust Class Diagram

In this section, we show how using the trust class diagrams to represent an agent involved in a trust relationship coupled with the notation for the trust relationship, we can model more than one trust relationship between two agents. For the purpose of elucidating this, we again make use of the example of two trust relationships between East Logistics and West Warehouse as outlined in Section 13.3.5.

The steps involved in *modeling* trust relationships between two agents in *multiple different contexts* using a trust class diagrams are as follows:

(1) For the trusting agent, draw its trust class diagram and label it with the identity of the trusting agent.

In our example, since East Logistics is the trusting entity, we draw the trust class diagram of the trusting agent and label it as East Logistics.

(2) For the trusted agent, draw its trust class diagram and label it with the identity of the trusted agent.

In our example, since West Warehouse is the trusted agent, we draw the trust class diagram of the trusted agent and label it as West Warehouse.

(3) Connect the trust class diagrams of the trusted agent and the trust class diagram of the trusting agent with the trust relationship notation, for each trust relationship that exists between them. In other words, there will be as many trust relationship notations connecting the trust class diagrams of the trusting agent and the trust class diagram of the trusted agent as there are trust relationships between the two of them that need to be modeled.

(4) Annotate each of the trust relationship notations between the trusting agent and the trusted agent with the corresponding trust tuple that conveys more meaningful information about the trust relationship.

In our example, since East Logistics and West Warehouse are involved in two trust relationships, we connect the trust case diagrams of both the entities by two trust relationship notations.

Let us assume that the upper trust relationship notation is used to model the trust relationship in the context of 'storing the goods' and the bottom one is used to model the trust relationship in the context of 'delivering goods'.

We then annotate each of the trust relationship notations with the corresponding tuples as shown in Figure 13.17.

13.6.5 Modeling Multiagent Trust Relationships

In this section, we show how multiagent trust relationships between agents can be modeled and represented using trust class diagrams. For the purpose of elucidation, we take the example of the

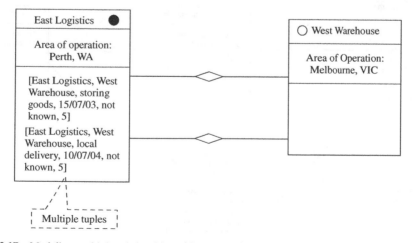

Figure 13.17   Modeling multiple relationships with same agents (see the two tuple representation in the third section of East Logistics class). Note that the relationship is represented by trust tuples, and there are multiple trust relationships between the same agents

complex web of trust relationships between East Logistics, West Warehouse, North Field and South Field given in Section 13.5.4.

The steps involved in modeling such multiagent trust relationships using trust class diagrams between agents are as follows:

(1) Determine those agents that act as trusting agents *only*. Draw the trust class of the trusting agent (see Figure 13.18) for each agent that acts only as the trusting agent and label them.

From the earlier five distinct trust relationships, we can see that East Logistics does not play the role of a trusted agent in any scenario. Hence, we draw a trust class diagram for the trusting agent East Logistics and label it as East Logistics.

(2) Determine those agents that act as trusted agents *only*. Draw the trust class of the trusted agent (Figure 13.18) for each agent that acts only as the trusted agent and label them.

From the five distinct trust relationships given earlier, we can see that North Field does not play the role of a trusted agent in any scenario. Hence, we draw a trust class diagram for the trusted agent for North Field and label it as North Field.

(3) Determine those agents that play the role of trusting agents and trusted agents. Draw their trust class diagrams (Figure 13.18) and label them with their respective identities.

From the earlier five distinct trust relationships, we can see that West Warehouse and South Field play the role of a trusting agent as well as a trusted agent. In some scenarios, they play the role of a trusting agent and in some others they play the role of a trusted agent; for example, West Warehouse plays the role of a trusted agent in scenarios A and B, whereas it plays the role of a trusting agent in scenario C. Similarly, South Field plays the role of a trusted agent in scenario D, whereas it plays the role of a trusting agent in scenario E. For West Warehouse and South Field, draw the trust class diagram for representing both a trusted agent and a trusting agent and label them with their respective identities. Finally, we need to ensure that the notation for representing West Warehouse as a trusted agent (with a clear circle) faces those agents for whom it acts as a trusted agent (East Logistics in this case, as can be seen in scenarios A and B) and the notation for representing West Warehouse as a trusting agent faces (with a solid circle) those agents for whom it acts as a trusting agent (North Field in this case, as can be seen in scenario C).

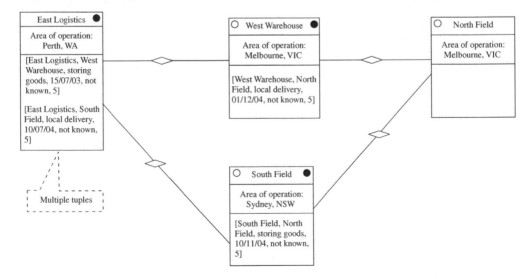

Figure 13.18 A class diagram for modeling multiple agents and multitrust relationships

Similarly, we need to ensure that the notation for representing South Field as a trusted agent faces (the clear circle) those agents for whom it acts as a trusted agent (East Logistics in this case, as can be seen in scenario D) and the notation for representing South Field as a trusting agent faces (with a solid circle) those agents for whom it acts as a trusting agent (North Field in this case, as can be seen in scenario E).

(4) For each of the trust relationships outlined in the five scenarios, connect the trust class diagram of the trusting agent with the trust class diagram of the trusted agent with a trust relationship notation and annotate them as the trust tuple for that trust relationship.

13.7 Trust Transition Diagrams

State transition diagrams are used to model the dynamic nature of a class [2, 3]. Class diagrams, on the contrary, are used to model the static nature of a class [2, 3]. State transition diagrams can be used to capture the various states of the class and how the transitions between these states takes place [2–4].

We discussed in Chapter 5 that as a result of various factors trust between two agents is dynamic. Moreover, trust relationships and trustworthiness because of these factors is dynamic, as discussed in Chapter 3 and Chapter 6. Over a period, the trust (hence the trustworthiness) that an agent has in another agent may change and in this period the only factor that may change in the trust tuple is the trustworthiness. The trust tuple is used to model/represent attributes of the single trust relationship, and for a given trust relationship, all the other attributes apart from the trustworthiness will remain the same. Hence, as can be clearly seen, if we need to model the dynamic nature of trust, we need to model and trace the change in trustworthiness over the duration of time.

In this section, we show how we can model the dynamic nature of trust between two agents using modifications to the *state transition diagrams* in UML. We make semantic modifications to the existing notations of *state* and *transition* used in state transition diagrams and syntactic modifications to the notation of trust tuple that we introduced in Section 13.3.5. We term our proposed method for modeling the dynamic nature of trust as the *'trust transition diagram'*.

13.7.1 Trust State and Trust Transition

As mentioned in the above section, we need to make semantic modifications to the notations of state and the notation of transition to model the dynamic nature of trust. We now explain the modified semantics for a state and a transition.

13.7.1.1 State

We make use of the state symbol to show the values of all the attributes of the trust relationship. We call this 'trust state'. In all further discussions, we refer to this as trust state. We define a trust state as *a notation that is used to represent the values of all the attributes of the trust relationship.*
We propose the following rules for the trust state:

- There should be one and only one *'trust start state'* in a trust transition diagram. The notation for showing the trust transition diagram start state is consistent with that of the start state in UML. The trust start state denotes the foremost status of the trust relationship.
- There should be one and only one *'trust end state'* in a trust transition diagram. The notation for showing the trust end diagram is consistent with that of the end state in UML. The trust end state denotes the very last *state* of the trust relationship.
- If, while charting the state transition diagram, the end state is not known, then there is no need to include the end state.
- Between the trust start state and trust end state there may be many trust states.

- Each trust state, including the start state and the trust end state should be labelled by the modified version of the trust tuple. We explain in the following section what we mean by modified trust tuple.
- Each trust state should be annotated by one and only one modified trust tuple.

13.7.1.2 Transition

We make use of an arrow symbol to show the shift from one trust state to another trust state. We refer to the *arrow symbol* to show the shift from one trust state to another state for a given trust relationship as 'trust transition'. We define a trust transition as *a notation used to describe the shift between two trust states of a trust relationship*.

We propose the following rules for trust transitions:

- Each trust state, except the trust start state and the trust end state, should have one and only one incoming and outgoing transition. There cannot be more than one incoming and outgoing trust transition for any trust *state*.
- The trust start state cannot have an incoming state but should have an outgoing state.
- The trust end state cannot have an outgoing state but should have an incoming state.
- Between any two given trust states there cannot be more than one trust transition.
- Each of these transitions is annotated by the reason that caused this transition (in other words, the trigger for the change in the *state*) in the *state* of the trust relationship to take place.
- The trigger for trust transition should be specified in natural language.
- The reason for trust transition should be prefixed by a '/'. We call this symbol '/' the trigger for trust transition.

The following points may be noted for trust transition diagrams:

- A given trust transition diagram can model the dynamic nature of only one trust relationship.
- A given trust relationship may be represented by more than one trust transition diagram.
- There is a one-to-many correspondence between trust relationships and trust transition diagrams.

13.7.2 Syntactic Modification of Trust Tuple

Apart from the above-mentioned changes to the semantics associated with the state and transition in UML, we need to change the syntax of the trust tuple that we had introduced in Section 13.3.5. In order to model the dynamic nature of Trust, we need to represent the exact time at which the *state* of the trust relationship changes. As we can see, the trust tuple only models the time at which the trust relationship started and when it ceased to exist. It cannot model the current *state* of the trust relationship. In order to do so, we modify the trust tuple and include a field in the time stamp to represent the time factor at the time the trustworthiness of a trust relationship changed.

The format of the modified trust tuple is described in the following text:

[Trusting agent, trusted agent, context, start time, end time, trustworthiness value, time spot]

The first six elements of the modified trust tuple are exactly the same as the trust tuple that we introduced in Section 13.3.5 and they carry the same semantics as those explained in that section. The only difference is the last element, the '*time spot*'.

The time spot is used to capture and represent the exact time at which the *state* of the trust relationship changes. We propose the following format for time stamp:

$$HH:MM:SS \parallel DD/MM/YYYY.$$

where, HH denotes the hour the trustworthiness of the trust relationship changed,
MM denotes the minute when the trustworthiness of the trust relationship changed,

SS denotes the seconds when the trustworthiness of the trust relationship changed,
DD stands for the day when the trustworthiness of the trust relationship changed,
MM stands for the month when the trustworthiness of the trust relationship changed and,
YYYY stands for the year when the trustworthiness of the trust relationship changed.

We now show through the following examples how the dynamic nature of trust relationships between two agents can be modeled using trust transition diagrams.

Let us assume that East Logistics had interacted with West Warehouse and used its warehouse space. Let us assume that the interaction had taken place on 22/10/2004. Subsequently, East Logistics had assigned West Warehouse a trustworthiness value of '4'.

At a later date, that is, on 22/01/2005, East Logistics again interacted with West Warehouse in the same context and assigned a Trustworthiness value of 5 to West Warehouse.

We can see from the above description that initially East Logistics had assigned West Warehouse a trustworthiness value of '4', when the trust relationship between them was established on 22/10/2004, in the context of 'storing goods in the warehouse'. As can be additionally seen from the above description, on 22/01/2005, East Logistics again interacted with West Warehouse in the same context and assigned a trustworthiness value of '5' this time. As a result of this, the value of the trustworthiness attribute in the trust tuple of this trust relationship has changed. We now show how we can model this change in the trustworthiness value, along with the time at which the change in trustworthiness took place, using trust transition diagrams explained earlier. When we refer to trust tuple while modeling the dynamic nature of trust, we refer to the modified trust tuple introduced. The steps involved are as follows:

(1) Determine the total number of states of the trust relationship, including the initial value of the initial state. For each of these states, determine the trust tuple.

In this case, we can see that there are two states. The first state is the one when the trust relationship was initially established on 22/10/2004. Apart from that, the second state is the one when the value of the trust relationship changed on 22/10/2005. The trust tuple for the initial state would be follows:

[East Logistics, West Warehouse, storing the goods, 22/10/2004, not known, 4, n/a]

Since the time at which the trust relationship was established is not mentioned, we did not include it in the trust tuple. Had the time been specified as well, we would have appended it with the date at which the trust relationship was established with a '||'.

On 22/01/2005, the trustworthiness value of the trust relationship between East Logistics and West Warehouse had changed. The new trust tuple of this trust relationship, taking into account the changed trustworthiness value, would be as follows:

[East Logistics, West Warehouse, storing the goods, 22/10/2004, not known, 5, 22/01/05]

(2) Draw the symbol of the start state and annotate it with the trust tuple that represents the start of the trust relationship.

In our case, we draw the start state symbol and label it with the trust tuple that describes the attributes of the trust relationship between East Logistics and West Warehouse when it started, that is, the following trust tuple,

[East Logistics, West Warehouse, storing the goods, 22/10/2004, not known, 4, n/a]

(3) For each additional state, draw a rectangle diagram to represent an additional trust state in the trust relationship and annotate it with the trust tuple. If an end state is mentioned, draw the end state and annotate it with the trust tuple as well. In case the end state has not been mentioned, do not include the end state.

In our case, we draw the state symbol and label it with the trust tuple that describes the attributes of the trust relationship between East Logistics and West Warehouse when it started, that is, the following trust tuple,

[East Logistics, West Warehouse, storing the goods, 22/10/2004, not known, 5, 22/01/2005]

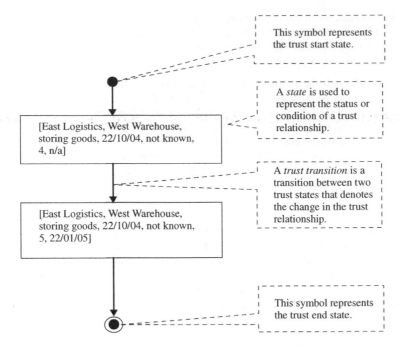

Figure 13.19   Dynamic trust modeling using trust transition diagrams

Since in our case the end state of the trust relationship is not known, we cannot draw an end state.

(4) In chronological order, connect the states using the trust transition diagrams.

In our case, since we have just two states, and chronologically the start state appears first and then the second state appears, we connect the start state to the second state using the trust transition diagram, as shown in Figure 13.19

13.7.3 Modeling the Dynamics of Trust

Figure 13.19 shows the trust transition diagram that models the dynamic changes of the trust relationship between Alice and Bob.

13.8 Conclusion

In order to help in the design and development of a trust and reputation system, which is a tool to help build business intelligence and consumer confidence, we have introduced a pictorial or graphical modeling language. This language will help designers and end-users to conceptualize the complex trust and reputation relationships, communicate with business providers or customers, and assist the automation of the development of trust and reputation system components and databases. We also introduced a detailed notion of a trust tuple along with the rules for representing this tuple. In addition, we presented trust relationship diagrams and trust property diagrams that model the trust relationships. We developed trust context diagrams based on use case diagrams. We detailed trust transition diagrams for modeling the dynamic nature of trust relationships. We explained these diagrams and provided detailed examples of how to use them. We also explained how trust relationships between multiple agents and multiple trust relationships between the same agents in different contexts may be modeled.

We also described the importance of and need for modeling trust. We introduced a new set of trust representation notations and introduced the representation of trust tuple. We illustrated with examples how trust can be modeled using trust relationship diagrams, trust case diagrams, trust class diagrams, and trust transition diagrams. For modeling trust, we modified the semantics associated with actor, class, association relationships, state and transitions. Additionally, we illustrated with examples how trust between two agents in single, multiple contexts and multiple agents can be modeled in trust relationships, trust case diagrams and class diagrams. Finally, we demonstrated how the dynamic nature of trust relationships can be modeled using trust transition diagrams.

In summary, in this chapter we have learnt the following:

- notation systems to include trusting agent, untrusted or unknown agent and malicious agent and relationship notations;
- the trust tuple representation and trust case notation;
- diagrams to represent trust relationships, trust cases, trust classes and trust state transitions;
- the significance and purpose of modeling;
- Modeling for single trust relationships, multiple trust relationships, multiagent and multitrust relationships, multirelationships between the same agents and the dynamic nature of trust.

In the next chapter, we review our previous chapters and studies, and conclude how trust and reputation help build business intelligence and consumer confidence.

References

[1] Ullman J.D. 1989 *Principles of Database and Knowledge-Base Systems*, Vol. 1, Computer Science Press, NY.
[2] Booch G., Rumbaugh J. & Jacobson I. 1999 *The Unified Modeling Language User*, Addison Wesley, USA.
[3] Fowler M. & Scott K. 2000 *UML Distilled: A Brief Guide to the standard Object Modeling Language*, 2nd Ed, Addison Wesley Longman.
[4] Dillon T.S. & Tan P.L. 1993 *Object Oriented Conceptual Modeling*, Prentice Hall, UK.

14

The Vision of Trust and Reputation Technology

14.1 Introduction

Business Intelligence (BI) has been defined in many ways. Some believe that BI is about a focus on the bottom line, some consider it as a focus on organizational direction and strategic planning, some define it as the way of using technologies to help in decision making, and some definitions focus on using organizational data through data mining to find out what customers want. Some regard business intelligence as being inclusive of issues such as security and privacy, and so on.

In this chapter, we give a review of existing definitions and propose a more advanced BI definition. We give an overview of 40 years of BI technology in the application development paradigm. We also give comparisons and contrasts between new-age technology such as trust and reputation systems and the existing well-known BI tools such as Enterprise Resource Planning (ERP), Customer Relationship Management (CRM), and so on.

We clearly list the new things that the trust and reputation systems can do to help BI and consumer confidence. We describe why trust and reputation is a science, why it is a methodology and why it is a technology and tool for BI. Finally, we give an overview of future research and development in this new class of technologies.

In the last few decades, the use of IT has progressed from provision of infrastructure for handling data, storage of data, querying data, monitoring, accounting and audit systems to the automation of processes previously manually carried out by human beings, to providing decision support and more recently to the provision of BI. BI is the new frontier for IT and business interactions. Hence, it is important that we review and understand its nature and the different elements that go into its making.

BI can be defined in many ways from many different perspectives. In this chapter, we give an advanced definition of BI. We shall see how trust and reputation systems can help build BI and consumer confidence. They are different from existing BI applications and tools. They are unique and categorized as a new-age technology. We also outline how they are reshaping e-businesses and why they are able to provide customer assurance and quality of service assessments.

14.2 Business Intelligence

BI moves away from the traditional concentration of business on using data purely for repetitive calculations, monitoring and controlling to that of obtaining knowledge in a form that is suitable for supporting and enabling business decisions from marketing, sales, relationship formation, fraud

Trust and Reputation for Service-Oriented Environments   Elizabeth Chang, Tharam Dillon and Farookh Hussain
© 2006 John Wiley & Sons, Ltd

detection up to major strategic decisions. In order to understand the nature of BI, we will initially begin by reviewing some of the existing definitions and notions of BI.

14.2.1 The Classical Definition of BI

In this section, we will discuss some of the definitions of *BI* given by some well-known organizations.

We thought it useful to look at the definitions not just of researchers but also those put forward by major IT companies that provide business systems of one kind or another. Such a review of definitions and meaning ascribed to the idea of BI will understandably not be comprehensive because of the limitations of space. Hence, we have tried to provide samples that touch on the different threads ascribed to this idea. Among the companies considered are IBM, Accuracast, Siebel, Cognos and Oracle. Samples of such definitions are given in the following text.

'*Business Intelligence is a concept of applying a set of technologies to turn data into meaningful information. With Business Intelligence Applications, large amounts of data originating in many different formats (spreadsheets, relationship databases, web logs) can be consolidated and presented to key business analysts.., and armed with timely, intelligent information that is easily understood, and the business analyst is enabled to affect change and develop strategies to drive higher profits*' [1].

Bergerou [2] citing Accuracast defined BI as '*the process for increasing the competitive advantage of a company by intelligent use of available data in decision-making. Business Intelligence consists of sourcing the data, filtering out unimportant information, analyzing the data, assessing the situation, developing solutions, analyzing risks and then supporting the decisions made*'.

Siebel [3] defines BI as '*a solution suite that integrates data from multiple enterprise sources and transforms it into key insights that enable executives, managers, and front-line employees to take actions that lead to dramatic improvements in business performance*'. Siebel further considers that the next generation of BI '*comprises a mission-critical architecture that scales to handle the largest data volumes and delivers critical information to tens of thousands of concurrent users across the enterprise*'.

Cognos [4] defined BI to be event driven. '*Event Driven BI monitors three classes of events in operational and Business Intelligence content – notification, performance and operation events – looking for key changes. Having detected changes, event-driven BI then notifies and alerts decision-makers, keeping them informed and up-to-minute. This personalized information can be pushed to decision makers no matter where they are, enabling them to make timely and effective decisions*'.

Moss and Hoberman [5] described BI as '*the processes, technologies, and tools needed to turn data into information, information into knowledge and knowledge into plans that drive profitable business action. BI encompasses data warehousing, business analytics tools and content/knowledge management*'.

14.2.2 The Advanced Definition of BI

We see from the above definitions that BI refers to businesses understanding their customers, knowing their needs and wants, studying their purchasing behaviour, identifying potential services that are in demand, understanding market conditions and reacting quickly, targeting new businesses, and it is also about learning what we do not know.

Three different views illustrate this: Menninger [6] states that managing the business is about '*known unknown*'. IBM [1] notes '*..gather, consolidate, cleanse and analyse data for purpose of understanding and acting on the key metrics that drive profitability in an enterprise*'. In Oracle [7], we find the view of BI as '*. . . Collect data about your business, for analysis and prediction, timely, easily and decision making capability for an organization at all levels, simplified administration, scalable, reliable and performance*'.

Taking these factors into consideration, we offer a more comprehensive definition of BI as follows:

Business Intelligence is accurate, timely and critical data, information and knowledge that supports strategic and operational decision making and risk assessment in uncertain and dynamic business environments. The source of the data, information and knowledge are both internal organizationally collected as well as externally supplied by partners, customers or third parties as a result of their own choice.

Data is defined as a set of facts about the corporation and its business. *Information* is an abstraction of data, which provides semantics about the data with defined meaning, context and value. *Knowledge* is a high-level representation and abstraction that permits one to reason, carry out pattern recognition, classification, planning or other high-level intelligent tasks and includes representations of uncertainty.

Data can be in the form of figures on sales, buyers, suppliers, inventories, budgets, and so on. *Information* can be in the form of customer demand, cooperation, competition, feedback, best products, or sales patterns, and so on. *Knowledge* can be considered as an abstraction of data and information. Knowledge can be obtained directly from experts or experiences and can also be derived from data mining of the corporate data sources that provides strategic advice on market trends, profit/loss projections, productivity measurements, quality of service, product reputation and bottom-line predictions. This knowledge can also be used to bring about an increase in consumer confidence and business value.

14.2.3 40 years of BI Development

Figure 14.1 is our view of the 40 years of BI development, together with its associated technology and tools.

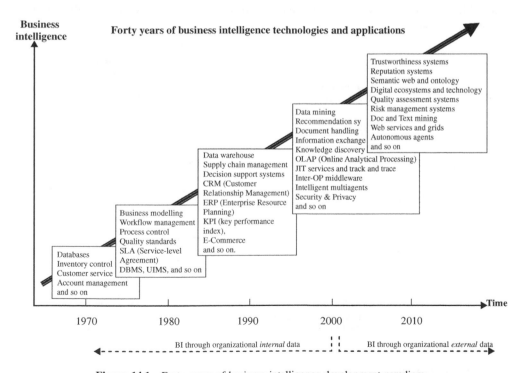

Figure 14.1   Forty years of business intelligence development paradigm

The notion of BI has evolved over the last thirty years and is likely to evolve further. This evolution of the notion of BI together with associated techniques is illustrated in Figure 14.1. Currently, BI is largely focused on the ideas of data mining, recommender systems and knowledge discovery techniques. These represent very important aspects of BI. However, they are circumscribed by the feature that they conduct this search for knowledge within organizational databases that represent either of the following categories:

(a) useful information about different aspects and units and individuals within the organizations;
(b) useful information that the organization itself has collected about transactions with other organizations and customers.

What they do not allow for is a collaborative notion of intelligence that utilizes

(a) knowledge from organizations outside of itself;
(b) data and information that are provided by customers and other organizations as a result of their own choice;
(c) information arising from the open nature of interactions and the Internet.

Once their new dimensions of interactions and knowledge are added, we see that BI in the future will include, among other things, trust and reputation systems, knowledge sharing, ontologies and ontology-based search engines and internal and external holistic risk management. This is illustrated by the projected notions of BI given in Figure 14.1.

14.3 Traditional IT and New-age Digital Ecosystems and Technology

14.3.1 Enabled Push and Pull Systems and Technologies

Figure 14.2 describes how enabled *push and pull* BI technologies help to advance BI and build consumer confidence.

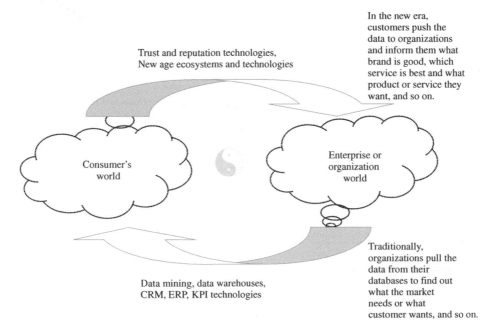

Figure 14.2 Enabled push and pull systems and technologies provide strategic input to BI

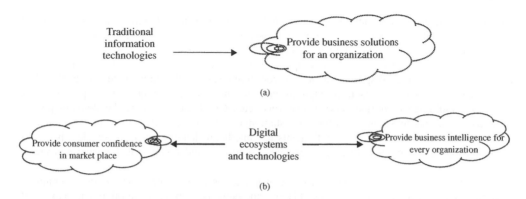

Figure 14.3 (a) The sole benefit of traditional information technologies. (b) The new-age digital ecosystems and technologies provide benefit to both customers and organization

Classical BI tools and technologies include data mining, which can help provide information about customer purchasing patterns or fraud detection by correlations between the data within the database; pattern analysis of this data warehousing, which can help provide ease of data query and access from different databases for optimal data look-up, view, analysis and report generation; CRM, which can help in market analysis; and ERP systems, which can help provide data integration, data transfer and transformation, resource scheduling and monitoring, budget, productivities and sales reporting, and so on. All these systems can provide corporate governance, process control, performance measurement, budget reporting, marketing forecasts, and so on, from the organizational databases, which may contain peta byte heterogeneous data.

Trust and reputation technologies represent the next generation of BI tools. So far, these technologies are still in their infancy. However, it is our vision that these technologies will be widely adopted in every industry, private or public, small or large, in the near future.

The new-age technologies such as trust and reputation (represented as Yin) together with existing organizational systems (represented as Yang) are not only complementary but also fundamental to each other. BI can be built from internal organizational data sources (traditional BI methods) and also can come from external sources, such as data from customers and end-users. This is well illustrated in Figure 14.2.

Within organizational data sources, the Yin and Yang pair analysis can also be defined. Cognos [4] distinguishes between the internal organization data as *deep data* (data that gives an in-depth view of the business or process) or *time critical data* (information that must be received quickly and used, among other things, for risk analysis). As a Yin and Yang pair, they are essential to each other.

As shown in Figure 14.3 (a and b), a contrast can be seen between traditional BI technologies and next generation BI technologies. Unlike traditional information technology, which only addresses the organization and business needs and provides the benefit to the sole or individual organization, the trust and reputation system provides benefit to both organizations and customers.

14.3.2 Digital Ecosystems and Technologies

A *Digital Ecosystem* is a self-organising digital infrastructure aimed at creating a digital environment for networked organisations that supports the cooperation, the knowledge sharing, the development of open and adaptive technologies [11] and evolutionary domain knowledge rich models.

Digital ecosystem transcends the traditional notion of rigorous defined e-business environments, such as centralised (client-server) or distributed (such as peer to peer) models moving rather to *agents*-based paradigm that is *loosely* coupled, *domain*-specific and has *demand* driven interactive communities which offer cost-effective digital services and value-creating activities that *attract* agents to participate and benefit from them. It is a "business model innovation in the Digital Economy" [11]. It captures the essence of the classical complex ecological community in nature, where biological organisms in analogy with economic organisms (such as business entities) or digital organisms (such as software or database applications) form a dynamic and interrelated complex ecosystem where biological species 'create and conserve resources that humans find valuable' [11].

An ecosystem advances itself through its core technology development. Examples such as ontologies, multi-agents, digital service discovery, etc.. Ontologies organize the domain-specific knowledge, and that tells you about the ecosystem and its agents' (species) functions, personality, who they are, and what they do. Multi-agents are intelligent leading software species that have strong reasoning capabilities which can manage, coordinate and collaborate between ecosystem agents.

A *Digital ecosystem* can be specifically developed for a particular domain or community, examples of digital business ecosystems [11] are "the network of buyers, suppliers and makers of related products or services" [11] or digital health ecosystem as a network of medical researchers, doctors, patients, etc, where species of different autonomous agents occupy the digital ecosystem.

Digital Ecosystems have the following significance

- Agents in ecosystem are autonomous entities that participate in the community for the benefit of themselves and the community
- They encompass loosely coupled relationships between agents or agent-families
- They are domain specific. All Agents come from a common background and engage in the community
- A demand-driven participation and collaboration with others in the community on Agent's own motivation, for the Agent's own benefit or profit
- Agents in ecosystems share a commonly agreed vocabulary
- Agents communicate knowledge through commonly shared ontology
- Human Agents are able to design a Digital Ecosystem Agent so that they can work together and coordinate with each other
- They are an open, social and collaborative learning environments
- Agents remedy problems through collaborative effort, sub-tasking, coordinated actions, shared intelligence and skills
- Agents provide the ecosystem with dynamism, efficiency, and stability
- Agents in an ecosystem are proactive, adaptive, and responsive
- Collaboration within an ecosystem can be carried out on-line or off-line
- Each ecosystem agent can be a client (when querying) or a server (when queried) with respect to the digital resources

Digital ecosystem technologies enable transparency and knowledge sharing that benefit not only the service providers, and customers, but also the environment itself. The innovation ecosystem initiative is part of the European Union *i2010 initiative* [11]. It integrate advanced social and economic sciences, system theory, self-organization of complex systems, epistemology and computer science [11] for new-age businesses and intelligence.

14.4 Trust and Reputation – An example of Digital Ecosystem and Technology

14.4.1 A Science and an Art

Trust and reputation systems and technology constitute a science, a methodology, a tool, a BI application, core organizational assets, complementary technology and new agile enterprise applications. These are briefly discussed in the following text.

14.4.1.1 A Science and an Art

Trust and reputation is a science. It provides a set of metrics, criteria, measurement scales, algorithms, methods, techniques and tools to help organizations and business providers by analysis and understanding of the complexity of dynamic consumer markets from customer submitted data.

Trust and reputation is an ART in that it provides a concrete business model based on science, social science, and business fundamentals. It also provides detailed concept definitions and ontology, theories and postulates, as well as scientific formulae together with a systematic process to guide the assigning of trustworthiness values and aggregating reputation values coupled with modelling notation systems.

14.4.1.2 A Tool

Trust and reputation is a technology and tool for BI. Specifically, it helps build BI and consumer confidence automatically with low costs in terms of IT and labour. Data collection and analysis is undertaken through the use of the tool itself, not by personnel from the organization.

14.4.1.3 A BI Application

Trust and reputation technologies are the next generation of business intelligence applications that cover diverse raw customer information, customer opinions, customer preferences, market trends and forecasting, competitors' movement, and so on, regardless of the region or continent, that every intelligent organization is interested in.

These sets of BI applications are the result of the Internet and the global reach that every organization has for competitive advantage and low-cost technology adoption. These applications enable us to give direct advice on business outputs such as services and products that directly reflect business strategies, goals, processes and economic outcomes.

14.4.1.4 A Core Asset of Every Organization

Trust and reputation technologies provide low-cost, real-time information for simple business indicators and market appraisals for an organization. Data is created through compilation of information from thousands or millions of users or customers, and not by the organization itself. This additional data set is a valuable asset for business. This enables organizations to have repositories of multiple data sources, including back-end databases (such as CRM and ERP), providing them with information to make multiple business direction planning decisions and developments.

14.4.1.5 Complementary Technology

Trust and reputation technology is a complementary technology to the existing technologies used within organizations, such as CRM, ERP or accounting systems. It provides information directly from the perspective of the consumer or customer. It does not duplicate any existing tools, technologies or systems that an organization already uses or has. As seen in Figure 14.2, new trust and reputation systems provide Yin and the existing system provides Yang. Yin and Yang need each other and they both provide a positive impact on corporate governance, business administration and production control and quality assessment for any organization.

Trust and reputation systems do not duplicate classical Online Analytical Processing (OLAP) systems. OLAP data sets are generally pulled from existing organizational databases, as these systems have web-based e-commerce systems. They can provide information such as sales through the Internet versus sales through telephone, fax and traditional personal contact methods. However, an e-commerce application is not a trust or reputation system.

14.4.1.6 New-age Agile Enterprise Application

Trust and reputation technologies are the next generation BI methodology, technology and tool that provide long-term data source and vital information for any business organization. This is because it enables an organization to have a better and more targeted market and consumer demand analysis from the data submitted by customers to make strategic decisions for business. This cutting-edge, innovative, low-cost technology is for every new-age agile enterprise that is ready to shift and move to a new BI and marketing paradigm.

14.4.2 BI and Consumer Confidence

Trust and reputation systems, a new-age digital ecosystem and technology, will help to build BI and customer confidence in the following ways.

14.4.2.1 Optimum Best Price for Goods and Services

One of the key uses of trust and reputation technologies for organizations is direct customer feedback or opinion or ratings which assist in enabling an organization to determine the best price for products to capture a significant market share. For example, if a similar product with similar functionalities and the same brand reputation was to gain a competitive advantage, a slightly lower price for a certain period, it would attract customers and increase sales. This is also true for the services industry and organizations.

14.4.2.2 Consumer Confidence with a Snap Shot of Rating Information

Most trust and reputation systems use visual representations such as *stars* or *numbers* for easy comprehension. These visual systems are friendly and require little knowledge and every customer can read and understand them without training. It is also easy to comprehend a trustworthiness rating. They represent the quality assurance for the customer about the product and services.

14.4.2.3 A Snap Shot of Supply and Demand for Businesses

Most trust and reputation systems provide a simple star rating visual representation for a snap shot of the statistics of various customer feedback, opinions, suggestions, ratings and quality measures about the organization's products and services for simple comprehension. This provides senior executives with an information resource for bottom-line analysis, strategic planning, profit and loss analysis, risk analysis, and so on.

14.4.2.4 Real-time and Just-in-Time

The trust and reputation systems provide real-time analysis and reports for both customers and businesses about products and services, as well as the trustworthiness of the quality of services, providers, partners or online brokers.

   With the support of the Internet, the organization can receive customer feedback with no time delay. A customer can receive just-in-time news, information on new products or services or survey and feedback details.

14.4.2.5 Monitoring and Prediction

Trust and reputation systems depend on the power of the algorithms and metrics used. These can provide senior executives with near-accurate market analysis, sales performance, consumer confidence, purchasing pattern, potential business risks, and so on. It also provides trustworthiness predictions for both consumers and businesses. Prediction and forecasting is one of the compulsory business processes that every organization has to undertake. Not all the ICT systems perform predictions. However, prediction is one of the two primary functions of trust and reputation systems. The two primary functions are *trustworthiness measurement* and *trustworthiness prediction*. The prediction provides the likelihood of changes in trust or quality of services as time passes.

14.4.2.6 Validation of Assumptions and Beliefs

Many senior business executives have their own assumptions and beliefs about what customers want and where they think they can make a lot of money from their products. Sometimes, when business does not generate profits, senior executives do not consider the possibility of error in their own beliefs and blame other issues. With trust and reputation systems, these assumptions and beliefs can be validated, proven right or wrong and assist managers to know what they do not know. Thus, they provide concrete information for business strategy improvement and redevelopment.

14.4.3 Other Areas of Applications

Trust and reputation systems will be an essential feature of many applications and these include the areas mentioned in the following text.

14.4.3.1 Quality of Service Assessment

The proposed trustworthiness technology and reputation systems can also apply to any of the quality of service assessments provided by giving a set of well-defined criteria.

14.4.3.2 Quality of Education Service and E-Learning Assessment

Education is a special type of service. The quality of education can be measured and assessed by using the same principles and methods as used in business for trust and reputation. For example, in the e-learning environment, remote students can get a degree such as an MBA without seeing the lecturers or tutors or any staff from a university. Therefore, the quality of students and their study can be measured in the same way as we measure a business provider or a member of a virtual community. To see how this is done, a prototype system can be viewed at www.ceebi.curtin.edu.au/e-Learner.

14.4.3.3 Quality of Public Funded Research & Development Assessment

With well-defined criteria and key performance indicators (KPI), it is also possible to utilize the trust and reputation principles and systems to assess public funded research. With trustworthiness systems, every thing can be measured and evaluated, such as the quality of applications, quality of applicants, quality of reviewers, quality of review, quality of selection panels and quality of expert panels. To see how this is done, a prototype system can be viewed at www.ceebi.curtin.edu.au/Quality.

14.4.3.4 Quality of Logistics Network Service Assessment

The same principles apply to a logistics network service measurement. Logistic services can be divided into local delivery, international delivery, interstate delivery, warehouse, distribution centres, export and import, and so on. Price and just-in-time service are very important to the customer. However, users are often overcharged. Sometimes there is an unpredictable difference between a quotation and the actual charges. With a well-defined set of criteria, a logistics service can be measured and published, thus creating an open quality service white board and achieving customer protection.

14.5 Future Research and Development

14.5.1 Data Adequacy and Accountability

For consumers and business providers to rely on trust and reputation systems, the technology itself must ensure information accuracy, security and privacy. Otherwise, systems could mislead customers and business and result in wasted time. More seriously, in the business environment, it could result in social or economic disaster. Therefore, many trust and reputation systems are designed with transparency, which expose their metrics, criteria, and calculation algorithms and data usage methods. This is vital for the next generation of ICTs so that there are no hidden agendas when talking to vulnerable customers.

14.5.2 Maintaining the Trust

This book addressed the question of how to establish the trust and assign trust values. However, one outstanding issue still not addressed is how to maintain the trust. This is important between business and consumers and between partners in partnerships. It is also relevant to all areas of social and economic environments.

14.5.3 Computational Strength and Performance

Many existing trust and reputation systems use only simple algorithms. For example, regarding reputation value aggregation, as presented in Chapter 9, 10, 11, there are simple as well as more sophisticated aggregation solutions. The more sophisticated solutions offer more accurate correlations of what customers really want, where simple formulae may not consider cheating, such as self-recommendations and a single agent with multiple IDs giving multiple recommendations, malicious agent recommendations and un-trusted agent recommendations.

More sophisticated algorithms not only define the above attacks on the integrity of trust and reputation systems, but also consider whether an opinion is first hand, second hand or third hand, the trustworthiness of recommendation agents and the trustworthiness of their opinions. However, this creates challenges for computational performance.

In this book, we have suggested a number of computation methods for measurement and prediction. There are also a few algorithms that have been developed over the last three years. So far, most of the algorithms are not yet tested with real examples, and give performance indicators and compression.

Trust and reputation systems have to have the capacity to handle millions of users' feedback and opinions, which requires a scalable database design with optimal query, power of data processing and calculation. These are the key challenges in trust and reputation system design and implementation.

The reputation system may sometimes require network optimization for maximum performance. This is especially the case when an agent is making a reputation query, in a given, simple time slot. There may be thousands of agent replies, followed by the process of checking the trustworthiness of recommenders. A lot of network bandwidth is required. Therefore, a careful analysis of a network load is crucial and the design of the reputation query for maximum performance is critical for successfully designed trust and reputation systems.

14.5.4 Ontology Drive and Agent-based System

A back-end office database can be controlled and slowly expanded in size. However, with front-end databases such as trust and reputation systems, the size of data may not easily be controlled. They could grow many times faster than back-end databases. With quantum hard disk space available today, the storage of data is not a big problem. However, indexing and searching is a growing concern of any web-based application. Therefore, the use of ontology modelling of business domain concepts and relationships, the development of business rules, and building of agent-based search engines are keys to distinguish leading edge developments from the old methodologies used in system development.

14.5.5 Data Mining Capability

Trust and reputation technologies not only calculate using data provided by the customers but can also carry out some data mining functions, such as the integration of recommendation systems, carrying out data correlation and pattern discovery, and are able to analyse the relationship between customer clusters, customer regions, sales and purchasing data and can provide strategic data for senior managers.

14.5.6 Data and Information Security

In order to keep data safe, track and trace facilities should be added to allow for monitoring the access of data and databases. This is to ensure that the data and information are accurate and securely stored. There should be an optimum system so that no dishonest information or malicious attacks on the data can occur.

14.5.7 Integration with Existing Systems

Integration functions may be added to allow system outputs to be integrated with other systems such as ERP or CRM or accounting systems. The purpose of integration is to help generate an integrated report easily and frequently, as this is what senior executives need to enable strategic decision making. In order to do this, a flexible software component interface should be developed that allows the integration of new systems with legacy systems.

14.6 The Vision and Conclusion

It is a long-term strategy to adopt trust and reputation methodologies and technology within the organization, especially within a networked environment. There is a global opportunity to take advantage of a low-cost BI tool for maximizing business chances, improving customer relationships, improving and targeting business strategies. Every dynamic new-age agile enterprise should prepare for this.

It is our vision that trust and reputation technologies will eventually transform what is normally believed to be a very risky environment into a disciplined and trusted business environment, because it helps change the organization and individual behaviour in a positive and professional way. It exposes dishonest services, unfair trading, biased recommendations, discriminatory actions, fraudulent behaviours, and untrue advertising. It will help build BI through disseminating customer feedback, buyer recommendations, third-party opinions, inference-free data and information sources, thus providing a guide for enhancing customer relationships, a method for learning consumer behaviours and a tool for capturing market reaction on products and services. It also encourages virtual collaboration and competition.

Through the support of trust and reputation technologies, the potential efficient, effective, just-in-time, low-cost and convenient service platform, as part of social life for ordinary users or customers and organizations, will become a reality. The further that technology develops, the further it will shape professional business behaviour in the networked environment. The further it increases consumer confidence, the further it will attract the population to use the Internet, increase sales, reputation of sellers, providers and the quality of products and services.

This inter-science and inter-disciplinary technology integrates computer science, social science, software engineering, ICT, business, commerce and psychology, having the ongoing potential to bring benefit and positive impacts in the networked economy. The Internet has enabled businesses and their agents to extend their reach to wide-scale communities. Customers and users can reach beyond their usual operations at very low cost. Such trust and reputation technologies will not only have an impact on the individual behaviour of businesses, customers, or agents in the service-oriented networks but will also have a strong impact in the virtual community as to what they think of the Internet and e-business.

It is envisioned by Bakos and Dellarocas [8] that a substantial fraction of economic transactions are likely to render online reputation systems into powerful quality assurance institutions in the social, economical and perhaps political environments. As such, technology trends deserve careful study and attention in order to promote honest trading over the threats of commercial litigation.

As it is no longer the case that online services are once-only transactions, the unlawful behaviour of an agent in a service network will eventually be widely and publicly known, thus helping to promote lawful trading environments.

'It is also a first time in human history, individuals can make their personal thoughts, reactions and opinions easily accessible to the global community of Internet users' [9].

References

[1] IBM 2005 *e-Server i5 and DB2: Business Intelligence Concepts*, http://dw.ittoolbox.com/documents/document.asp?i=2188 also see http://www.ibm.com/us/.

[2] Bergerou J. 2005 *AccuraCast Gaining Business Intelligence*, http://www.accuracast.com/ also see http://bi.ittoolbox.com/documents/document.asp?i=2914.

[3] Siebel Business Analytics. 2005 http://www.siebel.com/business-intelligence-platform/solutions-description.shtm.

[4] Cognos 2004 *Event-Driven Business Intelligence*, http://www.cognos.com/ also see, http://bi.ittoolbox.com/documents/document.asp?i=3002.

[5] Moss L. & Hoberman S. 2005 *The Importance of Data Modelling as a Foundation for Business Insight*, http://www.teradata.com/t/ also see http://dw.ittoolbox.com/documents/document.asp?i=2334.

[6] Menninger D. 2005 *An Insider's Guide to Business Intelligence in the Insurance Industry*, http://bi.ittoolbox.com.

[7] Oracle 2005 *Oracle Business Intelligence Reporting and Analysis*, http://www.oracle.com/index.html also see http://oracle.ittoolbox.com/documents/document.asp?i=3457.

[8] Bakos Y. & Dellarocas C. 2002 http://ccs.mit.edu/dell/.

[9] Dellarocas, 2003, http://ccs.mit.edu/dell/.

Index

Printed and bound in the UK by
CPI Antony Rowe, Eastbourne

Printed and bound by CPI Group (UK) Ltd, Croydon, CR0 4YY

27/10/2024

14580150-0002